Please renew or return items by the date
shown on your receipt

www.hertsdirect.org/libraries

Renewals and   0300 123 4049
enquiries:

Textphone for hearing   0300 123 4041
or speech impaired

RHS

GOOD PLANT GUIDE

LONDON, NEW YORK,
MUNICH, MELBOURNE, DELHI

Project Editor Shashwati Tia Sarkar
Project Art Editor Clare Marshall
Managing Editor Penny Warren
Pre-Production Producer Sarah Isle
Senior Producer Alex Bell
Publisher Mary Ling
Art Director Jane Bull

DK India
Project Editor Janashree Singha
Project Art Editor Vikas Sachdeva
Senior Editor Nidhilekha Mathur
DTP Designer Manish Chandra Upreti

RHS Editors
Simon Maughan, Rae-Spencer Jones

First Edition 1998
Project Editor Simon Maughan, **Editor** Tracie Lee
Art Editor Ursula Dawson, **Managing Editor** Louise Abbott

First edition published in Great Britain in 1998 by Dorling Kindersley
Publishers Limited, 80 Strand, London WC2R ORL

This revised edition published in 2014

Copyright © 1998, 2000, 2004, 2011, 2014 Dorling Kindersley Limited, London

A Penguin Random House Company

2 4 6 8 10 9 7 5 3 1
001–255752–March/2014
A CIP catalogue record for this book is available from
the British Library.

ISBN 978-1-4093-4986-0

Printed and bound by South China, China.

To find out more about RHS membership, visit our website
www.rhs.org.uk or call 0845 062 1111

Discover more at
www.dk.com

Contents

INTRODUCTION

Gardeners today have plenty of choice when buying plants. Not only are plant breeders constantly producing new cultivars, but plants are now more widely available, in garden centres and nurseries and also in superstores – and in less traditional plant-buying situations, information and advice may not be readily available. It is perhaps no wonder that gardeners sometimes find choosing the right plant a bewildering business.

The *RHS Good Plant Guide* was conceived to help gardeners select outstanding and reliable plants for their garden, whatever their level of expertise and experience. Because all of the plants it features are recommended by the Royal Horticultural Society – the majority holding its coveted Award of Garden Merit (see pp.10–12) – even the novice gardener can choose plants from its pages with confidence.

However, these recommendations refer only to the plants themselves, and not to those who grow them. It is impossible to guarantee the quality of plants offered for sale

Ever-widening plant choice Exotic flowers, like this blue poppy (*Meconopsis grandis*), may tempt the gardener, but check the conditions they require before buying.

by any nursery, garden centre or store, and here gardeners must exercise a degree of common sense and good judgment, in order to ensure that the individual plants they select and take home will thrive. The following pages include guidance on recognising healthy, well-grown plants and bulbs, and also show how to get them home safely and get them off to a good start.

CHOOSING FOR YOUR GARDEN

Selecting plants appropriate to your site and soil is essential, and each plant in the *A–Z of Plants* has its preferences indicated. Well-prepared soil, feeding and, in dry conditions, watering in the early stages are also important for the well-being of most plants. Entries in the *A–Z* include basic care for the plant concerned, along with hints and tips on pruning and winter protection where hardiness is borderline. By following this advice, the plants you choose using the *RHS Good Plant Guide* should perform well, and fulfil all the expectations conferred on them by the RHS recommendation.

USING THE GUIDE

The *RHS Good Plant Guide* presents a selection of over 1,000 plants to help you make the best choices for your garden. They are listed in an A–Z format with detailed entries and are illustrated with photographs.

You can use it to find details about a plant, whether you are reading its name on a label or have noted it down from a magazine article, plant catalogue, or website. It will tell you what type of plant it is, how it grows, what its ornamental features are, which conditions it prefers and where it looks best, as well as how to care for it. For quick reference, symbols summarise its main requirements.

The Planting Guide section lists some choice plants to suit specific requirements, such as plants to grow for scented flowers, or for berries, or if you go for themed plantings, a range of plants to attract birds into your garden. Page references are given for plants with entries and portraits elsewhere.

Show stoppers (facing page) Flower shows such as Chelsea provide opportunities to see a wide range of inspirational plants.

SYMBOLS USED IN THE GUIDE

♀ The RHS Award of Garden Merit

Soil Moisture Preferences/Tolerances
◊ Well-drained soil
◐ Moist soil
● Wet soil

Sun/Shade Preferences/Tolerances
☼ Full sun
☀ Partial shade: either dappled shade or shade for part of the day
❀ Full shade

NB Where two symbols from the same category appear, the plant is suitable for a range of conditions.

Hardiness Ratings
❦ Plants requiring a warm, permanently frost-free climate, which must be grown in heated glasshouses or as house plants in the UK. In practice some can be grown outside in summer as bedding or in containers.

❀ Plants requiring unheated glass (or the protection of an unheated glasshouse): some may be placed outdoors in summer.

❀❀ Plants hardy outside in some regions or in favoured sites, or those that, in temperate countries, are usually grown outside in summer but need frost-free cover over winter (e.g. dahlias).

❀❀❀ Fully hardy plants.

THE PLANTS IN THE GUIDE

The range of plants – over 1,000 – featured in the *RHS Good Plant Guide* has been carefully chosen to provide the best selection across all the different types of plants, and all the situations for which plants may be required. All should be readily available from sources in the United Kingdom, although gardeners in other countries may have to do a little more research to track down some specimens. The majority hold the Royal Horticultural Society's Award of Garden Merit, the trophy symbol for which is now a recognised "kitemark" of quality throughout the UK and even further afield, seen in nursery catalogues, on plant labels, and in a variety of gardening publications.

ABOUT THE AGM

The RHS Award of Garden Merit, re-instituted in 1992, recognises plants of outstanding excellence, whether grown in the open or under glass. Its remit is broad, encompassing all ornamental plants as well as fruit and vegetables. Any type of plant can be entered for the award, be it the most minuscule alpine plant or a majestic tree.

Besides being the highest accolade the Society can give a plant, the AGM system is primarily designed to be of practical use to ordinary gardeners, helping them in making a choice from the many thousands of plants currently available. When choosing an AGM plant, the gardener can be certain that it fulfils the following criteria:

***Geranium x riversleaianum* 'Russell Prichard'** Versatile, easy to grow plants earn their AGMs for sheer garden value.

***Camellia x williamsii* 'Brigadoon'** Choice forms of flowering plants are recognised by the award scheme.

***Hydrangea macrophylla* 'Altona'** The best selections of many garden stalwarts are recommended.

• It should be excellent for ornamental use, either in the open garden or under glass

• It should be of good constitution

• It should be available in the horticultural trade, or at least be available for propagation

• It should not be particularly susceptible to any pest or disease

• It should not require any highly specialised care, other than providing the appropriate conditions for the type of plant or individual plant concerned (for example, lime-free soil if required)

• It should not be subject to an unreasonable degree of reversion in its vegetative or floral characteristics. To explain this point simply, many plants with unusual characteristics differing from the species, such as double flowers or variegated leaves, have often been propagated from a single plant – a natural mutation or "sport" – that has appeared spontaneously. Plants bred from these "sports" – especially when raised from seed – are liable to have only the normal leaf colour or flower form. It takes several generations of careful and controlled propagation for the special feature to be stable enough for plants to be recognised and registered with a distinct cultivar name (which is usually chosen by the breeder) and allowed to be offered for sale. Many never breed reliably enough from seed with the special feature, and can only be reproduced by vegetative methods such as taking cuttings.

HOW PLANTS GAIN AWARDS

An AGM award is made either following plant trials, usually conducted at the Society's display garden at Wisley in Surrey, but also

Prunus laurocerasus '**Otto Luyken**' This evergreen is handsome year-round both as a shrub and as a hedge.

Anemone blanda '**White Splendour**' Floriferous cultivars for every site and season gain the RHS award.

Cornus alba '**Spaethii**' Plants that have variegated foliage should not be prone to excessive reversion.

at other locations where expertise in growing a particular genus is to be found, or on the recommendation of a panel of experts. The AGM is not graded, so there is no attempt to differentiate the good from the better or best, but in those groups in which there are many cultivars, standards are necessarily high so that the resulting selection offers helpful guidance to the gardener.

OTHER PLANTS IN THE GUIDE
In some categories of plants, notably annuals (particularly those for summer bedding), new cultivars appear so rapidly, often superseding others offered for sale, that the trialling process needed for AGM assessment is hard-pressed to keep up with developments. In order to give a wider choice in these somewhat under-represented categories, the *RHS Good Plant Guide* includes, in its *Planting Guide* section, other selected cultivars which have proved their reliability over the years. Some have been grown at the RHS gardens at Wisley and around the UK; some are recommended by other bodies, such as the respected Fleuroselect organisation, an association based

Putting plants to the test Dahlia trials at the RHS gardens at Wisley, helping to select the best forms for gardeners.

in the Netherlands that specialises in recommending cultivars and seed series of both annuals and biennials from top breeders, and whose awards are recognised throughout Europe.

FINDING PLANTS BY NAME
The botanical or "Latin" names for plants are used throughout the Guide, simply because these are the names that gardeners will find on plant tags and in nursery lists: these names are also international, transcending any language barriers.

To the uninitiated, plant nomenclature may seem confusing, and sometimes plant names appear to be very similar. It is important to recognise that the AGM is given only to specific plants within a genus – and not, for example, to another cultivar within the same species, however similar that plant might look to its awarded relative. Some plants have synonyms – older or alternative names – by which they may sometimes be referred to: the *RHS Good Plant Guide* gives popular recognised synonyms for a number of plants. A brief guide to plant nomenclature can be found on pp.24–25. Many common names for plants or plant types that feature in the *Guide* are also given in the A–Z entries.

SHOPPING FOR GOOD PLANTS

The first step in ensuring that your garden will be full of healthy plants is to choose and buy carefully.

WHERE TO BUY

Plants can be found for sale in such a variety of situations today that there are no hard and fast rules to be applied. Generally, only buy plants where you feel reasonably sure that the plant offered is actually what it says it is, that it has been well-grown, and is not going to bring any pests and diseases into your garden.

PLANTS BY POST

Buying plants online, or by mail order, is one of the easiest ways to obtain particular plants that you are keen to acquire. It gives you a huge choice, far greater than in most garden centres, and also gives you access to specialist nurseries that concentrate on certain plant groups, which may not be open to visitors. With a buyer's guide such as *The RHS Plant Finder,* you can track down almost any specimen from the comfort of your own home, and provided that someone will be at home on the day it arrives, it should be delivered in perfect health.

Nurseries are usually happy to replace any plant damaged in transit. Usually, the plants will arrive with their roots wrapped or "balled" in a little compost, and are best planted out immediately.

WHEN TO BUY

Trees, shrubs, and roses that are available as bare-rooted specimens are invariably deciduous. Woody plants with roots balled and wrapped may be deciduous or evergreen. All are available only in the dormant season; they are ideally purchased in late autumn or early spring, and should be planted as soon as possible.

Container-grown plants are offered for sale year round. They can be planted all year too, but the traditional planting times of spring and autumn, when conditions are not too extreme, are still the best. No plants should be put into freezing or baked, dry soil, but kept in their containers until conditions are more favourable.

In frost-prone climates, only fully hardy plants should be planted in winter. Spring is the time to plant plants of borderline hardiness, and to buy bedding. Beware buying summer bedding too early, however tempting it looks after the dark days of winter. Bedding sold in early and mid-spring is intended to be bought by people with greenhouses and cold frames, who can grow the plants on under cover until all danger of frost has passed.

YOUR SHOPPING LIST

Whether you are stocking a new garden from scratch, replacing a casualty, or simply wanting to add extra touches with some plants for a shady corner, say, or for a patio tub,

Well-grown plants for sale Healthy, clearly labelled plants in neat, well-kept surroundings are a good indication of excellent nursery care.

informed choice is the key to buying the right plants, and to getting the very best and healthiest specimens.

There are many factors to consider about your garden (see pp.16–17) and, equally, buying tips that will help you make the best choice from the selection of plants on offer (see pp.18–19). Many gardeners prefer to set out with specific plants in mind. However, impulse buys are one of gardening's great pleasures – and with the *RHS Good Plant Guide* to hand, you will be able to snap up just the right plants for your garden, and avoid expensive mistakes.

CHOOSING THE RIGHT PLANTS

Although there are ways to get around many of plants' climatic and soil requirements, you will avoid extra work and expense by choosing plants well-suited by the conditions your garden offers.

HARDINESS

Most of us have gardens that experience some degree of frost in winter. Optimism is no substitute for action if plants susceptible to frost damage are to survive. Preferably, move them into a greenhouse for the winter, but a frost-free shed or cool utility room can also be used. In cold gardens, make full use of the ever-widening choice of plants offered as annual summer bedding. Reserve warm spots for shrubs and perennials of borderline hardiness, and protect them in winter with a

Extending the range
Container growing may be the answer for gardeners who covet plants with special soil needs, like these acid-loving azaleas, that their own gardens cannot meet.

thick mulch such as straw around the base, kept dry with plastic sheeting or bubble polythene.

SUN AND SHADE

It is well worth "mapping" your garden to see how much sun areas receive at different times of the day and year. For good growth and the best display of features such as coloured foliage, always match plants' sun and shade requirements.

YOUR SOIL

Most plants tolerate soil that falls short of perfect, and many survive in conditions that are very far from ideal. It is always preferable – and far more labour-saving – to choose plants that suit your soil, rather than manipulate your soil to suit plants, by adding, say, quantities of peat or lime. The effect never lasts, and is now also considered environmentally inadvisable in the case of peat. It is, however, helpful to improve the soil before planting. This enables plants to establish quicker, so they need less watering and are less likely to struggle and die. Use garden compost, manure, and mulches to help improve the soil structure but, as these are low in nutrients, add fertilizer too, particularly on poor sandy soils.

Determining your soil acidity or its reverse, alkalinity, measured by units known as pH values, is important. Most plants tolerate a broad range of pH values around neutral, but some groups have specifically evolved to be suited by soil that is either definitely acid or definitely alkaline. There is no point ignoring these soil preferences and trying to grow, for example, azaleas in a chalky soil; they will not thrive. Choose plants that will enjoy your soil or, if you really covet some specialised plant groups, grow them in containers, in a potting mix that meets their needs.

DRAINAGE

The expression "well-drained yet moisture-retentive" is one of the most widely used yet seemingly baffling expressions in gardening. However, it is not such a contradiction in terms as it seems. What it applies to is a soil that in composition is not dominated by pebbles or grains of sand, which cannot hold water, nor clay which binds up and retains water in a solid gluey mass. It is the presence of a dark, spongy organic material known as humus that enables soil to hold moisture *and* air simultaneously, so that they are available to plants. Working in organic matter rich in humus (see p.21) is the most useful way to improve soil texture, either across the whole garden or as plants are planted.

CHOOSING A HEALTHY PLANT

Try to resist buying plants that are in poor condition, even if it is the sole example on offer of a plant you really crave. Nursing a pathetic specimen could take a whole growing season, during which time the plant will scarcely reward you, and with container-grown plants so widely available it is likely that, later in the season, you will find a healthier specimen that you can plant straight away. In practice, however, there are few gardeners that have never taken pity on a neglected plant – but you must always harden your heart to those with signs of pest infestation or disease. You risk importing problems that could spread to other plants.

WHAT TO BUY

Plants offered for sale should be clearly labelled, healthy, undamaged, and free from pests and diseases. Inspect the plant thoroughly for signs of neglect. With root-balled plants, check that the root-ball is firm and evenly moist, with the netting or hessian wrapping intact. Containerised plants should always have been raised in containers. Do not buy plants that appear to have been hastily uprooted from a nursery bed or field and potted up. There should be no sign of roots at the surface.

Look under the pot for protruding roots, a sign that the plant is pot-bound (has been in its container for too long). Always lift the pot to make sure roots have not penetrated the standing area, another sign of a root-bound specimen. Crowded roots do not penetrate the surrounding soil readily after planting, and the plant will not establish well.

Leaves are glossy and healthy

Early pruning has produced an attractive shape

Roots are well-grown but not crowded

Good specimen This well-grown skimmia has a substantial root ball that is well in proportion to the bushy top growth.

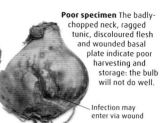

Healthy bulb A sound neck, tunic (the papery outer covering) and basal plate, where the roots will grow, all recommend this bulb to the gardener.

Sound, firm basal plate

Poor specimen The badly-chopped neck, ragged tunic, discoloured flesh and wounded basal plate indicate poor harvesting and storage: the bulb will not do well.

Infection may enter via wound

HEALTHY GROWTH

Strong top growth is another key factor. Avoid plants with pale or yellowing foliage, damaged shoot tips or etiolated growth (soft, pale and weak-looking shoots). Look at the compost surface: it should not be covered in weeds or moss.

With herbaceous plants, small healthy specimens are cheaper than large ones and will soon grow away once planted. Groups of plants often look better than single specimens. With all types of herbaceous plants, including annuals and bedding, pick out the stockiest, bushiest specimens, with (if appropriate) plenty of flower buds, rather than open flowers.

PLANT SHAPE AND FORM

When buying trees and shrubs, which are to make a long-lasting contribution to your garden, consider their shape as well as their good health. Choose plants with a well-balanced framework, and with top growth that is not far too large for the root-ball or pot. A stocky, multi-stemmed plant usually makes the best well-rounded shrub, while young trees with a single trunk should have just that – and a straight, sound one too – right from the start. A one-sided shrub may be acceptable for wall-training, but will put on more growth from the base, and cover space better, if its shoot tips are pruned after planting.

BUYING BULBS

Buy bulbs (see above) as if they were onions you were intending to cook with and eat – reject any that are soft, discoloured, diseased or damaged. While it is not ideal, there is no harm in buying bulbs that have begun to sprout a little.

PREPARING THE SOIL

While the best plants for your garden are those suited to the growing conditions available, most soils can be improved to extend the range of plants that can be grown.

IDENTIFYING YOUR SOIL TYPE

Investigating your soil to discover what its qualities are is one of the most useful things you can do to ensure your plants thrive. Generally, the soil will either be sandy or have a clay texture, or somewhere between the two. If your soil has a light, loose texture and drains rapidly, it is probably sandy. Sandy soil has a rough, gritty feel when rubbed and makes a characteristic scraping sound against the blade of a spade. Although easy to dig, it is low in fertility. Clay soil is heavy, sticky, cold, easy to mould when wet, and can become waterlogged. It is difficult to work, but is often very fertile. A good mix of the two – a "medium loam" – is ideal.

To further establish which plants will be best suited to your soil, you can discover its pH value (to what degree it is acid or alkaline) and the levels of nutrients it contains using simple testing kits that are available at most garden centres.

CLEARING WEEDS

In a new garden, or if you are planning to plant a previously uncultivated area, your first task will be to clear away any debris and weeds. Pre-planting clearance of all weeds is essential, since they will compete with your plants for light, moisture, and nutrients. Where the ground is infested with perennial weeds, an alternative to laborious hand-clearing is to spray it over in the season before planting with a systemic herbicide that will kill their roots. This is most effective in early summer when the weeds are growing strongly. Although this may mean delaying planting, your patience will be rewarded later.

Isolated perennial weeds can be forked out carefully, removing all root fragments. Annual weeds may be hoed or sprayed off immediately before planting.

WORKING THE SOIL

Digging or forking over the soil helps to break down compacted areas and increase aeration, which encourages good plant growth. The deeper you dig the better, but never bring poor-quality subsoil up to the surface: plants

Digging with ease The correct tools and technique make digging more comfortable, and safer for your back. Test the weight and height of a spade before buying, and always keep a straight back when using it.

Forking over Drive the fork into the ground, then lift and turn the fork over to break up and aerate soil. Spread a layer of organic matter over the soil first so that it is incorporated into the soil as you work.

need all the nourishment that the darker, more nutritious topsoil can give them. If you need to dig deep, remove the topsoil, fork over the subsoil, then replace the topsoil.

Dig heavy, clay soils in late autumn. The weathering effects of frost and cold over the winter will help improve the soil's texture by breaking it down into smaller pieces. Avoid digging when the soil is wet, as this will damage the soil structure. It will save work if you incorporate soil conditioners as you go.

ADDING SOIL CONDITIONERS

Digging and forking will improve soil texture to some extent, but to really bring a soil to life, the addition

of a soil conditioner is invaluable. The drainage, aeration, fertility, and moisture-holding properties of most soil types can be improved simply by adding well-rotted organic matter, rich in humus. This is best done during the season before planting, to allow the soil to settle. Garden compost and leafmould enhance the moisture retention of a sandy soil and improve nutrient levels. Applied regularly, they also improve the structure of clay soil. Clay soils can be further opened up by the addition of horticultural gravel to a depth of at least 30cm (12in). Organic mulches of bark, cocoa shells, or chippings are also eventually broken down into the soil.

PLANTING OUT

Whatever you are planting, make sure that you give your plants a really promising start. Careful planting saves both time and money, and well-chosen plants positioned in optimum conditions will perform well and should resist attack from pests and diseases.

PRE-PLANTING PLANNING

Before planting, check the potential heights and spreads of plants to ensure that you leave the correct distances between them. When designing plant groups, consider different plants' season of interest, and their appearance in winter.

WHEN TO PLANT

The best seasons for planting are autumn and spring. Autumn planting allows plants to establish quickly before the onset of winter, since the soil is still warm and moist enough to permit root growth. Spring planting is better in cold areas for plants that are not fully hardy, or dislike wet winter conditions.

Perennials can be planted at any time of the year, except during extreme conditions. They should grow rapidly and usually perform well within their first year. To reduce stress on a perennial when planting during dry or hot weather, prune off

Basic planting Soak plants well in a bowl of water. Position the plant so that the root ball surface is flush with soil level, then fill round the sides.

Settling the soil Backfill the hole, gently firming the soil to ensure good contact with the roots. Water in well, then add a layer of mulch over the root area.

its flowers and the largest leaves before planting. In full, hot sun, shade the plant with garden netting.

PLANTING TECHNIQUES

For container-grown or root-balled plants, dig a hole about twice the size of the root-ball. If necessary, water the hole well to ensure that the soil is thoroughly moist. Carefully remove the plant from its pot and gently tease out the roots with your fingers. Check that the plant is at the correct depth in its hole, then backfill the hole with a mix of compost, fertilizer, and soil. Firm the soil, and water thoroughly to settle it around the roots. Apply a mulch around, but not touching, the plant base to aid moisture retention and suppress weeds.

For bare-rooted plants, the planting technique is essentially the same, but it is vital that the roots never dry out before replanting. Dig holes in advance, and if there is any delay in planting, heel the plants in or store in moist garden compost or sand until conditions are suitable. Bare-rooted plants need a planting hole wide enough to accommodate their roots when fully spread. Ensure the depth allows the final soil level to be the same as it was in the pot or nursery. After planting, tread the soil gently to firm.

WALL SHRUBS AND CLIMBERS

Always erect supports before planting. Plant wall shrubs and climbers at least 25cm (10in) from walls and fences, so that the roots are not in a rain shadow, and lean the plant slightly inward toward the wall. Fan out the main shoots and attach them firmly to their support. Shrubs and non-clinging climbers will need further tying-in as they grow; the shoots of twining climbers may also need gentle guidance.

ANNUALS AND BEDDING

In order to ensure that these plants look their best, it is essential that they are well-planted. A moist soil and regular deadheading are keys to success. Bedding plants need regular feeding throughout the growing season, especially in containers, but many hardy annuals flower best in soil that is not enriched. Many annuals are available as seedling "plugs", which should be grown on under cover. When planted, the well-developed root system suffers little damage, ensuring rapid growth.

PLANTING BULBS

In general, bulbs can be planted at between one and three times their own depth. Plant small bulbs quite shallowly; those of bigger plants like large tulips and lilies more deeply.

UNDERSTANDING PLANT NAMES

Throughout the *RHS Good Plant Guide*, all plants are listed by their current botanical names. The basic unit of plant classification is the species, with a two-part name correctly given in italic text: the first part is the genus, and the second part is the species name or "epithet".

GENUS

A group of one or more plants that share a wide range of characteristics, such as *Chrysanthemum* or *Rosa,* is known as a genus. A genus name is quite like a family surname, because it is shared by a group of individuals that are all closely related. Hybrid genera (crosses between plants from two genera, such as x *Halimiocistus*), are denoted by a multiplication sign before the genus name.

SPECIES

A group of plants capable of breeding together to produce similar offspring are known as a species. In a two-part botanical name, the species epithet distinguishes a species from other plants in the same genus; rather like a Christian or given name. A species epithet usually refers to a particular feature of that species, like *alpinum* or *tricolor*, or it may refer to the person who first discovered the plant.

SUBSPECIES, VARIETY AND FORMA

Naturally occurring variants of a species – subspecies, variety, or forma – are given an additional name in italics, prefixed by the abbreviations "subsp.", "var." or "f.". These are minor subdivisions of a

Genus/Species *Malus floribunda* is of the same genus as apples (*Malus domestica*); its species name means "mass of flowers".

Variety *Dictamnus albus* var. *purpureus* has purplish flowers instead of the pink-white of the species.

Hybrid The multiplication sign after the genus name in *Osmanthus* x *burkwoodii* denotes a species hybrid.

species, differing slightly in their botanical structure or appearance.

HYBRIDS

If different species within the same genus are cultivated together, they may cross-breed, giving rise to hybrids sharing attributes of both parents. This process is exploited by gardeners who wish to combine the valued characteristics of two distinct plants. The new hybrid is then increased by propagation. An example is *Camellia* x *williamsii*, which has the parents *C. japonica* and *C. saluensis*.

CULTIVARS

Variations of a species that are selected or artificially raised are given a vernacular name. This appears in single quotation marks after the species name. Some cultivars are also registered with

trademark names, often used commercially instead of the valid cultivar name. If the parentage is obscure or complex, the cultivar name may directly follow the generic name – *Dahlia* 'Conway'. In a few cases, particularly the roses, the plant is known by a popular selling name, which is not the correct cultivar name; here, the popular name comes before the cultivar name, as in *Rosa* BONICA 'Meidomonac'.

GROUPS AND SERIES

Several very similar cultivars may, for convenience, be classified in named Groups or Series that denote their similarities. Sometimes, they can be a deliberate mixture of cultivars, of the same character but with flowers in different colours.

Cultivar Species parentage may not be given if it is complex or unknown, as with *Osteospermum* 'Buttermilk'.

Cultivar of species *Ophiopogon planiscapus* 'Nigrescens' is an unusual form cultivated for its black leaves.

Seed series *Antirrhinum majus* Sonnet Series is a mixture of bright-coloured cultivars for summer bedding.

THE
A–Z OF
PLANTS

Attractive and reliable plants for every garden, and for every part of the garden, can be found in this section, the majority carrying the recommendation of the RHS Award of Garden Merit. However much their ornamental features and season of interest appeal, always check their hardiness, eventual size, and site and soil requirements before you buy.

Abelia 'Edward Goucher'

This semi-evergreen shrub with arching branches bears glossy, dark green leaves which are bronze when young. Trumpet-shaped, lilac-pink flowers appear from summer to autumn. Like most abelias, it is suitable for a sunny border.

CULTIVATION: *Grow in well-drained, fertile soil, in sun with shelter from cold winds. Remove dead or damaged growth in spring, cutting some of the older stems back to the ground after flowering to promote new growth.*

☼ ◊ ♀ ❀ ❀ ❀ ‡1.5m (5ft) ↔2m (6ft)

Abelia floribunda

An evergreen shrub with arching shoots, from which tubular, bright pink-red flowers hang in profuse clusters in early summer. The leaves are oval and glossy dark green. Ideal for a sunny border, but may not survive cold winters except in a very sheltered position.

CULTIVATION: *Grow in well-drained, fertile soil, in full sun with shelter from cold, drying winds. Prune older growth back after flowering, removing any dead or damaged growth in spring.*

☼ ◊ ♀ ❀ ❀ ‡3m (10ft) ↔4m (12ft)

Abelia x *grandiflora*

A rounded, semi-evergreen shrub bearing arching branches. Cultivated for its attractive, glossy dark green leaves and profusion of fragrant, pink-tinged white flowers which are borne from midsummer to autumn. Suitable for a sunny border.

CULTIVATION: *Grow in well-drained, fertile soil, in full sun with shelter from cold, drying winds. Prune back older growth after flowering, and remove damaged growth in spring.*

☼ ◊ ❋ ❋ ❋   ‡3m (10ft) ↔4m (12ft)

Abutilon megapotamicum

The trailing abutilon is a semi-evergreen shrub bearing pendulous, bell-shaped, red and yellow flowers from summer to autumn. The oval leaves are bright green and heart-shaped at the base. In cold climates, train against a warm wall or grow in a conservatory.

CULTIVATION: *Best in well-drained, moderately fertile soil, in full sun. Remove any wayward shoots during late winter or early spring.*

☼ ◊ ♀ ❋ ❋   ‡↔2m (6ft)

Abutilon vitifolium 'Veronica Tennant'

A fast-growing, upright, deciduous shrub that may attain the stature of a small, bushy tree. Masses of large, bowl-shaped flowers hang from the stout, grey-felted shoots in early summer. The softly hairy green leaves are sharply toothed. May not survive in cold areas. 'Tennant's White', with pure white flowers, is also recommended.

CULTIVATION: *Grow in well-drained, moderately fertile soil, in full sun. Prune young plants after flowering to encourage a good shape; do not prune established plants.*

☀ ◊ ♀ ✿ ✿   ‡5m (15ft) ↔2.5m (8ft)

Acacia baileyana 'Purpurea'

The Cootamundra wattle is a shrubby small tree bearing feathery, purple-tinted foliage. Like most acacias, it is grown primarily for its small yellow pompon flowers that cover the plant from winter to spring. 'Purpurea' has tinted foliage, the darker colour particularly noticeable in new growth.

CULTIVATION: *Grow in well-drained, neutral to acid soil in full sun. Trim back after flowering, if necessary. Under glass, grow in a soil-based compost.*

☀ ◊ ♀ ✿   ‡to 8m (25ft) ↔to 6m (20ft)

Acaena microphylla

This summer-flowering, mat-forming perennial bears heads of small, dull red flowers with spiny bracts which develop into decorative burrs. The finely divided, mid-green leaves are bronze-tinged when young and evergreen through all but the hardest winters. Good for a rock garden, trough, or raised bed.

CULTIVATION: *Grow in well-drained soil, in full sun or partial shade. Pull out rooted stems around the main plant to restrict spread.*

☼ ◐ ◊ ♀ ❀ ❀ ❀
‡5cm (2in) ↔ 60cm (24in)

Acanthus spinosus

Bear's breeches is a striking, architectural perennial bearing long, arching, dark green leaves which have deeply cut and spiny edges. From late spring to midsummer, pure white, two-lipped flowers with purple bracts are borne on tall, sturdy stems; they are good for cutting and drying. Grow in a spacious border.

CULTIVATION: *Best in deep, well-drained, fertile soil, in full sun or partial shade. Provide plenty of space to display its architectural merits.*

☼ ◐ ◊ ❀ ❀ ❀
‡1.5m (5ft) ↔ 60cm (24in)

Acer griseum

The paper-bark maple is a slow-growing, spreading, deciduous tree valued for its peeling orange-brown bark. The dark green leaves, divided into three leaflets, turn orange to red and scarlet in autumn. Tiny yellow flowers are carried in hanging clusters during early or mid-spring, followed by brown, winged fruits.

CULTIVATION: *Grow in moist but well-drained, fertile soil, in sun or partial shade. In winter only, remove shoots that obscure the bark on the trunk and lower parts of the main branches.*

☼ ☼ ◊ ♀ ❀ ❀ ❀   ‡↔10m (30ft)

Acer grosseri var. *hersii*

This variety of the snake-bark maple with boldly green- and white-streaked bark is a spreading to upright, deciduous tree. The three-lobed, triangular, bright green leaves turn orange or yellow in autumn. Hanging clusters of tiny, pale yellow flowers appear in spring, followed by pink-brown, winged fruits.

CULTIVATION: *Grow in moist but well-drained, fertile soil, in full sun or partial shade. Shelter from cold winds. Remove shoots that obscure the bark on the trunk and main branches in winter.*

☼ ☼ ◊ ❀ ❀ ❀   ‡↔15m (50ft)

Acer japonicum 'Aconitifolium'

A deciduous, bushy tree or large shrub bearing deeply lobed, mid-green leaves which turn brilliant dark red in autumn. It is very free-flowering, producing upright clusters of conspicuous, reddish-purple flowers in mid-spring, followed by brown, winged fruits. 'Vitifolium' is similar, not so free-flowering but with fine autumn colour.

CULTIVATION: *Grow in moist but well-drained, fertile soil, in partial shade. In areas with severe winters, mulch around the base in autumn. Remove badly placed shoots in winter only.*

☀ ◊ ♀ ❋❋❋   ‡5m (15ft) ↔6m (20ft)

Acer negundo 'Flamingo'

Round-headed, deciduous tree with pink-margined, oval leaflets which turn white in summer. With regular pruning, it can be grown as a shrub; this also produces larger leaves with an intensified colour. Flowers are tiny and inconspicuous. Plain-leaved var. *violaceum* has glaucous shoots and long tassels of violet flowers.

CULTIVATION: *Grow in any moist but well-drained, fertile soil, in full sun or partial shade. For larger leaves and a shrubby habit, cut back to a framework every 1 or 2 years in winter. Remove any branches with all-green leaves.*

☀ ☀ ◊ ❋❋❋
‡15m (50ft) ↔10m (30ft)

Japanese Maples (*Acer palmatum*)

Cultivars of *Acer palmatum*, the Japanese maple, are mostly small, round-headed, deciduous shrubs, although some, such as 'Sango-kaku', will grow into small trees. They are valued for their delicate and colourful foliage, which often gives a beautiful display in autumn. The leaves of 'Butterfly', for example, are variegated grey-green, white, and pink, and those of 'Osakazuki' turn a brilliant red before they fall. In mid-spring, hanging clusters of small, reddish-purple flowers are produced, followed by winged fruits later in the season. Japanese maples are excellent for gardens of any size.

CULTIVATION: *Grow in moist but well-drained, fertile soil, in sun or partial shade. Restrict pruning and training to young plants only; remove badly placed or crossing shoots in winter to develop a well-spaced network of branches. Keep pruning to a minimum on established plants.*

☼ ☀ ◊ ❀ ❀ ❀

↕↔5m (15ft)

↕3m (10ft) ↔1.5m (5ft)

↕2m (6ft) ↔3m (10ft)

↕2m (6ft) ↔3m (10ft)

1 *Acer palmatum* 'Bloodgood' ♀ **2** *A. palmatum* 'Butterfly' **3** *A. palmatum* 'Chitose-yama' ♀ **4** *A. palmatum* 'Garnet' ♀

‡5m (15ft) ↔ 4m (12ft)

MORE CHOICES

'Burgundy Lace' Very
deeply cut red-purple
leaves, 4m (12ft) tall
and widely spreading.

var. *coreanum* 'Korean
Gem' Green leaves
turning crimson/
scarlet in autumn.

var. *dissectum*

var. *dissectum*
'Crimson Queen'

var. *dissectum*
'Inabe-Shidare'
Deeply divided
purple-red leaves.

'Seiryû'

‡↔ 6m (30ft)

‡↔ 1.5m (5ft)

‡6m (20ft) ↔ 5m (15ft)

5 *A. palmatum* 'Linearilobum' **6** *A. palmatum* 'Osakazuki' ♀ **7** *A. palmatum* 'Red
Pygmy' ♀ **8** *A. palmatum* 'Sango-kaku' (*syn.* 'Senkaki') ♀

Acer pensylvanicum 'Erythrocladum'

This striped maple is an upright, deciduous tree. Its leaves are bright green, turning clear yellow in autumn. In winter, brilliant pink or red young shoots make a fiery display; they become orange-red with white stripes as they mature. Hanging clusters of small, greenish-yellow flowers in spring are followed by winged fruits. Best grown as a specimen tree.

CULTIVATION: *Grow in fertile, moist but well-drained soil in sun or partial shade. Remove crossing or damaged shoots from late autumn to midwinter only.*

☼ ☀ ◊ ❀❀❀
↕12m (40ft) ↔10m (30ft)

Acer platanoides 'Crimson King'

This cultivar of *Acer platanoides* is a large, spreading, deciduous tree. It is grown for its dark red-purple leaves, which deepen to dark purple as they mature. The foliage is preceded in spring by clusters of small, red-tinged yellow flowers. These develop into winged fruits. A colourful specimen tree.

CULTIVATION: *Grow in fertile, moist but well-drained soil, in sun or partial shade. Prune in late autumn to midwinter only, to remove any crossing, crowded, or unhealthy growth.*

☼ ☀ ◊ ♀ ❀❀❀
↕25m (80ft) ↔15m (50ft)

Acer platanoides 'Drummondii'

This is a much smaller, more spreading deciduous tree than the species, and is as broad as it is tall. Its leaves have a wide, pale green to cream margin, and colour well in autumn. In spring, clusters of yellow flowers are seen; from these later develop winged fruits. A bright specimen tree for a medium-sized garden.

CULTIVATION: *Grow in fertile, moist but well-drained soil, in sun or partial shade. Prune in late autumn to midwinter only, to remove any crossing, crowded, or unhealthy growth.*

☼ ☀ ◊ ❀ ❀ ❀   ‡↔10–12m (30–40ft)

Acer pseudoplatanus 'Brilliantissimum'

This small, slow-growing cultivar of sycamore is a spreading, deciduous tree with a dense head. It bears colourful, five-lobed leaves, which turn from salmon-pink to yellow then dark green as they mature. Hanging clusters of tiny, yellow-green flowers appear in spring, followed by winged fruit. An attractive maple for smaller gardens.

CULTIVATION: *Grow in any soil, in sun or partial shade. Tolerates exposed sites. Prune young plants in winter to develop well-spaced branches and a clear trunk.*

☼ ☀ ◊ ♀ ❀ ❀ ❀   ‡6m (20ft) ↔8m (25ft)

Acer rubrum 'October Glory'

This cultivar of the red, or swamp, maple is a round-headed to open-crowned, deciduous tree with glossy dark green foliage, turning bright red in early autumn. The upright clusters of tiny red flowers in spring are followed by winged fruits. A fine specimen tree for a large garden.

CULTIVATION: *Grow in any fertile, moist but well-drained soil, in full sun or partial shade, although best autumn colour is seen in acid soil. Between late autumn and midwinter, remove any crossing or congested branches.*

☼ ☀ ◊ ♀ ❀ ❀ ❀
↕20m (70ft) ↔10m (30ft)

Acer tataricum subsp. *ginnala*

The Amur maple is a rounded, bushy, deciduous tree with slender, arching branches and glossy bright green leaves, very deeply lobed; in autumn, these become a deep, rich bronze-red. In spring, it bears upright clusters of cream flowers, from which develop red winged fruits. Best as a specimen tree.

CULTIVATION: *Grow in any fertile, moist but well-drained soil, in full sun or in partial shade. Restrict pruning to between late autumn and midwinter to remove any crossing or congested branches, if necessary.*

☼ ◊ ❀ ❀ ❀   ↕10m (30ft) ↔8m (25ft)

Achillea ageratifolia

This small yarrow is a fast-growing, creeping perennial, forming mats of silvery, hairy leaf rosettes above which, in summer, small white flowerheads stand on upright stems. Grow at the front of a sunny border or in a rock or scree garden, or in planting spaces in a paved area.

CULTIVATION: *Grow in any moderately fertile, free-draining soil, in sun. Divide plants that have spread too widely or become straggly in spring or autumn. Deadhead to encourage further flowers.*

☼ ◊ ♀ ❀ ❀ ❀
‡5–8cm (2–3in) ↔ to 45cm (18in)

Achillea 'Coronation Gold'

This cultivar of yarrow is a clump-forming perennial that bears large, flat heads of golden-yellow flowers from midsummer to early autumn. The luxuriant, evergreen, fern-like leaves are silver-grey, complementing the flower colour (contact may aggravate skin allergies). Excellent for a mixed or herbaceous border, and for cutting and drying.

CULTIVATION: *Grow in moist but well-drained soil in an open site in full sun. Divide large or congested clumps to maintain vigour.*

☼ ◊ ♦ ♀ ❀ ❀ ❀
‡75–90cm (30–36in) ↔ 45cm (18in)

Achillea filipendulina 'Gold Plate'

A strong-growing, clump-forming and upright, evergreen perennial with grey-green leaves. Flat-headed clusters of bright golden-yellow flowers are borne on strong stems from early summer to early autumn; they make good cut flowers. Grow in a mixed or herbaceous border. Contact with the foliage may aggravate skin allergies.

CULTIVATION: *Grow in moist but well-drained soil in an open, sunny site. To maintain good performance, divide clumps when large and congested.*

☼ ◊ ◊ ♥ ✿ ✿ ✿
‡1.2m (4ft) ↔ 45cm (18in)

Achillea 'Heidi'

This charming rose-pink yarrow is shorter than most and makes a useful perennial where there may not be enough space for larger varieties. Flat, dish-like flowerheads appear through midsummer and are attractive to many flying insects, including butterflies. The plant's smaller stature means that the flower stems need less support than other yarrows.

CULTIVATION: *Grow in any moist but well-drained soil in sun. Remove old flowerheads to promote further flowering. Divide large clumps in spring or autumn.*

☼ ◊ ◊ ♥ ✿ ✿ ✿
‡45cm (18in) ↔ 45cm (18in) or more

Achillea 'Moonshine'

A clump-forming, evergreen perennial with narrow, feathery, grey-green leaves. Light yellow flowerheads with slightly darker centres appear from early summer to early autumn in flattish clusters; they dry well for arrangements. Excellent for mixed borders and for informal, wild, or cottage-style plantings. May not survive over winter in cold climates without protection.

CULTIVATION: *Grow in well-drained soil in an open, sunny site. Divide every 2 or 3 years in spring to maintain vigour. Protect from cold with a winter mulch.*

☼ ◊ ♥ ❀❀   ↕↔60cm (24in)

Achillea 'Paprika' (Galaxy Series)

A robust yarrow yielding rich red flowerheads, dotted with white centres, which look striking in a mixed border. Like many yarrows, this is a tough plant that can tolerate periods of drought and hot weather. Achilleas make good choices for a wildlife garden because insects love them. They also make very good cut flowers.

CULTIVATION: *Grow in moist but well-drained soil in sun, ideally through a support. Remove old flowerheads to promote reflowering. Divide large clumps in spring or autumn.*

☼ ◊ ◗ ❀❀❀
↕60cm (24in) ↔60cm (24in) or more

Aconitum 'Bressingham Spire'

A compact perennial producing very upright spikes of hooded, deep violet flowers from midsummer to early autumn. The leaves are deeply divided and glossy dark green. Ideal for woodland or borders in partial or dappled shade. For lavender flowers on a slightly taller plant, look for *A. carmichaelii* 'Kelmscott'. All parts of these plants are poisonous.

CULTIVATION: *Best in cool, moist, fertile soil, in partial shade, but will tolerate most soils and full sun. The tallest stems may need staking.*

☼ ☀ ◊ ♀ ❀❀❀
‡90–100cm (36–39in) ↔30cm (12in)

Actaea simplex 'Brunette'

A useful late-season perennial with dark purple-brown foliage that is ideal for a shady corner of the garden. In late summer and autumn, flowerheads resembling white bottlebrushes rise above the divided foliage. When the plants are arranged in clumps, this will create an effective display. Place among ferns and other shade-loving plants.

CULTIVATION: *Grow in moist, fertile soil in partial shade. Keep the soil enriched with organic matter. Divide large clumps in spring.*

☼ ◊ ♀ ❀❀❀
‡1.2m (4ft) ↔60cm (24in) or more

Actinidia kolomikta

A vigorous, deciduous climber with large, deep green leaves that are purple-tinged when young and develop vivid splashes of white and pink as they mature. Small, fragrant white flowers appear in early summer. Female plants produce small, egg-shaped, yellow-green fruits, but only if a male plant is grown nearby. Train against a wall or up into a tree.

CULTIVATION: *Best in well-drained, fertile soil. For best fruiting, grow in full sun with protection from strong winds. Tie in new shoots as they develop, and remove badly placed shoots in summer.*

☼ ◊ ♀ ❁ ❁ ❁   ↕5m (15ft)

Adiantum pedatum

A deciduous relative of the maidenhair fern bearing long, mid-green fronds up to 35cm (14in) tall. These have glossy dark brown or black stalks which emerge from creeping rhizomes. There are no flowers. *A. aleuticum* is very similar and also recommended; var. *subpumilum* is a dwarf form of this fern, only 15cm (6in) tall. Grow in a shady border or light woodland.

CULTIVATION: *Best in cool, moist soil, in deep or partial shade. Remove old or damaged fronds in early spring. Divide and replant rhizomes every few years.*

☼ ☀ ◊ ❁ ❁ ❁   ↕↔30–40cm (12–16in)

Adiantum venustum

The Himalayan maidenhair fern has black-stalked, triangular, mid-green fronds, beautifully divided into many small leaflets. The new foliage is bright bronze-pink when it emerges in late winter to early spring from creeping rhizomes. It is evergreen above –10°C (14°F). There are no flowers. Decorative ground cover for a woodland garden or shady border.

CULTIVATION: *Grow in moderately fertile, moist but well-drained soil in partial shade. Remove old or damaged fronds in spring, and divide the rhizomes every few years in early spring.*

☀ ◊ ◊ ♀ ✿✿✿
‡15cm (6in) ↔ indefinite

Aeonium haworthii

A succulent subshrub with slender branches, each crowned by a neat rosette of bluish-green, fleshy leaves with red margins. Clusters of pale yellow to pinkish-white flowers are seen in spring. It is a popular pot plant for conservatories and porches.

CULTIVATION: *Under glass, grow in standard cactus compost in filtered light, and allow the compost to dry out between waterings. Outdoors, grow in moderately fertile, well-drained soil in partial shade. Minimum temperature 10°C (50°F).*

☀ ◊ ♀ ❀
‡↔60cm (24in)

Aeonium 'Zwartkop'

An upright, succulent subshrub
with few branches, each tipped by a
rosette of black-purple leaves. Large,
pyramid-shaped clusters of bright
yellow flowers appear in late spring.
Makes an unusually coloured pot
plant. *A. arboreum* 'Atropurpureum'
is similarly striking.

CULTIVATION: *Grow in standard cactus
compost in filtered light under glass,
allowing the compost to dry out between
waterings. Outdoors, grow in reasonably
fertile, well-drained soil in partial shade.
Minimum temperature 10˚C (50˚F).*

☼ ◊ ♀☂   ↕↔to 2m (6ft)

Aesculus x *carnea* 'Briotii'

This cultivar of red horse chestnut
is a spreading tree, admired in early
summer for its large, upright cones
of dark rose-red flowers. The leaves,
dark green, are divided into 5–7
leaflets. The flowers are followed
by spiny fruits. For large gardens;
for a smaller tree, look for the
sunrise horse chestnut, *A.* x *neglecta*
'Erythroblastos', to 10m (30ft), or the
red buckeye, *A. pavia*, to 5m (15ft).

CULTIVATION: *Grow in deep, fertile, moist
but well-drained soil in full sun or partial
shade. Remove dead, diseased, or crossing
branches during winter.*

☼ ☼ ◊ ◊ ❀❀❀
↕20m (70ft) ↔15m (50ft)

Aesculus parviflora

A large, thicket-forming, deciduous shrub, closely related to the horse chestnut, that bears large-lobed, dark green leaves. The foliage is bronze when young, turning yellow in autumn. Upright white flowerheads, up to 30cm (12in) tall, appear in midsummer, followed by smooth-skinned fruits.

CULTIVATION: *Grow in moist but well-drained, fertile soil, in sun or partial shade; it will not grow in wet ground. If necessary, restrict spread by pruning stems to the ground after leaf fall.*

☼ ☀ ◊ ◊ ♀ ❀ ❀ ❀
↕3m (10ft) ↔5m (15ft)

Aethionema 'Warley Rose'

A short-lived, evergreen or semi-evergreen, compact shrub bearing clusters of bright pink, cross-shaped flowers in late spring and early summer. The small, narrow leaves are blue-grey. For flowers of a much paler pink, look for *A. grandiflorum*, very similar if a little taller. Aethionemas are ideal for a rock garden or on a wall.

CULTIVATION: *Best in well-drained, fertile, alkaline soil, but tolerates poor, acid soils. Choose a site in full sun.*

☼ ◊ ♀ ❀ ❀ ❀   ↕↔15–20cm (6–8in)

Agapanthus campanulatus subsp. *patens*

A vigorous, clump-forming perennial bearing round heads of bell-shaped, light blue flowers on upright stems during late summer and early autumn. The strap-shaped, greyish-green leaves are deciduous. Useful in borders or large containers as a late-flowering perennial. This and 'Loch Hope', with deep blue flowers, are the hardiest of the AGM-recommended agapanthus.

CULTIVATION: *Grow in moist but well-drained, fertile soil or compost, in full sun. Water freely when in growth, and sparingly in winter.*

☼ ◐ ♀ ❀❀
‡to 45cm (18in) ↔30cm (12in)

Agapanthus 'Loch Hope'

'Loch Hope' is a reliably sturdy cultivar, and with its large heads of deep blue flowers it makes an excellent late-summer perennial. Established clumps are impressive in flower, but can take up a lot of space. The lush green, strap-shaped foliage dies back in winter. Agapanthus grow well in pots in soil-based compost.

CULTIVATION: *Grow in moist but well-drained, fertile soil in full sun. Divide large clumps in spring. Protect with a mulch in cold winters.*

☼ ◐ ◐ ♀ ❀❀
‡to1.5m (5ft) ↔60cm (24in) or more

Agave victoriae-reginae

A frost-tender, succulent perennial bearing basal rosettes of triangular, dark green leaves with white marks. The central leaves curve inward, each tipped with a brown spine. Upright spikes of creamy-white flowers appear in summer. A good specimen plant: in frost-prone areas, grow in containers for summer display, taking it under cover for winter shelter.

CULTIVATION: *Best in sharply drained, moderately fertile, slightly acid soil, or standard cactus compost. Site in full sun. Minimum temperature 2˚C (36˚F).*

☼ ◊ ♀ 🏠   ↕↔to 50cm (20in)

Ageratina altissima 'Chocolate'

Formerly classified under *Eupatorium*, white snakeroot is a clump-forming perennial that is good in shade, even dry shade, and bears clusters of small white flowers in late summer. The toothed leaves are an attractive deep brown colour through most of the spring and summer, gradually greening with age, before the whole plant dies back over winter.

CULTIVATION: *Best in reasonably moist soil in partial shade. Divide plants in spring, as necessary.*

☼ ◊ ❄❄❄
↕1m (3ft) ↔60cm (24in) or more

Ajuga reptans 'Atropurpurea'

An excellent evergreen perennial for ground cover, spreading freely over the soil surface by means of rooting stems. Dark blue flowers are borne in whorls along the upright stems during late spring and early summer. The glossy leaves are deep bronze-purple. *A. reptans* 'Catlin's Giant' has similarly coloured leaves. Invaluable for border edging under shrubs and robust perennials.

CULTIVATION: *Best in moist but well-drained, fertile soil, but tolerates most soils. Site in sun or partial shade.*

☼ ☀ ◊ ◊ ❄❄❄
‡15cm (6in) ↔1m (3ft)

Alchemilla mollis

Lady's mantle is a drought-tolerant, clump-forming, tallish ground cover perennial that produces sprays of tiny, bright greenish-yellow flowers from early summer to early autumn; these are ideal for cutting, and dry well for winter arrangements. The pale green leaves are rounded with crinkled edges. It looks well in a wildflower garden; for a rock garden, *A. erythropoda* is similar but smaller, with blue-tinged leaves.

CULTIVATION: *Grow in any moist but well-drained, humus-rich soil, in an open, sunny site. Deadhead soon after flowering as it self-seeds very freely.*

☼ ◊ ◊ ♀ ❄❄❄
‡60cm (24in) ↔75cm (30in)

Tall Ornamental Onions (*Allium*)

Tall onions grown for garden display are bulbous perennials from the *Allium* genus; their attractive flowerheads look excellent in a mixed border, especially grouped together. The tiny summer flowers are usually massed into dense, rounded, or hemispherical heads – like those of *A. giganteum* – or they may hang loosely, like the yellow flowers of *A. flavum*. When crushed, the strap-shaped leaves release a pungent aroma; they are often withered by flowering time. The seedheads tend to dry out intact, standing well into autumn and continuing to look attractive. Some alliums self-seed and will naturalize.

CULTIVATION: *Grow in fertile, well-drained soil in full sun to simulate their dry native habitats. Plant bulbs 5–10cm (2–4in) deep in autumn; divide and replant older clumps at the same time or in spring. In climates with cold winters, provide a thick winter mulch for* A. cristophii *and* A. caeruleum.

☀ ◊ ❀❀❀

‡1m (3ft) ↔15cm (6in)

MORE CHOICES

A carinatum subsp. *pulchellum* Purple flowers, 30–45cm (12–18in) tall.

A. cernuum 'Hidcote' Nodding pink flowers.

A. hollandicum Purplish-pink flowers, 1m (3ft) tall, very similar to 'Purple Sensation'.

‡60cm (24in) ↔2.5cm (1in)

1 *Allium* 'Beau Regard' ♀ **2** *A. caeruleum* ♀

‡to 35cm (14in) ↔ 5cm (2in)

‡30–60cm (12–24in) ↔ 18cm (7in)

‡1.5m–2m (5–6ft) ↔ 15cm (6in)

‡1.2m (4ft) ↔ 15cm (6in)

‡80cm (32in) ↔ 20cm (8in)

‡1m (3ft) ↔ 7cm (3in)

‡60cm (24in) ↔ 10cm (4in)

3 *A. cristophii* ♀ **4** *A. flavum* ♀ **5** *A. giganteum* ♀ **6** *A.* 'Gladiator' ♀ **7** *A.* 'Globemaster' ♀
8 *A. hollandicum* 'Purple Sensation' ♀ **9** *A. sphaerocephalon*

Small Ornamental Onions (*Allium*)

Shorter alliums are summer-flowering, bulbous perennials for the front of a border or rock garden. They form clumps as they establish; some, such as *A. moly*, will self-seed. The flowers are borne in clustered heads which may be large or small; those of *A. karataviense* can be 8cm (3in) across despite its small stature. Flower colours range from bright gold to purple, blue, and pale pink. The dry seedheads are attractive too, lasting well into winter. The strap-shaped leaves are often withered by flowering time. Those of ornamental chives, such as *A. schoenopraesum* 'Pink Perfection' and the very similar 'Black Isle Blush', are edible.

CULTIVATION: *Best in well-drained, humus-rich soil in full sun. Plant bulbs 5–10cm (2–4in) deep in autumn; divide and replant old or crowded clumps in autumn or spring. Provide a thick, dry winter mulch for* A. karataviense *where winters are cold.*

☼ ◊ ✿✿✿

‡10–25cm (4–10in) ↔10cm (4in)

‡15–25cm (6–10in) ↔5cm (2in)

‡5–20cm (2–8in) ↔3cm (1¼in)

‡30–60cm (12–24in) ↔5cm (2in)

1 *Allium karataviense* ♀ **2** *A. moly* **3** *A. oreophilum* **4** *A. schoenoprasum* 'Pink Perfection'

Alnus glutinosa 'Imperialis'

This attractive cultivar of the common alder is a broadly conical tree with deeply dissected, lobed, mid-green leaves. Groups of yellow-brown catkins are seen in late winter, followed by small oval cones in summer. This is a beautiful foliage tree, particularly good close to water as it tolerates poor, wet soil.

CULTIVATION: *Thrives in any moderately fertile, moist but not waterlogged soil, in full sun. Prune after leaf fall, if necessary, to remove any damaged or crossing branches.*

☼ ◐ ♀ ✿ ✿ ✿   ‡25m (80ft) ↔5m (15ft)

Aloe vera

A fast-growing tender succulent that is justifiably popular as a house plant. It forms clumps of thick, lance-shaped, bright green fleshy leaves. The clear gel inside the leaves can be used to soothe mild burns and skin irritations. Grow in a pot on the kitchen windowsill in cooler months. In summer, it will thrive in a well-lit spot outside, if kept well watered.

CULTIVATION: *In containers, grow in a well-drained, soil-based potting compost. Repot each year with fresh compost and divide clumps as necessary.*

☼ ◐ ♀ ⌂   ‡60cm (24in) ↔indefinite

Alonsoa warscewiczii

This species of mask flower is a compact, bushy perennial, grown for its bright scarlet, sometimes white, flowers. These are on display from summer to autumn amid the dark green leaves. Useful as summer bedding or in a mixed border, it also provides good cut flowers.

CULTIVATION: *Outdoors, grow in any fertile, well-drained soil in full sun, or in loam-based potting compost if grown in a container. Water moderately.*

☼ ◊ ❀❀
‡45–60cm (18–24in) ↔30cm (12in)

Alstroemeria 'Apollo'

This is a good border perennial with yellow-centred, white, lily-like flowers that appear from midsummer. The flowers are patterned with deep red flecks, and carried over a period of many weeks above the strap-shaped green leaves. They are ideal for cutting, but wear gloves when gathering as the sap can irritate the skin. Dwarf varieties are good for containers.

CULTIVATION: *Grow in light, well-drained soil in full sun. In cold areas, plant deeply in spring and mulch in winter to protect the crown from frost.*

☼ ◊ ♀ ❀❀   ‡1m (3ft) ↔90cm (36in)

Amelanchier x *grandiflora* 'Ballerina'

A spreading, deciduous tree grown for its profusion of white spring flowers and colourful autumn foliage. When young, the glossy leaves are tinted bronze, becoming mid-green in summer, then red and purple in autumn. The sweet, juicy fruits are red at first, ripening to purplish-black in summer. They can be eaten if cooked, and are attractive to birds.

CULTIVATION: *Grow in moist but well-drained, fertile, neutral to acid soil, in full sun or partial shade. Allow shape to develop naturally; only minimal pruning is necessary, in winter.*

☼ ☀ ◊ ◑ ✤✤✤   ‡6m (20ft) ↔8m (25ft)

Amelanchier lamarckii

A many-stemmed, upright, deciduous shrub bearing leaves that are bronze when young, maturing to dark green in summer, then brilliant red and orange in autumn. Hanging clusters of white flowers are produced in spring. The ripe, purple-black fruits that follow are edible when cooked, and are attractive to birds. Also known as *A. canadensis*.

CULTIVATION: *Grow in moist but well-drained, humus-rich, neutral to acid soil, in sun or partial shade. Develops its shape naturally with only minimal pruning when dormant in winter.*

☼ ☀ ◊ ◑ ♀ ✤✤✤
‡10m (30ft) ↔12m (40ft)

Amsonia hubrichtii

The star-shaped, light blue, flowers of Arkansas blue star create a cool early summer accent, and its distinctive, feathery, light green foliage turns burnished gold shades in autumn. This clump-forming perennial attracts butterflies, and is ideal for wildflower and woodland gardens.

CULTIVATION: *Grow in any moist but well-drained soil in full sun or dappled shade. Will tolerate some drought.*

☼ ☀ ◊ ❀❀❀
‡to 1m (3ft) ↔ to 1.2m (4ft)

Anaphalis triplinervis 'Sommerschnee'

A clump-forming perennial carrying pale grey-green, white-woolly leaves. Tiny yellow flowerheads, surrounded by brilliant white bracts, appear during midsummer in dense clusters; they are excellent for cutting and drying. Provides good foliage contrast in borders that are too moist for the majority of other grey-leaved plants.

CULTIVATION: *Grow in any reasonably well-drained, moderately fertile soil that does not dry out in summer. Choose a position in full sun or partial shade.*

☼ ☀ ◊ ♀ ❀❀❀
‡50cm (20in) ↔ 45–60cm (18–24in)

Anchusa azurea 'Loddon Royalist'

An upright, clump-forming perennial that is much-valued in herbaceous borders for its spikes of intensely dark blue flowers. These are borne on branching stems in early summer, above the lance-shaped and hairy, mid-green leaves which are arranged at the base of the stems. For a rock garden, *A. cespitosa* looks similar in miniature, only 5–10cm (2–4in) tall.

CULTIVATION: *Grow in deep, moist but well-drained, fertile soil, in sun. Often short-lived, but easily propagated by root cuttings. If growth is vigorous, staking may be necessary.*

☼ ◊ ❀❀❀   ↕90cm (36in) ↔60m (24in)

Andromeda polifolia 'Compacta'

This plant is known as bog rosemary for its leathery, linear leaves that bear a striking resemblance to the herb rosemary. Unlike its namesake, however, this plant loves moist conditions and bears hanging clusters of pink bell-shaped flowers in spring and early summer. This compact form would suit a shaded part of a rock garden, in acid soil.

CULTIVATION: *Grow in moist, acid soil enriched with plenty of organic matter in full sun or partial shade. Trim plants in spring, as necessary.*

☼ ☼ ◊ ◊ ♀ ❀❀❀
↕to 30cm (12in) ↔to 20cm (8in)

Androsace carnea subsp. *laggeri*

An evergreen, cushion-forming perennial that bears small clusters of tiny, cup-shaped, deep pink flowers with yellow eyes, in late spring. The pointed, mid-green leaves are arranged in tight rosettes. Rock jasmines grow wild in alpine turf and rock crevices, making them ideal for rock gardens or troughs. *A. sempervivoides* has scented flowers, ideal for a raised bed.

CULTIVATION: *Grow in moist but sharply drained, gritty soil, in full sun. Provide a top-dressing of grit or gravel to keep the stems and leaves dry.*

☼ ◊ ♀ ❀❀❀
‡5cm (2in) ↔8–15cm (3–6in)

Anemone blanda 'White Splendour'

A spreading, spring-flowering perennial that soon forms clumps of stems growing from knobbly tubers. The solitary, upright, flattish white flowers, with pink-tinged undersides, are borne above oval, dark green leaves divided into delicately lobed leaflets. Excellent for naturalizing in sunny or shaded sites with good drainage. Mix it with 'Radar', with white-centred magenta flowers, or 'Ingramii', with deep blue flowers.

CULTIVATION: *Grow in well-drained soil that is rich in humus. Choose a position in full sun or partial shade.*

☼ ☀ ◊ ♀ ❀❀❀   ↕↔15cm (6in)

Anemone hupehensis 'Hadspen Abundance'

Upright, woody-based, late-flowering border perennial that spreads by shoots growing from the roots. Reddish-pink flowers, with petal margins that gradually fade to white, are borne on branched stems during mid- and late summer. The deeply divided, long-stalked, dark green leaves are oval and sharply toothed. *A. hupehensis* 'Prinz Heinrich' is similar, but spreads more vigorously.

CULTIVATION: *Grow in moist, fertile, humus-rich soil, in sun or partial shade. Provide a mulch in cold areas.*

☼ ☀ ◊ ◑ ♀ ✿✿✿
‡60–90cm (24–36in) ↔40cm (16in)

Anemone x *hybrida* 'Honorine Jobert'

This upright, woody-based perennial with branched, wiry stems is an invaluable long-flowering choice for late summer to mid-autumn, when single, cupped white flowers, with pink-tinged undersides and golden-yellow stamens, are borne above divided, mid-green leaves. (For pure white flowers without a hint of pink, look for 'Géante des Blanches'.) It can be invasive.

CULTIVATION: *Grow in moist but well-drained, moderately fertile, humus-rich soil, in sun or partial shade.*

☼ ☀ ◊ ◑ ♀ ✿✿✿
‡1.2–1.5m (4–5ft) ↔indefinite

Anemone nemorosa 'Robinsoniana'

A vigorous, carpeting perennial that produces masses of large, star-shaped, pale lavender-blue flowers on maroon stems from spring to early summer, above deeply-divided, mid-green leaves which die down in midsummer. Excellent for underplanting or a woodland garden; it naturalizes with ease. 'Allenii' has deeper blue flowers; choose 'Vestal' for white flowers.

CULTIVATION: *Grow in loose, moist but well-drained soil that is rich in humus, in light dappled shade.*

☀ ◊ ♀ ❀❀❀
‡8–15cm (3–6in) ↔30cm (12in) or more

Anemone ranunculoides

This spring-flowering, spreading perennial is excellent for naturalizing in damp woodland gardens. The large, solitary, buttercup-like yellow flowers are borne above the "ruffs" of short-stalked, rounded, deeply lobed, fresh green leaves.

CULTIVATION: *Grow in moist but well-drained, humus-rich soil, in semi-shade or dappled sunlight. Tolerates drier conditions when dormant in summer.*

☀ ◊ ◑ ♀ ❀❀❀
‡5–10cm (2–4in) ↔to 45cm (18in)

Antennaria microphylla

A mat-forming, semi-evergreen perennial carrying densely white-hairy, spoon-shaped, grey-green leaves. In late spring and early summer, heads of small, fluffy, rose-pink flowers are borne on short stems. Use in a rock garden, as low ground cover at the front of a border or in crevices in walls or paving. The flowerheads dry well for decoration.

CULTIVATION: *Best in well-drained soil that is no more than moderately fertile. Choose a position in full sun.*

☼ ◊ ❀❀❀
‡5cm (2in) ↔ to 45cm (18in)

Anthemis punctata subsp. *cupaniana*

A mat-forming, evergreen perennial that produces a flush of small but long-lasting, daisy-like flowerheads in early summer, and a few blooms later on. The white flowers with yellow centres are borne singly on short stems, amid dense, finely cut, silvery-grey foliage which turns dull grey-green in winter. Excellent for border edges.

CULTIVATION: *Grow in well-drained soil, in a sheltered, sunny position. Cut back after flowering to maintain vigour.*

☼ ◊ ♀ ❀❀ ‡30cm (12in) ↔45cm (18in)

Aquilegia canadensis

A lovely woodland or border plant with an airy habit, bearing nodding red and yellow, spurred flowers in early summer. The nectar-loaded spurs will lure larger insects like bumblebees and butterflies. The plant is likely to cross-pollinate with other aquilegias, so remove colour variants if necessary.

CULTIVATION: *Grow in any moist but well-drained soil in dappled shade. Propagate by seed.*

☼ ◊ ◊ ♀ ❀❀❀
‡to 90cm (36in) ↔ to 30cm (12in)

Aquilegia vulgaris 'Nivea'

An upright, vigorous, clump-forming perennial, sometimes sold as MUNSTEAD WHITE, bearing leafy clusters of nodding, short-spurred, pure white flowers in late spring and early summer. Each greyish-green leaf is deeply divided into lobed leaflets. Attractive in light woodland or in a herbaceous border; plant with soft blue *Aquilegia* 'Hensol Harebell' for a luminous mix in light shade.

CULTIVATION: *Best in moist but well-drained, fertile soil. Choose a position in full sun or partial shade.*

☼ ☼ ◊ ♀ ❀❀❀
‡90cm (36in) ↔ 45cm (18in)

Aquilegia vulgaris 'Nora Barlow'

This upright, vigorous perennial is much-valued for its leafy clusters of funnel-shaped, double-pompon flowers. These are pink and white with pale green petal tips, and appear from late spring to early summer. The greyish-green leaves are deeply divided into narrow lobes. Suits herbaceous borders and cottage garden-style plantings.

CULTIVATION: *Grow in moist but well-drained, fertile soil. Position in an open, sunny site.*

☼ ◊ ❀❀❀
‡90cm (36in) ↔45cm (18in)

Arabis procurrens 'Variegata'

A mat-forming, evergreen or semi-evergreen perennial bearing loose clusters of cross-shaped white flowers on tall, slender stems during late spring. The narrow, mid-green leaves, arranged into flattened rosettes, have creamy-white margins and are sometimes pink-tinged. Useful in a rock garden. May not survive in areas with cold winters.

CULTIVATION: *Grow in any well-drained soil, in full sun. Remove completely any stems with plain green leaves.*

☼ ◊ ♀ ❀❀
‡5-8cm (2-3in) ↔30-40cm (12-16in)

Aralia elata 'Variegata'

The variegated Japanese angelica tree is deciduous, with a beautiful, exotic appearance. The large leaves are divided into leaflets irregularly edged with creamy white. Flat clusters of small white flowers appear in late summer, followed by round black fruits. Suitable for a shady border or wooded streambank in a large garden. Less prone to suckering than the plain-leaved *Aralia elata*.

CULTIVATION: *Grow in fertile, humus-rich, moist soil, in sun or part shade. Remove any branches with all-green foliage in late winter. Needs a sheltered site; strong winds can damage leaves.*

☀ ◊ ♡ ❈ ❈ ❈   ↕↔5m (15ft)

Araucaria heterophylla

The Norfolk Island pine is a grand, cone-shaped conifer, valued for its geometrical shape and unusual branches of whorled foliage. The leaves are tough, light green, and scale-like. There are no flowers. An excellent, fast-growing, gale-tolerant tree for coastal sites; in cold climates, it can be grown as a conservatory plant.

CULTIVATION: *Grow in moderately fertile, moist but well-drained soil in an open site with shelter from cold drying winds. Tolerates partial shade when young.*

☀ ◊ ◊ ♡☺
↕25–45m (80–150ft) ↔6–8m (20–25ft)

Arbutus x *andrachnoides*

This strawberry tree is a broad,
sometimes shrubby tree, with
peeling, red-brown bark. The
mid-green leaves are finely toothed
and glossy. Clusters of small white
flowers are borne from autumn to
spring, only rarely followed by fruits.
Excellent for a large shrub border,
or as a specimen tree.

CULTIVATION: *Grow in fertile, well-drained
soil rich in organic matter, in a sheltered
but sunny site. Protect from cold winds,
even when mature, and keep pruning to
a minimum, in winter if necessary. It
tolerates alkaline soil.*

☼ ◊ ♀ ❀ ❀ ❀   ↕↔8m (25ft)

Arbutus unedo

The strawberry tree is a spreading,
evergreen tree with attractive, rough,
shredding, red-brown bark. Hanging
clusters of small, urn-shaped white
flowers, which are sometimes
pink-tinged, open during autumn as
the previous season's strawberry-like
red fruits ripen. The glossy deep
green leaves are shallowly toothed.
Excellent for a large shrub border,
with shelter from wind.

CULTIVATION: *Best in well-drained, fertile,
humus-rich, acid soil. Tolerates slightly
alkaline conditions. Choose a sheltered site
in sun. Prune low branches in spring, but
keep to a minimum.*

☼ ◊ ♀ ❀ ❀ ❀   ↕↔8m (25ft)

Arctostaphylos uva-ursi

Bearberry is a hardy, evergreen low-growing shrub with glossy small leaves. Clusters of small pink or white, nectar-rich flowers appear in summer, followed by red berries and bronze foliage. Use as ground cover in rock gardens or at the front of the border. 'Massachusetts' offers disease-resistant foliage.

CULTIVATION: *Does best in open areas with well-drained acid soil, and full sun. Tolerates drought and coastal conditions.*

☼ ☀ ◊ ❀❀❀
‡10cm (4in) ↔to 50cm (20in)

Arenaria montana

This sandwort is a low-growing, spreading, vigorous, evergreen perennial freely bearing shallowly cup-shaped white flowers in early summer. The small, narrowly lance-shaped, greyish-green leaves on wiry stems form loose mats. Easily grown in wall or paving crevices, or in a rock garden.

CULTIVATION: *Grow in sandy, moist but sharply drained, poor soil, in full sun. Must have adequate moisture.*

☼ ◊ ♀ ❀❀❀
‡2–5cm (³/₄–2in) ↔30cm (12in)

Argyranthemum 'Jamaica Primrose'

A bushy, evergreen perennial that bears daisy-like, primrose-yellow flowerheads with darker yellow centres throughout summer above fern-like, greyish-green leaves. In frost-prone areas, grow as summer bedding or in containers, bringing under cover for the winter. 'Cornish Gold' – shorter, with flowers of a deep yellow – is also recommended.

CULTIVATION: *Grow in well-drained, fairly fertile soil or compost, in a warm, sunny site. Mulch outdoor plants well. Pinch out shoot tips to encourage bushiness. Min. temp. 2°C (36°F).*

☼ ◊ ♀ ☙   ‡1.1m (3½ft) ↔1m (3ft)

Argyranthemum 'Vancouver'

This compact, summer-flowering, evergreen subshrub is valued for its double, daisy-like pink flowerheads with rose-pink centres and fern-like, grey-green leaves. Suits a mixed or herbaceous border; in frost-prone areas, grow as summer bedding or in containers that can be sheltered in frost-free conditions over winter.

CULTIVATION: *Grow in well-drained, fairly fertile soil, in sun. Apply a deep, dry mulch to outdoor plants. Pinch out growing tips to encourage bushiness. Minimum temperature 2°C (36°F).*

☼ ◊ ♀ ☙   ‡90cm (36in) ↔80cm (32in)

Armeria juniperifolia

This tiny, hummock-forming, evergreen subshrub bears small, purplish-pink to white flowers which are carried in short-stemmed, spherical clusters during late spring. The small, linear, grey-green leaves are hairy and spine-tipped, and are arranged in loose rosettes. Native to mountain pastures and rock crevices, it is ideal for a rock garden or trough. Also known as *A. caespitosa*.

CULTIVATION: *Grow in well-drained, poor to moderately fertile soil, in an open position in full sun.*

☼ ◊ ♀ ✿✿✿
↕5–8cm (2–3in) ↔ to 15cm (6in)

Armeria juniperifolia 'Bevan's Variety'

A compact, cushion-forming, evergreen subshrub that bears small, deep rose-pink flowers. These are carried in short-stemmed, rounded clusters during late spring, over the loose rosettes of small and narrow, pointed, grey-green leaves. Suits a rock garden or trough; for the front of the border, look for *Armeria* 'Bee's Ruby', a similar perennial plant growing to 30cm (12in) tall, with chive-like flowerheads.

CULTIVATION: *Grow in well-drained, poor to moderately fertile soil. Choose an open site in full sun.*

☼ ◊ ♀ ✿✿✿
↕to 5cm (2in) ↔ to 15cm (6in)

Artemisia ludoviciana 'Silver Queen'

An upright, bushy, clump-forming, semi-evergreen perennial bearing narrow, downy leaves, sometimes jaggedly toothed; silvery-white when young, they become greener with age. White-woolly plumes of brown-yellow flowers are borne from midsummer to autumn. Indispensable in a silver-themed border. 'Valerie Finnis' is also recommended; its leaves have more deeply cut edges.

CULTIVATION: *Grow in well-drained soil, in an open, sunny site. Cut back in spring for best foliage effect.*

☼ ◊ ❀❀❀
‡75cm (30in) ↔ 60cm (24in) or more

Artemisia 'Powis Castle'

A vigorous, shrubby, woody-based perennial forming a dense, billowing clump of finely cut, aromatic, silver-grey leaves. Sprays of insignificant, yellow-tinged silver flowerheads are borne in late summer. Excellent in a rock garden or border. May not survive in areas with cold winters.

CULTIVATION: *Grow in well-drained, fertile soil, in full sun. Will die back in heavy, poorly drained soils and may be short-lived. Cut to the base in autumn to maintain a compact habit.*

☼ ◊ ♀ ❀❀
‡60cm (24in) ↔ 90cm (36in)

Arum italicum subsp. *italicum* 'Marmoratum'

An unusual plant that is at its best from autumn to early spring. It bears cream-veined, deep green, glossy leaves and spears of poisonous berries. The fruits first appear in autumn, ripening from green to red as winter progresses. As the last berries fade, fresh foliage begins to emerge above the ground. Plant among early bulbs, such as snowdrops.

CULTIVATION: *Plant tubers in autumn or spring in moist but well-drained, humus-rich soil. Needs an open, sunny site to flower and fruit well.*

☼ ☀ ◌ ◗ ♀ ❀❀❀
↕↔30cm (12in)

Aruncus dioicus

Goatsbeard forms arresting clumps of rich green foliage from which loose and arching clusters of creamy or greenish white flowers emerge in the first half of summer. It is a graceful woodland plant that suits damp soils in shady areas, such as a pond edge. It will tolerate full sun as long as the soil remains moist. Plants may self seed generously.

CULTIVATION: *Grow in moist, fertile soil in sun or shade. Cut spent flower stems back hard in autumn.*

☼ ☀ ☀ ◌ ◗ ♀ ❀❀❀
↕2m (6ft) ↔1.2m (4ft)

Asplenium scolopendrium

The hart's tongue fern has irregular crowns of shuttlecock-like, tongue-shaped, leathery, bright green fronds, to 40cm (16in) long. Heart-shaped at the bases, they often have wavy margins, markedly so in the cultivar 'Crispum Bolton's Nobile'. On the undersides of mature fronds, rust-coloured spore cases are arranged in a herringbone pattern. There are no flowers. Good in alkaline soil.

CULTIVATION: *Grow in moist but well-drained, humus-rich, preferably alkaline soil with added grit, in partial shade.*

☼ ◊ ◊ ♀ ❀❀❀
‡45–70cm (18–28in) ↔ 60cm (24in)

Aster alpinus

This spreading, clump-forming perennial is grown for its mass of daisy-like, purplish-blue or pinkish-purple flowerheads with deep yellow centres. These are borne on upright stems in early and midsummer, above short-stalked, narrow, mid-green leaves. A low-growing aster, it is suitable for the front of a border or in a rock garden. Several outstanding cultivars are available.

CULTIVATION: *Grow in well-drained, moderately fertile soil, in sun. Mulch annually after cutting back in autumn.*

☼ ◊ ♀ ❀❀❀
‡to 25cm (10in) ↔ 45cm (18in)

Aster amellus 'King George'

A clump-forming, bushy perennial bearing loose clusters of large, daisy-like, violet-blue flowerheads with yellow centres which open from late summer to autumn. The rough, mid-green leaves are hairy and lance-shaped. An invaluable late-flowering border plant; other recommended cultivars include 'Framfieldii' (lavender-blue flowers), 'Jacqueline Genebrier' (bright red-purple) and 'Veilchenkönigen' (deep purple).

CULTIVATION: *Grow in open, well-drained, moderately fertile soil, in full sun. Thrives in alkaline conditions.*

☼ ◊ ♀ ✽✽✽   ↕↔45cm (18in)

Aster 'Andenken an Alma Pötschke'

This vigorous, upright, clump-forming perennial carries sprays of large, daisy-like, bright salmon-pink flowerheads with yellow centres from late summer to mid-autumn, on stiff stems above rough, stem-clasping, mid-green leaves. Good for cutting, or in late-flowering displays. *A. novae-angliae* 'Harrington's Pink' is very similar, with paler flowers.

CULTIVATION: *Grow in moist but well-drained, fertile, well-cultivated soil, in sun or semi-shade. Divide and replant every third year to maintain vigour and flower quality. May need staking.*

☼ ☀ ◊ ✽✽✽   ↕1.2m (4ft) ↔60cm (24in)

Aster x *frikartii* '**Mönch**'

This upright, bushy perennial provides a continuous show of long-lasting, daisy-like, lavender-blue flowerheads with orange centres during late summer and early autumn. The dark green leaves are rough-textured and oblong. A useful plant for adding cool tones to a late summer or autumn display, as is the very similar 'Wunder von Stäfa'.

CULTIVATION: *Best in well-drained, moderately fertile soil. Position in an open, sunny site. Mulch annually after cutting back in late autumn.*

☼ ◊ ♀ ❀❀❀
‡70cm (28in) ↔ 35–40cm (14–16in)

Aster lateriflorus var. *horizontalis*

A clump-forming, freely branching perennial bearing clusters of daisy-like, sometimes pink-tinged white flowerheads with darker pink centres, from midsummer to mid-autumn. The slender, hairy stems bear small, lance-shaped, mid-green leaves. An invaluable late-flowerer for a mixed border.

CULTIVATION: *Grow in moist but well-drained, moderately fertile soil, in partial shade. Keep moist in summer.*

☼ ◊ ♀ ❀❀❀
‡60cm (24in) ↔ 30cm (12in)

Aster 'Little Carlow'

A clump-forming, upright perennial that produces large clusters of daisy-like, violet-blue flowers with yellow centres, in early autumn. The dark green leaves are oval to heart-shaped and toothed. Valuable for autumn displays; the flowers also dry very well. Related to *A. cordifolius* 'Chieftain' (mauve flowers) and 'Sweet Lavender', also recommended.

CULTIVATION: *Best in moist, moderately fertile soil, in partial shade, but tolerates well-drained soil, in full sun. Mulch annually after cutting back in late autumn. May need staking.*

☼ ☼ ◊ ◊ ♀ ❀ ❀ ❀
‡90cm (36in) ↔45cm (18in)

Astilbe x *crispa* 'Perkeo'

A summer-flowering, clump-forming perennial, low-growing compared with other astilbes, that bears small, upright plumes of tiny, star-shaped, deep pink flowers. The stiff, finely cut, crinkled, dark green leaves are bronze-tinted when young. Suitable for a border or rock garden; the flowers colour best in light shade. 'Bronce Elegans' is another good, compact astilbe where space is limited.

CULTIVATION: *Grow in reasonably moist, fertile soil that is rich in organic matter. Choose a position in partial shade.*

☼ ◊ ♀ ❀ ❀ ❀
‡15–20cm (6–8in) ↔15cm (6in)

Astilbe 'Fanal'

A leafy, clump-forming perennial grown for its long-lasting, tapering, feathery heads of tiny, dark crimson flowers in early summer; they later turn brown, keeping their shape well into winter. The dark green leaves, borne on strong stems, are divided into several leaflets. Grow in a damp border or woodland garden, or use for waterside plantings. 'Brautschleier' is similar, with creamy-white flower plumes.

CULTIVATION: *Grow in moist, fertile, preferably humus-rich soil. Choose a position in full sun or partial shade.*

☼ ◑ ◊ ♀ ❀❀❀
‡60cm (24in) ↔ 45cm (18in)

Astilbe 'Sprite'

A summer-flowering, leafy, clump-forming dwarf perennial that is suitable for waterside plantings. The feathery, tapering plumes of tiny, star-shaped, shell-pink flowers arch elegantly over a mass of broad, mid-green leaves composed of many narrow leaflets. *Astilbe* 'Deutschland' has the same arching, as opposed to upright, flower plumes in cream.

CULTIVATION: *Grow in reliably moist, fertile soil that is rich in organic matter. Choose a site in a partial shade.*

☼ ◊ ♀ ❀❀❀
‡50cm (20in) ↔ to 1m (3ft)

Astrantia major '**Sunningdale Variegated**'

A clump-forming perennial bearing attractive, deeply lobed, basal leaves which have unevenly variegated, creamy-yellow margins. From early summer, domes of tiny, green or pink, often deep purple-red flowers with star-shaped collars of pale-pink bracts, are carried on wiry stems. Thrives in a moist border, woodland garden, or on a stream bank.

CULTIVATION: *Grow in any moist but well-drained, fertile soil. Needs full sun to obtain the best leaf colouring.*

☼ ◊ ♀ ✿ ✿ ✿
‡30–90cm (12–36in) ↔45cm (18in)

Astrantia maxima

Sometimes known as Hattie's pincushion, this mat-forming perennial produces domed, rose-pink flowerheads with star-shaped collars of papery, greenish-pink bracts, on tall stems during summer and autumn. The mid-green leaves are divided into three toothed lobes. Flowers are good for cutting and drying, for use in cottage-style arrangements.

CULTIVATION: *Grow in any moist, fertile, preferably humus-rich soil, in sun or semi-shade. Tolerates drier conditions.*

☼ ☀ ◊ ◊ ♀ ✿ ✿ ✿
‡60cm (24in) ↔30cm (12in)

Athyrium filix-femina

The lady fern has much-divided, light green, deciduous fronds which are borne like upright shuttlecocks, about 1m (3ft) long, arching outward with age. Frond dissection is very varied, and the stalks are sometimes red-brown. There are no flowers. Useful for shaded sites, like a woodland garden. Its cultivars 'Frizelliae' and 'Vernoniae' have unusual, distinctive fronds.

CULTIVATION: *Grow in moist, fertile, neutral to acid soil enriched with leaf mould or garden compost. Choose a shaded, sheltered site.*

☼ ◊ ♀ ❀❀❀
‡to 1.2m (4ft) ↔ 60–90cm (24–36in)

Aubrieta 'Red Cascade' (Cascade Series)

In common with other aubrietas, this variety forms a spreading mat of mid-green foliage that bursts into flower each spring. Aubrietas are good for ground cover and look particularly effective spilling over dry stone walls and raised beds in full sun. They are also well suited to rock gardens. The smallish, four-petalled flowers are bright red.

CULTIVATION: *Grow in moderately fertile, well-drained, preferably neutral to alkaline soil in full sun. Trim back after flowering to keep compact.*

☼ ◊ ♀ ❀❀❀
‡15cm (6in) ↔ 1m (3ft) or more

Aucuba japonica 'Crotonifolia' (female)

This variegated form of spotted laurel is a rounded, evergreen shrub with large, glossy dark green leaves boldly speckled with golden-yellow. Upright clusters of small purplish flowers are borne in mid-spring, followed by red berries in autumn. Ideal for dense, semi-formal hedging.

CULTIVATION: *Grow in any but waterlogged soil, in full sun for best foliage colour, or in shade. Provide shelter in cold areas. Plant with male cultivars to ensure good fruiting. Tolerates light pruning at any time; cut back in spring to promote bushiness.*

☼ ☀ ◊ ◑ ♀ ✽✽✽   ↕↔3m (10ft)

Aurinia saxatilis

An evergreen perennial that forms dense clusters of bright yellow flowers in late spring which give rise to its common name, gold dust. The flowers of its cultivar 'Citrinus', also recommended, are a more lemony yellow. The oval, hairy, grey-green leaves are arranged in clumps. Ideal for rock gardens, walls, and banks. Also sold as *Alyssum saxatilis*.

CULTIVATION: *Grow in moderately fertile soil that is reliably well-drained, in a sunny site. Cut back after flowering to maintain compactness.*

☼ ◊ ♀ ✽✽✽
↕20cm (8in) ↔to 30cm (12in)

Ballota pseudodictamnus

An evergreen subshrub that forms
mounds of rounded, yellow-grey-
green leaves on upright, white-woolly
stems. Whorls of small, white or
pinkish-white flowers, each enclosed
by a pale green funnel, are produced
in late spring and early summer.
May not survive in areas with cold,
wet winters.

CULTIVATION: *Grow in poor, very well-
drained soil, in full sun with protection
from excessive winter wet. Cut back in
early spring to keep plants compact.*

☼ ◊ ♀ ❀ ❀
‡45cm (18in) ↔ 60cm (24in)

Baptisia australis

Blue false indigo is a gently
spreading, upright perennial with
a long season of interest. The bright
blue-green leaves, on grey-green
stems, are divided into three oval
leaflets. Spikes of indigo-blue flowers,
often flecked white or cream, open
throughout early summer. The dark
grey seed pods can be dried for
winter decoration.

CULTIVATION: *Grow in deep, moist but
well-drained, fertile, preferably neutral
to acid soil, in full sun. Once planted, it
is best left undisturbed.*

☼ ◊ ♀ ❀ ❀ ❀   ‡1.5m (5ft) ↔ 60cm (24in)

Begonias with Decorative Foliage

These perennial begonias are typically grown as annuals for their large, usually asymmetrical, ornamental leaves which are available in a variety of colours. For example, there are lively leaves of 'Merry Christmas' outlined with emerald green, or there is the more subtle, dark green, metallic foliage of *B. metallica*. Some leaves are valued for their unusual patterns; *B. masoniana* is appropriately known as the iron-cross begonia. Under the right conditions, 'Thurstonii' may reach shrub-like proportions, but most, like 'Munchkin', are more compact. Grow as summer bedding, in a conservatory or as house plants.

CULTIVATION: *Grow in fertile, well-drained, neutral to acid soil or compost, in light dappled shade. Promote compact, leafy growth by pinching out shoot tips during the growing season. When in growth, feed regularly with a nitrogen-rich fertilizer. Minimum temperature 15°C (59°F).*

☀ ◊ ⌂

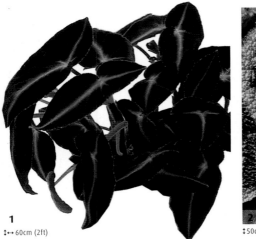

1
↕↔ 60cm (2ft)

↕50cm (20in) ↔ 45cm (18in)

1 *Begonia listada* ♀ **2** *B. masoniana* ♀

3 ‡25cm (8in) ↔ 30cm (12in)

‡90cm (36in) ↔ 60cm (24in)

‡20cm (8in) ↔ 25cm (10in)

6 ‡30cm (12in) ↔ 45cm (18in)

‡2m (6ft) ↔ 45cm (18in)

8 ‡20cm (8in) ↔ 25cm (10in)

3 *B.* 'Merry Christmas' **4** *B. metallica* ♀ **5** *B.* 'Munchkin' ♀
6 *B.* 'Silver Queen' ♀ **7** *B.* 'Thurstonii' ♀ **8** *B.* 'Tiger Paws' ♀

Flowering Begonias

Usually grown outdoors as annuals, these bold-flowered begonias are very variable in size and shape, offering a range of summer uses to the gardener: for specific information on growth habit, check the plant label or ask advice when buying. Upright or compact begonias, such as 'Pin Up' or the Olympia series, are ideal for summer bedding; for containers and hanging baskets, there are pendulous or trailing varieties like 'Illumination Orange'. Begonias can also be grown as house plants. The flowers also come in a wide variety of sizes and colours; they are either single or double, and appear in loose clusters throughout summer.

CULTIVATION: *Fertile, humus-rich, neutral to acid soil or compost with good drainage. Flowers are best in partial shade; they suffer in direct sun. When in growth, give a balanced fertilizer. Those of ✿ hardiness will not survive below 15°C (59°F).*

☼ ◑ ◊

1 ‡to 45cm (18in) ↔to 35cm (14in)

2 ‡↔to 40cm (16in)

3 ‡20cm (8in) ↔20–22cm (8–9in)

4 ‡60cm (24in) ↔30cm (12in)

1 *B.* 'Alfa Pink' ✿ **2** *Begonia* 'All Round Dark Rose Green Leaf' ✿
3 *B.* 'Expresso Scarlet' ✿ **4** *B.* 'Illumination Orange' ✿

‡to 35cm (14in) ↔ to 30cm (12in)

‡↔ to 20cm (8in)

‡75cm (30in) ↔ 60cm (24in)

‡↔ 30cm (12in)

‡60cm (24in) ↔ 45cm (18in)

‡25cm (10in) ↔ 20cm (8in)

‡75cm (30in) ↔ 45cm (18in)

5 *B.* 'Inferno Apple Blossom' ❀ 6 *B.* 'Irene Nuss' ♀❀ 7 *B.* 'Nonstop' ❀ 8 *B.* 'Olympia White' ❀ 9 *B.* 'Orange Rubra' ♀❀ 10 *B.* 'Pin Up' ❀ 11 *B.* sutherlandii ♀❀

Bellis perennis 'Pomponette'

This double-flowered form of the common daisy is usually grown as a biennial for spring bedding. Pink, red, or white flowerheads with quill-shaped petals appear from late winter to spring, above the dense clumps of spoon-shaped, bright green leaves. 'Dresden China' and 'Rob Roy' are other recommended bellis.

CULTIVATION: *Grow in well-drained, moderately fertile soil, in full sun or partial shade. Deadhead to prolong flowering and to prevent self-seeding.*

☼ ☼ ◊ ❀ ❀ ❀   ↕↔10–20cm (4–8in)

Berberis darwinii

The Darwin barberry is a vigorous, arching, evergreen shrub that carries masses of small, deep golden-orange flowers on spiny stems from mid- to late spring; these are followed by blue berries in autumn. The leaves are glossy dark green and spiny. Use as a vandal-resistant or barrier hedge.

CULTIVATION: *Grow in any but water-logged soil, in full sun or partial shade with shelter from cold, drying winds. Trim after flowering, if necessary.*

☼ ☼ ◊ ◊ ♀ ❀ ❀ ❀
↕3m (10ft) or more ↔3m (10ft)

Berberis x *ottawensis* 'Superba'

This spiny, rounded, deciduous, spring-flowering shrub bears clusters of small, pale yellow, red-tinged flowers which are followed by red berries in autumn. The red-purple leaves turn crimson before they fall. Effective as a specimen shrub or in a mixed border.

CULTIVATION: *Grow in almost any well-drained soil, preferably in full sun. Thin out dense growth in midwinter.*

☼ ◊ ◊ ❀ ❀ ❀   ↕↔2.5m (8ft)

Berberis x *stenophylla* 'Corallina Compacta'

While *Berberis stenophylla* is a large, arching shrub, ideal for informal hedging, this cultivar of it is tiny; a small, evergreen shrub bearing spine-tipped, deep green leaves on arching, spiny stems. Quantities of tiny, light orange flowers appear from mid-spring, followed by small, blue-black berries.

CULTIVATION: *Best in fertile, humus-rich soil that is reliably drained, in full sun. Cut back hard after flowering.*

☼ ◊ ♀ ❀ ❀ ❀   ↕↔to 30cm (12in)

Berberis thunbergii 'Bagatelle'

A very compact, spiny, spring-flowering, deciduous shrub with deep red-purple leaves which turn orange and red in autumn. The pale yellow flowers are followed by glossy red fruits. Good for a rock garden, but may not survive in areas with cold winters. *B. thunbergii* 'Atropurpurea Nana', or 'Crimson Pygmy', is another small purple-leaved berberis, to 60cm (24in) tall.

CULTIVATION: *Grow in well-drained soil, in full sun for best flower and foliage colour. Thin out dense, overcrowded growth in mid- to late winter.*

☼ ◊ ✿ ✿   ‡30cm (12in) ↔40cm (16in)

Berberis thunbergii 'Rose Glow'

A compact, spiny, deciduous shrub with reddish-purple leaves that gradually become flecked with white as the season progresses. Tiny, pale yellow flowers appear in mid-spring, followed by small red berries. Good as a barrier hedge.

CULTIVATION: *Grow in any but water-logged soil, in full sun or partial shade. Cut out any dead wood in summer.*

☼ ☼ ◊ ◊ ♀ ✿ ✿ ✿
‡2m (6ft) or more ↔2m (6ft)

Berberis verruculosa

A slow-growing, compact, spring-flowering barberry that makes a fine evergreen specimen shrub. The cup-shaped, golden-yellow flowers are carried amid the spine-tipped, glossy dark green leaves on spiny, arching stems. Oval to pear-shaped black berries develop in autumn.

CULTIVATION: *Best in well-drained, humus-rich, fertile soil, in full sun. Keep pruning to a minimum.*

☼ ◊ ♀ ❀❀❀   ↕↔1.5m (5ft)

Berberis wilsoniae

A very spiny, semi-evergreen, arching shrub forming dense mounds of grey-green foliage which turns red and orange in autumn. Clusters of pale yellow flowers in summer are followed by coral-pink to pinkish-red berries. Makes a good barrier hedge. Avoid seed-grown plants as they may be inferior hybrids.

CULTIVATION: *Grow in any well-drained soil, in sun or partial shade. Flowering and fruiting are best in full sun. Thin out dense growth in midwinter.*

☼ ◑ ◊ ❀❀❀   ↕1m (3ft) ↔2m (6ft)

Bergenia 'Ballawley'

This clump-forming, evergreen perennial, one of the first to flower in spring, bears bright crimson flowers which are carried on sturdy red stems. The leathery, oval leaves turn bronze-red in winter. Suits a woodland garden, or plant in groups to edge a mixed border. *B. cordifolia* 'Purpurea' has similarly coloured leaves, with deep magenta flowers.

CULTIVATION: *Grow in any well-drained soil, in full sun or light shade. Shelter from cold winds. Mulch in autumn.*

☼ ☀ ◊ ◑ ✿✿✿
‡to 60cm (24in) ↔60cm (24in)

Bergenia 'Silberlicht'

An early-flowering, clump-forming, evergreen perennial bearing clusters of cup-shaped white flowers, often flushed pink, in spring. (For pure white flowers on a similar plant, look for 'Bressingham White', or for deep pink, 'Morgenröte'.) The mid-green leaves are leathery, with toothed margins. Good underplanting for shrubs, which give it some winter shelter.

CULTIVATION: *Grow in any well-drained soil, in full sun or partial shade. Shelter from cold winds to avoid foliage scorch. Provide a mulch in autumn.*

☼ ☀ ◊ ◑ ♀ ✿✿✿
‡30cm (12in) ↔50cm (20in)

Betula nigra

The black birch is a tall, conical to spreading, deciduous tree with glossy, mid- to dark green, diamond-shaped leaves. It has shaggy, red-brown bark that peels in layers on young trees; on older specimens, the bark becomes blackish or grey-white and develops cracks. Yellow-brown male catkins are conspicuous in spring. Makes a fine specimen tree, but only for a large garden.

CULTIVATION: *Grow in moist but well-drained, moderately fertile soil, in full sun. Remove any damaged, diseased, or dead wood in late autumn.*

☼ ◊ ❁❁❁　　‡18m (60ft) ↔12m (40ft)

Betula pendula 'Youngii'

Young's weeping birch is a deciduous tree with an elegant, weeping habit. The yellow-brown male catkins appear in early spring before the triangular leaves; the foliage turns golden-yellow in autumn. An attractive tree for a small garden, more dome-shaped than 'Tristis' or 'Laciniata', other popular weeping birches, growing wider than it is tall.

CULTIVATION: *Any moist but well-drained soil, in an open, sunny site. Keep pruning to a minimum; remove any shoots on the trunk in late autumn.*

☼ ◊ ❁❁❁　　‡8m (25ft) ↔10m (30ft)

Betula utilis var. *jacquemontii*

The West Himalayan birch is an open, broadly conical, deciduous tree with smooth, peeling white bark. Catkins are a feature in early spring, and the dark green leaves turn rich golden-yellow in autumn. Plant where winter sun will light up the bark, particularly brilliantly white in the cultivars 'Silver Shadow', 'Jermyns', and 'Grayswood Ghost'.

CULTIVATION: *Grow in any moist but well-drained soil, in sun. Remove any damaged or dead wood from young trees in late autumn; once established, keep pruning to a minimum.*

☼ ◊ ❀❀❀ ‡15m (50ft) ↔7.5m (23ft)

Blechnum chilense

This striking evergreen fern forms large clumps of leathery, ribbed fronds that are very tall for a fern, in some cases up to 1.8m (6ft). It is an excellent fern for a lush look in a woodland setting. In time, the clumps will develop short trunks. In areas with hard winters, choose a sheltered site.

CULTIVATION: *Grow in a shady site in moist, neutral to acid soil enriched with organic matter. Tolerates some sun if soil is kept reliably moist.*

☼ ☀ ☀ ◊ ◊ ♀ ❀❀❀ ‡to 1.8m (6ft) ↔indefinite

Brachyglottis 'Sunshine'

A bushy, mound-forming, evergreen
shrub bearing oval leaves which are
silvery-grey when young, becoming
dark green with white-felted
undersides as they develop. Daisy-like
yellow flowers appear from early to
midsummer. Some gardeners prefer
it as a foliage plant, pinching
or snipping off the flower buds before
they open. Thrives in coastal sites.

CULTIVATION: *Grow in any well-drained
soil, in a sunny, sheltered site. Trim back
after flowering. Responds well to hard
pruning in spring.*

☼ ◊ ♀ ❀❀❀
‡1–1.5m (3–5ft) ↔2m (6ft) or more

Bracteantha
Bright Bikini Series

These straw flowers are upright
annuals or short-lived perennials
with papery, double flowers in red,
pink, orange, yellow, and white from
late spring to autumn. The leaves are
grey-green. Use to edge a border, or
grow in a window box; flowers are
long-lasting and cut and dry well. For
single colours rather than a mixture,
try 'Frosted Sulphur' (lemon-yellow),
'Silvery Rose', and 'Reeves Purple'.

CULTIVATION: *Grow in moist but well-
drained, moderately fertile soil. Choose
a position in full sun.*

☼ ◊ ❀❀
‡to 30cm (12in) ↔30cm (12in)

Brunnera macrophylla 'Hadspen Cream'

This clump-forming perennial with attractive foliage is ideal for ground cover in borders and among deciduous trees. In mid- and late spring, upright clusters of small, bright blue flowers appear above heart-shaped leaves, plain green in *Brunnera macrophylla*, but with irregular, creamy-white margins in this attractive cultivar.

CULTIVATION: *Grow in moist but well-drained, humus-rich soil. Choose a position that is cool and lightly shaded.*

☀ ◊ ◊ ♀ ❀❀❀
‡45cm (18in) ↔ 60cm (24in)

Buddleja alternifolia

A dense, deciduous shrub carrying slender, arching branches. Fragrant, lilac-purple flowers are produced in neat clusters during early summer among the narrow, grey-green leaves. Makes a good wall shrub, or can be trained with a single, clear trunk as a striking specimen tree. Attractive to beneficial insects.

CULTIVATION: *Best in chalky soil, but can be grown in any soil that is well-drained, in full sun. Cut stems back to strong buds after flowering; responds well to hard pruning in spring.*

☀ ◊ ♀ ❀❀❀   ‡↔4m (12ft)

Buddleja davidii

All cultivars of *B. davidii*, the butterfly bush, are fast-growing, deciduous shrubs with a wide range of flower colours. As the popular name suggests, the flowers attract butterflies and other beneficial garden insects in profusion. The long, arching shoots carry lance-shaped, mid- to grey-green leaves, up to 25cm (10in) long. Conical clusters of bright, fragrant flowers, usually about 30cm (12in) long, are borne at the end of arching stems from summer to autumn; those of 'Royal Red' are the largest, up to 50cm (20in) long. These shrubs respond well to hard pruning in spring, which keeps them a compact size for a small garden.

CULTIVATION: *Grow in well-drained, fertile soil, in sun. Restrict size and encourage better flowers by pruning back hard to a low framework each spring. To prevent self-seeding, cut spent flowerheads back to a pair of leaves or sideshoots; this may also result in a second blooming.*

☀ ◊ ❀❀❀

‡3m (10ft) ↔ 5m (15ft)   ‡3m (10ft) ↔ 5m (15ft)   ‡3m (10ft) ↔ 5m (15ft)

1 *B. davidii* 'Empire Blue' **2** *B. davidii* 'Royal Red' ♀ **3** *B. davidii* 'White Profusion' ♀

Buddleja globosa

The orange ball tree is a deciduous or semi-evergreen shrub bearing, unusually for a buddleja, round clusters of tiny, orange-yellow flowers which appear in early summer. The lance-shaped leaves are dark green with woolly undersides. A large shrub, it is prone to becoming bare at the base and does not respond well to pruning, so grow towards the back of a mixed border.

CULTIVATION: *Best on well-drained, chalky soil, in a sunny position with shelter from cold winds. Pruning should be kept to a minimum or the next year's flowers will be lost.*

☼ ◊ ♀ ❀❀❀   ↕↔5m (15ft)

Buddleja 'Lochinch'

A compact, deciduous shrub, very similar to a *Buddleja davidii* (see p.93), bearing long spikes of lilac-blue flowers from late summer to autumn. The leaves are downy and grey-green when young, becoming smooth and developing white-felted undersides as they mature. Very attractive to butterflies.

CULTIVATION: *Grow in any well-drained, moderately fertile soil, in sun. Cut back all stems close to the base each year, as the buds begin to swell in spring.*

☼ ◊ ♀ ❀❀   ↕2.5m (8ft) ↔3m (10ft)

Buxus sempervirens 'Elegantissima'

This variegated form of the common box is a rounded, dense, evergreen shrub bearing small and narrow, glossy bright green leaves edged with cream. The flowers are of little significance. Responding well to trimming, it is very good as an edging plant or for use as a low hedge. 'Latifolia Maculata' is also variegated, with yellow leaf markings.

CULTIVATION: *Grow in any well-drained soil, in sun or light shade. Trim in spring and summer; overgrown shrubs respond well to hard pruning in late spring.*

☼ ☀ ◊ ♀ ❀ ❀ ❀   ↕↔1.5m (5ft)

Buxus sempervirens 'Suffruticosa'

A very dense, slow-growing box carrying small, evergreen, glossy bright green leaves. Widely used as an edging plant or for clipping into precise shapes. During late spring or early summer, inconspicuous flowers are produced. Excellent as a hedge.

CULTIVATION: *Grow in any well-drained, fertile soil, in sun or semi-shade. The combination of dry soil and full sun can cause scorching. Trim hedges in summer; overgrown specimens can be hard-pruned in late spring.*

☼ ☀ ◊ ❀ ❀ ❀   ↕1m (3ft) ↔1.5m (5ft)

Calamagrostis brachytricha

Feather reed grass is a handsome plant that carries tall, upright stems of narrow flowering plumes in late summer. These open out and fade to silvery grey tinged with pink as the season progresses, lasting well into winter. This is a useful plant for both the summer and winter garden, and looks best when planted in drifts. It forms neat clumps of arching foliage.

CULTIVATION: *Grow in any fertile, moist but well-drained soil in sun or partial shade. Cut back old flowerheads in late winter.*

☼ ☀ ◊ ◑ ♀ ❀❀❀
↕to 1.5m (5ft) ↔60cm (24in)

Callicarpa bodinieri var. *giraldii* 'Profusion'

An upright, deciduous shrub grown mainly for its long-lasting autumn display of shiny, bead-like, deep violet berries. The large, pale green, tapering leaves are bronze when they emerge in spring, and pale pink flowers appear in summer. Brings a long season of interest to a shrub border; for maximum impact, plant in groups. May not survive in areas with cold winters.

CULTIVATION: *Grow in any well-drained, fertile soil, in full sun or dappled shade. Cut back about 1 in 5 stems to the base each year in early spring.*

☼ ☀ ◊ ♀ ❀❀
↕3m (10ft) ↔2.5m (8ft)

Callirhoe involucrata

Commonly called purple poppy mallow, this beautiful, ground-hugging perennial bears cup-shaped, silky magenta flowers in from late spring to summer. Ideal in a sunny rockery or as ground cover in informal gardens.

CULTIVATION: *Grow in well-drained sandy, loam soil in full sun. A long tap root makes transplanting difficult, but it self-sows in ideal conditions. Protect from winter wet.*

☼ ◊ ❀❀❀
‡to 30cm (12in) ↔ to 1m (3ft)

Callistemon citrinus '**Splendens**'

This attractive cultivar of the crimson bottlebrush is an evergreen shrub usually with arching branches. Dense spikes of brilliant red flowers appear in spring and summer, amid the grey-green, lemon-scented leaves which are bronze-red when young. Grow at the base of a sunny wall to give protection from winter cold.

CULTIVATION: *Best in well-drained, fertile, neutral to acid soil, in full sun. Pinch out tips of young plants to promote bushiness. Tolerates hard pruning in spring.*

☼ ◊ ♀ ❀❀
‡2–8m (6–25ft) ↔ 1.5–6m (5–20ft)

Calluna vulgaris

Cultivars of *C. vulgaris* are upright to spreading, fine-leaved heathers. They make excellent evergreen groundcover plants if weeds are suppressed before planting. Dense spikes of bell-shaped flowers appear from midsummer to late autumn, in shades of red, purple, pink, or white; 'Kinlochruel' is quite distinctive with its double white flowers in long clusters. Seasonal interest is extended into winter by cultivars with coloured foliage, such as 'Robert Chapman' and 'Beoley Gold'.

Heathers are very attractive to bees and other beneficial insects, and make good companions to dwarf conifers in an evergreen border.

CULTIVATION: *Best in well-drained, humus-rich, acid soil, in an open, sunny site, to recreate their native moorland habitats. Trim off flowered shoots in early spring with shears; remove overlong shoots wherever possible, cutting back to their point of origin below the flower cluster.*

☼ ◊ ❀ ❀ ❀

‡25cm (10in) ↔ 35cm (14in)

‡35cm (14in) ↔ to 75cm (30in)

‡25cm (10in) ↔ 40cm (16in)

‡25cm (10in) ↔ 65cm (26in)

1 *C. vulgaris* 'Beoley Gold' ♀ **2** *C. vulgaris* 'Darkness' ♀ **3** *C. vulgaris* 'Kinlochruel' ♀ **4** *C. vulgaris* 'Robert Chapman' ♀

Caltha palustris

The marsh marigold is a clump-forming, aquatic perennial which thrives in a bog garden or at the margins of a stream or pond. Cup-shaped, waxy, bright golden-yellow flowers appear on tall stems in spring, above the kidney-shaped, glossy green leaves. For double flowers, look for the cultivar 'Flore Pleno'.

CULTIVATION: *Best in boggy, rich soil, in an open, sunny site. Tolerates root restriction in aquatic planting baskets in water no deeper than 23cm (9in), but prefers shallower conditions.*

☼ ● ❀❀❀
‡10–40cm (4–16in) ↔45cm (18in)

Camellia 'Inspiration'

A dense, upright, evergreen shrub or small tree bearing masses of saucer-shaped, semi-double, deep pink flowers from midwinter to late spring. The dark green leaves are oval and leathery. Good for the back of a border or as a specimen shrub.

CULTIVATION: *Best in moist but well-drained, fertile, neutral to acid soil, in partial shade with shelter from cold, drying winds. Mulch around the base with shredded bark. After flowering, prune back young plants to encourage a bushy habit and a balanced shape.*

☼ ◊ ◗ ♀ ❀❀❀   ‡4m (12ft) ↔2m (6ft)

Camellia japonica

These long-lived and elegant, evergreen shrubs or small trees for gardens with acid soil are very popular in shrub borders, woodland gardens, or standing on their own in the open ground or in containers. The oval leaves are glossy and dark green; they serve to heighten the brilliance of the single to fully double flowers in spring. Single and semi-double flowers have a prominent central boss of yellow stamens. Most flowers are suitable for cutting, but blooms may be spoiled by late frosts.

CULTIVATION: *Grow in moist but well-drained, humus-rich, acid soil in partial shade. Choose a site sheltered from early morning sun, cold winds, and late-season frosts. Do not plant too deeply; the top of the root ball must be level with the firmed soil. Maintain a mulch 5–7cm (2–3in) deep of leaf mould or shredded bark. Little pruning is necessary, although moderate trimming of young plants can help to create a balanced shape.*

☀ ◐ ◊ ◊ ❀ ❀ ❀

1 ↕5m (15ft) ↔ 8m (25ft)

2 ↕9m (28ft) ↔ 8m (25ft)

3 ↕9m (28ft) ↔ 8m (25ft)

4 ↕2m (6ft) ↔ 1m (3ft)

5 ↕9m (28ft) ↔ 8m (25ft)

6 ↕9m (28ft) ↔ 8m (25ft)

1 *Camellia japonica* 'Adolphe Audusson' ♥ **2** 'Alexander Hunter' ♥ **3** 'Berenice Boddy' ♥ **4** 'Bob's Tinsie' ♥ **5** 'Coquettii' ♥ **6** 'Elegans'

MORE CHOICES

'Akashigata' Deep pink flowers.

'Bob Hope' Dark red.

'C. M. Hovey' Crimson/scarlet.

'Doctor Tinsley' Pinky-white.

'Grand Prix' Bright red with yellow stamens.

'Hagoromo' Pale pink.

'Masayoshi' White with red marbling.

'Miss Charleston' Ruby-red.

'Nuccio's Gem' White.

‡9m (28ft) ↔ 8m (25ft)

‡9m (28ft) ↔ 8m (25ft)

‡9m (28ft) ↔ 8m (25ft)

‡9m (28ft) ↔ 8m (25ft)

‡9m (28ft) ↔ 8m (25ft)

‡9m (28ft) ↔ 8m (25ft)

‡9m (28ft) ↔ 8m (25ft)

‡9m (28ft) ↔ 8m (25ft)

7 *Camellia japonica* 'Gloire de Nantes' ♀ **8** 'Guilio Nuccio' ♀ **9** 'Jupiter' (*syn.* 'Paul's Jupiter') ♀ **10** 'Lavinia Maggi' ♀ **11** 'Mrs D. W. Davis' ♀ **12** 'R. L. Wheeler' ♀ **13** 'Rubescens Major' **14** 'Tricolor' ♀

Camellia 'Lasca Beauty'

An open, upright shrub greatly valued for its very large, semi-double, pale pink flowers, which appear in mid-spring. They stand out against the dark green foliage. Grow in a cool greenhouse, and move outdoors to a partially shaded site in early summer.

CULTIVATION: *Grow in lime-free potting compost in bright filtered light. Water freely with soft water when in growth; more sparingly in winter. Apply a balanced fertilizer once in mid-spring and again in early summer.*

☀ ◊ ◊ ❀
‡2–5m (6–15ft) ↔1.5–3m (5–10ft)

Camellia 'Leonard Messel'

This spreading, evergreen shrub with oval, leathery, dark green leaves is one of the hardiest camellias available. It produces an abundance of large, flattish to cup-shaped, semi-double, clear pink flowers from early to late spring. Suits a shrub border.

CULTIVATION: *Best in moist but well-drained, fertile, neutral to acid soil. Position in semi-shade with shelter from cold, drying winds. Maintain a mulch of shredded bark or leaf mould around the base. Pruning is not necessary.*

☀ ◊ ◊ ♀ ❀❀❀   ‡4m (12ft) ↔3m (10ft)

Camellia 'Mandalay Queen'

This large, widely branching shrub bears deep rose-pink, semi-double flowers in spring. The broad, leathery leaves are dark green. Best grown in a cool greenhouse, but move it outdoors in summer.

CULTIVATION: *Best in lime-free potting compost with bright filtered light. Water freely with soft water when in growth, more sparingly in winter. Apply a balanced fertilizer in mid-spring and again in early summer.*

☼ ◊ ◊ ❄
↕ to 15m (50ft) ↔ 5m (15ft)

Camellia sasanqua 'Narumigata'

An upright shrub or small tree valued for its late display of fragrant, single-petalled white flowers in mid- to late autumn. The foliage is dark green. Makes a good hedge, though in very cold areas it may be better against a warm, sunny wall. *C. sasanqua* 'Crimson King' is a very similar shrub, with red flowers.

CULTIVATION: *Grow in moist but well-drained, humus-rich, acid soil, and maintain a thick mulch. Choose a site in full sun or partial shade, with shelter from cold winds. Tolerates hard pruning after flowering.*

☼ ☀ ◊ ◊ ♀ ❄❄
↕ to 6m (20ft) ↔ to 3m (10ft)

Camellia x williamsii

Cultivars of *C. x williamsii* are strong-growing, evergreen shrubs much-valued for their bright, lustrous foliage and the unsurpassed elegance of their rose-like flowers which range from pure white to crimson. Most flower in mid- and late spring, although 'Anticipation' and 'Mary Christian' begin to flower from late winter. Many are fully hardy and can be grown in any climate, although the flowers are susceptible to damage in hard frosts. They make handsome specimens for a cool conservatory or shrub border, and 'J.C. Williams', for example, will train against a cold wall. Avoid sites exposed to morning sun.

CULTIVATION: *Best in moist but well-drained, acid to neutral soil, in partial shade. Shelter from frost and cold winds, and mulch with shredded bark. Prune young plants after flowering to promote bushiness; wall-trained shrubs should be allowed to develop a strong central stem.*

☼ ◐ ◊ ♦ ❀ ❀ ❀

‡4m (12ft) ↔2m (6ft)

2
‡3m (10ft) ↔2.5m (8ft)

‡5m (15ft) ↔2.5m (8ft)

1 *C. x williamsii* 'Anticipation' ♀ **2** *C. x williamsii* 'Brigadoon' ♀
3 *C. x williamsii* 'Donation' ♀

4
‡↔4m (12ft)

5
‡4m (12ft) ↔2.5m (8ft)

6
‡4m (12ft) ↔2.5m (8ft)

‡4m (12ft) ↔2.5m (8ft)

‡4m (12ft) ↔2.5m (8ft)

‡↔3m (10ft)

4 *C.* x *williamsii* 'George Blandford' ♀ **5** *C.* x *williamsii* 'Joan Trehane' ♀
6 *C.* x *williamsii* 'J.C. Williams' ♀ **7** *C.* x *williamsii* 'Mary Christian'
8 *C.* x *williamsii* 'Saint Ewe' ♀ **9** *C.* x *williamsii* 'Water Lily' ♀

Campanula cochleariifolia

Fairies' thimbles is a low-growing, rosette-forming perennial bearing, in midsummer, abundant clusters of open, bell-shaped, mauve-blue or white flowers. The bright green leaves are heart-shaped. It spreads freely by means of creeping stems, and can be invasive. Particularly effective if allowed to colonize areas of gravel, paving crevices, or the tops of dry walls.

CULTIVATION: *Prefers moist but well-drained soil, in sun or partial shade. To restrict spread, pull up unwanted plants.*

☼ ☀ ◊ ❀ ❀ ❀
‡to 8cm (3in) ↔to 50cm (20in) or more

Campanula glomerata 'Superba'

A fast-growing, clump-forming perennial carrying dense heads of large, bell-shaped, purple-violet flowers in summer. The lance-shaped to oval, mid-green leaves are arranged in rosettes at the base of the plant and along the stems. Excellent in herbaceous borders or informal, cottage-style gardens.

CULTIVATION: *Best in moist but well-drained, neutral to alkaline soil, in sun or semi-shade. Cut back after flowering to encourage a second flush of flowers.*

☼ ☀ ◊ ◊ ♀ ❀ ❀ ❀
‡75cm (30in) ↔1m (3ft) or more

Campanula lactiflora 'Loddon Anna'

An upright, branching perennial producing sprays of large, nodding, bell-shaped, soft lilac-pink flowers from midsummer above mid-green leaves. Makes an excellent border perennial, but may need staking in an exposed site. Mixes well with the white-flowered 'Alba', or the deep purple 'Prichard's Variety'.

CULTIVATION: *Best in moist but well-drained, fertile soil, in full sun or partial shade. Trim after flowering to encourage a second, although less profuse, flush of flowers.*

☼ ☀ ◊ ◑ ♀ ❀❀❀
↕1.2–1.5m (4–5ft) ↔60cm (24in)

Campanula portenschlagiana

The Dalmatian bellflower is a robust, mound-forming, evergreen perennial with long, bell-shaped, deep purple flowers from mid- to late summer. The leaves are toothed and mid-green. Good in a rock garden or on a sunny bank. May become invasive. *Campanula poschkarskyana* can be used in a similar way; look for the recommended cultivar 'Stella'.

CULTIVATION: *Best in moist but well-drained soil, in sun or partial shade. Very vigorous, so plant away from smaller, less robust plants.*

☼ ☀ ◊ ♀ ❀❀❀
↕to 15cm (6in) ↔50cm (20in) or more

Campsis x *tagliabuana* 'Madame Galen'

Woody-stemmed climber that will cling with aerial roots against a wall, fence or pillar, or up into a tree. From late summer to autumn, clusters of trumpet-shaped, orange-red flowers open among narrow, toothed leaves. For yellow flowers, choose *C. radicans* f. *flava*.

CULTIVATION: *Prefers moist but well-drained, fertile soil, in a sunny, sheltered site. Tie in new growth until the allotted space is covered by a strong framework. Prune back hard each winter to promote bushiness.*

☀ ◌ ◖ ♥ ❀ ❀ ❀   ↕10m (30ft) or more

Cardiocrinum giganteum

The giant lily is a spectacular, summer-flowering, bulbous perennial with trumpet-shaped white flowers which are flushed with maroon-purple at the throats. The stems are stout, and the leaves broadly oval and glossy green. It needs careful siting and can take up to seven years to flower. Grow in woodland, or a sheltered border in shade.

CULTIVATION: *Best in deep, moist but well-drained, reliably cool, humus-rich soil, in semi-shade. Intolerant of hot or dry conditions. Slugs can be a problem.*

☀ ◌ ◖ ❀ ❀ ❀   ↕1.5–4m (5–12ft) ↔45cm (18in)

Carex elata 'Aurea'

Bowles' golden sedge is a colourful,
tussock-forming, deciduous perennial
for a moist border, bog garden or
the margins of a pond or stream.
The bright leaves are narrow and
golden-yellow. In spring and early
summer, small spikes of relatively
inconspicuous, dark brown flowers
are carried above the leaves. Often
sold as *C.* 'Bowles' Golden'.

CULTIVATION: *Grow in moist or wet,
reasonably fertile soil. Position in full
sun or partial shade.*

☼ ☀ ◊ ♦ ♀ ❀❀❀
↕ to 70cm (28in) ↔ 45cm (18in)

Carex oshimensis 'Evergold'

A very popular, evergreen, variegated
sedge, bright and densely tufted with
narrow, dark green, yellow-striped
leaves. Spikes of tiny, dark brown
flowers are borne in mid- and late
spring. Tolerates freer drainage than
many sedges and is suitable for a
mixed border.

CULTIVATION: *Needs moist but well
drained, fertile soil, in sun or partial
shade. Remove dead leaves in summer.*

☼ ☀ ◊ ♦ ♀ ❀❀❀
↕ 30cm (12in) ↔ 35cm (14in)

Carpenteria californica

The tree anemone is a summer-flowering, evergreen shrub bearing large, fragrant white flowers with showy yellow stamens. The glossy, dark green leaves are narrowly oval. It is suitable for wall-training, which overcomes its sometimes sprawling habit. In cold areas, it must be sited against a warm, sheltered wall.

CULTIVATION: *Grow in well-drained soil, in full sun with shelter from cold winds. In spring, remove branches that have become exhausted by flowering, cutting them back to their bases.*

☼ ◊ ❀❀ ‡2m (6ft) or more ↔2m (6ft)

Caryopteris x *clandonensis* 'Heavenly Blue'

A compact, upright, deciduous shrub grown for its clusters of intensely dark blue flowers which appear in late summer and early autumn. The irregularly toothed leaves are grey-green. In cold areas, position against a warm wall.

CULTIVATION: *Grow in well-drained, moderately fertile, light soil, in full sun. Prune all stems back hard to low buds in late spring. A woody framework will develop, which should not be cut into.*

☼ ◊ ❀❀ ‡↔1m (3ft)

Cassiope 'Edinburgh'

A heather-like, upright, evergreen shrub producing nodding, bell-shaped flowers in spring; these are white with small, greenish-brown outer petals. The scale-like, dark green leaves closely overlap along the stems. Suits a rock garden (not among limestone) or a peat bed. *C. lycopoides* has very similar flowers, but is mat-forming, almost prostrate, only 8cm (3in) tall.

CULTIVATION: *Grow in reliably moist, humus-rich, acid soil, in partial shade. Trim after flowering.*

☀ ◑ ♀ ✽✽✽   ↕↔to 25cm (10in)

Catalpa bignonioides 'Aurea'

This bright-leaved variety of the Indian bean tree is a superb foliage plant, best grown as a specimen tree or at the back of a mixed border. In a border, cut it back to a stump each year, and it will respond by producing luxuriant foliage. The yellowish green, deciduous, heart-shaped leaves are tinted bronze when young. White flower clusters in spring give rise to bean-like pods.

CULTIVATION: *Grow in fertile, moist but well-drained soil in sun.*

☀ ◌ ◑ ♀ ✽✽✽   ↕↔10m (30ft)

Ceanothus 'Autumnal Blue'

A vigorous, evergreen shrub that produces a profusion of tiny but vivid, rich sky-blue flowers from late summer to autumn. The leaves are broadly oval and glossy dark green. One of the hardiest of the evergreen ceanothus, it is suitable in an open border as well as for informal training on walls; best against a warm wall in cold areas.

CULTIVATION: *Grow in well-drained, moderately fertile soil, in full sun with shelter from cold winds. Tolerates lime. Tip-prune young plants in spring, and trim established plants after flowering.*

☼ ◊ ♀ ❀❀   ↕↔3m (10ft)

Ceanothus 'Blue Mound'

This mound-forming, late spring-flowering ceanothus is an evergreen shrub carrying masses of rich dark blue flowers. The leaves are finely toothed and glossy dark green. Ideal for ground cover, for cascading over banks or low walls, or in a large, sunny rock garden. *Ceanothus* 'Burkwoodii' and 'Italian Skies' can be used similarly. May not survive in areas with cold winters.

CULTIVATION: *Grow in well-drained, fertile soil, in full sun. Tip-prune young plants and trim established ones after flowering, in midsummer.*

☼ ◊ ♀ ❀❀   ↕1.5m (5ft) ↔2m (6ft)

Ceanothus x *delileanus* 'Gloire de Versailles'

This deciduous ceanothus is a fast-growing shrub. From midsummer to early autumn, large spikes of tiny, pale blue flowers are borne amid broadly oval, finely toothed, mid-green leaves. 'Topaze' is very similar, with dark blue flowers. They benefit from harder annual pruning than evergreen ceanothus.

CULTIVATION: *Grow in well-drained, fairly fertile, light soil, in sun. Tolerates lime. In spring, shorten the previous year's stems by half or more, or cut right back to a low framework.*

☼ ◊ ♀ ❀❀❀   ↕↔1.5m (5ft)

Ceanothus thyrsiflorus var. *repens*

This low and spreading ceanothus is a mound-forming, evergreen shrub, bearing rounded clusters of tiny blue flowers in late spring and early summer. The leaves are dark green and glossy. A good shrub to clothe a sunny or slightly shaded bank, but needs the protection of a warm, sunny site in cold areas.

CULTIVATION: *Best in light, well-drained, fertile soil, in sun or light shade. Trim back after flowering to keep compact.*

☼ ☀ ◊ ♀ ❀❀   ↕1m (3ft) ↔2.5m (8ft)

Ceratostigma plumbaginoides

A spreading, woody-based, sub-shrubby perennial bearing clusters of brilliant blue flowers in late summer. The oval, bright green leaves, carried on upright, slender red stems, become red-tinted in autumn. Good for a rock garden, and also suitable for groundcover.

CULTIVATION: *Grow in light, moist but well-drained, moderately fertile soil. Choose a sheltered site in full sun. Cut back stems to about 2.5–5cm (1–2in) from the ground in mid-spring.*

☼ ◊ ♀ ❀❀
‡to 45cm (18in) ↔30cm (12in) or more

Cercidiphyllum japonicum

The katsura tree is a fast-growing, spreading tree with small, heart-shaped, mid-green leaves. New leaves emerge bronze, but are perhaps seen at their best in autumn, when the colour turns to pale yellow, orange, red or pink before they fall; at this time, the tree also exudes a "burnt-sugar" scent. The best foliage colours are seen on trees in neutral to acid soils.

CULTIVATION: *Grow in fertile, moist but well-drained soil, with plenty of depth for the roots.*

☼ ☀ ◊ ◑ ♀ ❀❀❀
‡to 20m (70ft) ↔to 15m (50ft)

Cercis canadensis 'Forest Pansy'

The eastern redbud, or Judas tree, is remarkable for the pale pink flowers that adorn the bare tree in spring before the heart-shaped leaves emerge. This variety has red-purple foliage that turns purple and gold in autumn. Solo, the tree will develop a rounded shape, but it can be trained against a fence or wall or grown as a foliage plant in a mixed border.

CULTIVATION: *Tolerates most well-drained soils, particularly hot, dry sites, including those on chalk or limestone.*

☼ ☀ ◊ ♀ ✿✿✿   ↕↔5m (15ft)

Cercis siliquastrum

The Judas tree is a handsome, broadly spreading, deciduous tree which gradually develops a rounded crown. Clusters of pea-like, bright pink flowers appear on the previous year's wood either before or with the heart-shaped leaves in mid-spring. The foliage is bronze when young, maturing to dark blue-green, then to yellow in autumn. Flowering is best after a long, hot summer.

CULTIVATION: *Grow in deep, reliably well-drained, fertile soil, in full sun or light dappled shade. Prune young trees to shape in early summer, removing any frost-damaged growth.*

☼ ☀ ◊ ✿✿✿   ↕↔10m (30ft)

Chaenomeles speciosa 'Moerloosei'

A fast-growing and wide-spreading, deciduous shrub bearing large white flowers, flushed dark pink, in early spring. Tangled, spiny branches carry oval, glossy dark green leaves. The flowers are followed in autumn by apple-shaped, aromatic, yellow-green fruits. Use as a free-standing shrub or train against a wall.

CULTIVATION: *Grow in well-drained, moderately fertile soil, in full sun for best flowering, or light shade. If wall-trained, shorten sideshoots to 2 or 3 leaves in late spring. Free-standing shrubs require little pruning.*

☀ ☼ ◊ ♀ ❀❀❀ ‡2.5m (8ft) ↔5m (15ft)

Chaenomeles x *superba* 'Crimson and Gold'

This spreading, deciduous shrub bears masses of dark red flowers with conspicuous golden-yellow anthers from spring until summer. The dark green leaves appear on the spiny branches just after the first bloom of flowers; these are followed by yellow-green fruits. Useful as ground cover or low hedging, but may not survive in cold areas.

CULTIVATION: *Grow in well-drained, fertile soil, in sun. Trim lightly after flowering; shorten sideshoots to 2 or 3 leaves if grown against a wall.*

☀ ◊ ♀ ❀❀ ‡1m (3ft) ↔2m (6ft)

Lawson Cypresses (*Chamaecyparis lawsoniana*)

Cultivars of *C. lawsoniana* are popular evergreen conifers, available in many different shapes, sizes, and foliage colours. All have red-brown bark and dense crowns of branches that droop at the tips. The flattened sprays of dense, aromatic foliage, occasionally bearing small, rounded cones, make the larger types of Lawson cypress very suitable for thick hedging, such as bright blue-grey 'Pembury Blue', or golden-yellow 'Lanei Aurea'. Use compact cultivars in smaller gardens such as 'Ellwoodii' (to 3m/10ft) and 'Ellwood's Gold'; dwarf upright types such as 'Chilworth Silver' make eye-catching feature plants for containers, rock gardens, or borders.

CULTIVATION: *Grow in moist but well-drained soil, in sun. They tolerate chalky soil, but not exposed sites. Trim regularly from spring to autumn; do not cut into older wood. To train as formal hedges, pruning must begin on young plants.*

☼ ◊ ❀❀❀

‡1.5m (5ft) or more ↔60cm (24in)   ‡to 15m (50ft) ↔to 5m (15ft)   ‡to 15m (50ft) ↔2–5m (6–15ft)

1 *C. lawsoniana* 'Ellwood's Gold' ♀ **2** *C. lawsoniana* 'Lanei Aurea' ♀
3 *C. lawsoniana* 'Pembury Blue' ♀

Chamaecyparis nootkatensis 'Pendula'

This large and drooping conifer develops a gaunt, open crown as it matures. Hanging from the arching branches are evergreen sprays of dark green foliage with small, round cones which ripen in spring. Its unusual habit makes an interesting feature for a large garden.

CULTIVATION: *Best in full sun, in moist but well-drained, neutral to slightly acid soil; will also tolerate dry, chalky soil. Regular pruning is not required.*

☼ ◊ ◊ ❁❁❁
↕ to 30m (100ft) ↔ to 8m (25ft)

Chamaecyparis obtusa 'Nana Gracilis'

This dwarf form of Hinoki cypress is an evergreen, coniferous tree with a dense pyramidal habit. The aromatic, rich green foliage is carried in rounded, flattened sprays, bearing small cones which ripen to yellow-brown. Useful in a large rock garden, particularly to give an Oriental style. 'Nana Aurea' looks very similar but grows to only half the size.

CULTIVATION: *Grow in moist but well-drained, neutral to slightly acid soil, in full sun. Also tolerates dry, chalky soil. Regular pruning is not necessary.*

☼ ◊ ♀ ❁❁❁   ↕ 3m (10ft) ↔ 2m (6ft)

Chamaecyparis pisifera 'Boulevard'

A broad, evergreen, conifer that develops into a conical tree with an open crown. The soft, blue-green foliage is borne in flattened sprays with angular green cones, maturing to brown. Very neat and compact in habit; an excellent specimen tree for poorly drained, damp soil.

CULTIVATION: *Grow in reliably moist, preferably neutral to acid soil, in full sun. No regular pruning is required.*

☼ ◐ ♦ ♥ ❀❀❀
‡10m (30ft) ↔ to 5m (15ft)

Chamaerops humilis

The dwarf fan palm is a bushy foliage plant with several shaggy, fibrous stems topped by fan-like leaves with spiny stalks. Unspectacular flowers appear from spring. Owing to its tenderness – it will only survive short spells just below freezing – it is best grown as a specimen plant in a cool greenhouse or outdoors in a container, which can be moved under cover in winter.

CULTIVATION: *Grow in well-drained, reasonably fertile soil in sun. In a container, choose a soil-based potting compost and feed monthly in summer.*

☼ ☀ ◊ ♥ ❀
‡2–3m (6–10ft) ↔ 1–2m (3–6ft)

Chimonanthus praecox 'Grandiflorus'

Wintersweet is a vigorous, upright, deciduous shrub grown for the fragrant flowers borne on its bare branches in winter; on this cultivar they are large, cup-shaped, and deep yellow with maroon stripes inside. The leaves are mid-green. Suitable for a shrub border or for training against a warm, sunny wall.

CULTIVATION: *Grow in well-drained, fertile soil, in a sunny, sheltered site. Best left unpruned when young so that mature flowering wood can develop. Cut back flowered stems of wall-trained plants in spring.*

☼ ◊ ♀ ❀❀❀   ‡4m (12ft) ↔3m (10ft)

Chionodoxa luciliae

Glory of the snow is a small, bulbous perennial bearing star-shaped, clear blue flowers with white eyes, in early spring. The glossy green leaves are usually curved backwards. Grow in a sunny rock garden, or naturalize under deciduous trees. Sometimes referred to as *C. gigantea* of gardens. *C. forbesii* is very similar, with more erect leaves.

CULTIVATION: *Grow in any well-drained soil, with a position in full sun. Plant bulbs 8cm (3in) deep in autumn.*

☼ ◊ ♀ ❀❀❀
‡15cm (6in) ↔3cm (1¹/₄in)

Choisya ternata

Mexican orange blossom is a fast-growing, rounded, evergreen shrub valued for its attractive foliage and fragrant flowers. The aromatic, dark green leaves are divided into three. Clusters of star-shaped white flowers appear in spring. A fine, pollution-tolerant shrub for town gardens, but prone to frost damage in exposed sites. 'Aztec Pearl' is similar, its flowers perhaps not quite as fragrant.

CULTIVATION: *Grow in well-drained, fairly fertile soil, in full sun. Naturally forms a well-shaped bush without pruning. Cutting back flowered shoots encourages a second flush of flowers.*

☼ ◊ ♀ ❋ ❋ ❋   ‡↔2.5m (8ft)

Choisya ternata
SUNDANCE

This slower-growing, bright yellow-leaved variety of Mexican orange blossom is a compact, evergreen shrub. The aromatic leaves, divided into three leaflets, are a duller yellow-green if positioned in shade. Flowers are rare. Grow against a warm wall, with extra protection in cold climates.

CULTIVATION: *Best in well-drained, fertile soil, in full sun for the best leaf colour. Provide shelter from cold winds. Trim wayward shoots in summer, removing any frost-damaged shoots in spring.*

☼ ◊ ♀ ❋ ❋   ‡↔2.5m (8ft)

Garden Chrysanthemums

These upright, bushy perennials are a mainstay of the late border, with bright, showy flowerheads traditionally used for display and cutting. The lobed or feathery leaves are aromatic and bright green. They flower from late summer to mid-autumn, depending on the cultivar; very late-flowering cultivars are best raised under glass in frost-prone climates (see p.125). The flower form, although always many-petalled, varies from the daisy-like 'Pennine Alfie' to the blowsy, reflexed blooms of 'George Griffith'. Lift in autumn and store over winter in frost-free conditions. Plant out after any risk of frost has passed.

CULTIVATION: *Grow in moist but well-drained, neutral to slightly acid soil enriched with well-rotted manure, in a sunny, sheltered site. Stake tall flower stems. Apply a balanced fertilizer when in growth, until flower buds begin to show.*

☼ ◊ ◑ ✿ ✿

‡1.2m (4ft) ↔75cm (30in)

‡1.2m (4ft) ↔60–75cm (24–30in)

‡1m (3ft) ↔60–75cm (24–30in)

‡50cm (20in) ↔25cm (10in)

‡30–60cm (12–24in) ↔60cm (24in)

‡1.1m (3¹/₂ft) ↔60–75cm (24–30in)

1 *Chrysanthemum* 'Amber Enbee Wedding' **2** *C.* 'Amber Yvonne Arnaud'
3 *C.* 'Angora' **4** *C.* 'Bravo' ♀ **5** *C.* 'Bronze Fairie' **6** *C.* 'Cherry Nathalie'

‡1.1m (3¹/₂ft) ↔ 60–75cm (24–30in)

‡90cm (30in) ↔ 30cm (12in)

‡1.3–1.5m (4¹/₂–5ft) ↔ 75cm (30in)

‡1.2m (4ft) ↔ 75cm (30in)

‡90cm (30in) ↔ 30cm (12in)

‡60cm (24in) ↔ 30cm (12in)

‡90cm (30in) ↔ 30cm (12in)

‡1.2m (4ft) ↔ 75cm (30in)

‡1.2m (4ft) ↔ 75cm (30in)

7 _C._ 'Eastleigh' **8** _C._ 'Flo Cooper' **9** _C._ 'George Griffiths' ♀ **10** _C._ 'Madeleine'
11 _C._ 'Mancetta Bride' **12** _C._ 'Mavis' ♀ **13** _C._ 'Myss Madi' **14** _C._ 'Pennine Alfie'
15 _C._ 'Pennine Flute' ♀

‡70cm (28in) ↔ 30cm (10in)

‡65cm (25in) ↔ 30cm (12in)

‡1.2m (4ft) ↔ 60-75cm (24-30in)

‡1m (3ft) ↔ 45cm (18in)

‡1.2m (4ft) ↔ 75cm (30in)

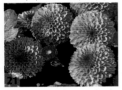

‡30-60cm (12-24in) ↔ 60cm (24in)

‡1.2m (4ft) ↔ 60-75cm (24-30in)

‡85cm (34in) ↔ 45cm (18in)

MORE CHOICES

'Margaret' Pink, with reflexed petals.
'Max Riley' Yellow.
'Pennine Signal' Scarlet.
'Yellow Pennine Oriel' Sprays of small yellow flowers.

‡1.2m (4ft) ↔ 60-75cm (24-30in)

‡1.2m (4ft) ↔ 60-75cm (24-30in)

16 *Chrysanthemum* 'Pennine Lace' **17** *C.* 'Pennine Marie' ♀ **18** *C.* 'Pennine Oriel' ♀
19 *C.* 'Primrose Allouise' ♀ **20** *C.* 'Purple Pennine Wine' **21** *C.* 'Salmon Fairie'
22 *C.* 'Salmon Margaret' **23** *C.* 'Southway Swan' **24** *C.* 'Wendy' **25** *C.* 'Yvonne Arnaud'

Late-Flowering Chrysanthemums

This group of herbaceous perennials comes into flower from autumn into winter, which means that the protection of a warm greenhouse during this time is essential in cold areas. None can tolerate frost, but they can be moved outdoors during summer. The showy flowerheads come in a wide range of shapes and sizes and are available in a blaze of golds, bronzes, yellows, oranges, pinks, and reds. Protection from the elements enables perfect blooms for exhibition to be nurtured. Grow late-flowering chrysanthemums in containers or in a greenhouse border.

CULTIVATION: *Best in a good loam-based potting compost, kept slightly moist at all times, in bright, filtered light. Stake plants as they grow, and give liquid fertilizer until flower buds begin to form. Ensure adequate ventilation and a min. temp. of 10°C (50°F). In frost-free areas, grow outdoors in fertile, moist but well-drained soil in full sun.*

☼ ◊ ◑ ❀

‡1.2m (4ft) ↔ 60cm (24in)

‡1.2–1.5m (4–5ft) ↔ 60cm (24in)

‡1.4m (4¹/₂ft) ↔ 60cm (24in)

MORE CHOICES

'Apricot Shoesmith Salmon'
'Bronze Cassandra'
'Dark Red Mayford Perfection'
'Pink Gin' Light purple.
'Rose Mayford Perfection'
'Rynoon' Light pink.

‡1.2m (4ft) ↔ 75–100cm (30–39in)

‡1.2m (4ft) ↔ 60–75cm (24–30in)

1 *Chrysanthemum* 'Beacon' ♀ **2** *C.* 'Golden Cassandra' ♀ **3** *C.* 'Roy Coopland' ♀
4 *C.* 'Satin Pink Gin' **5** *C.* 'Yellow John Hughes' ♀

Chusquea culeou

This is a graceful bamboo, which will form a dense, fountain-like thicket of whiskered, glossy, olive-green canes with papery white leaf sheaths. The narrow leaves are subtly chequered and mid-green. It is a perfect specimen for a spacious woodland garden. It also makes a good screening plant and is tolerant of coastal conditions.

CULTIVATION: *Grow in moist but well-drained, enriched soil in sun or partial shade.*

☼ ☼ ◊ ◊ ♀ ❀ ❀ ❀
↕ up to 6m (20ft) ↔ 2.5m (8ft) or more

Cistus x *aguilarii* 'Maculatus'

This fast-growing, evergreen shrub bears large, solitary white flowers for a few weeks in early and midsummer. At the centre of each flower is a mass of bright golden yellow stamens surrounded by five crimson blotches. The lance-shaped leaves are sticky, aromatic and bright green. Excellent on a sunny bank or in containers. May perish in cold areas.

CULTIVATION: *Grow in well-drained, poor to moderately fertile soil, in a sunny, sheltered site. Lime-tolerant. If necessary, trim lightly in early spring or after flowering, but do not prune hard.*

☼ ◊ ♀ ❀ ❀
↕ ↔ 1.2m (4ft)

Cistus x *dansereaui* 'Decumbens'

This evergreen rock rose makes effective shrubby ground cover. From early to midsummer, white, yellow-centred, crimson-blotched flowers appear over many weeks. The narrow leaves are glossy and dark green, borne on sticky shoots. Cistus prefers sunny sites and is suited to a rock garden, raised bed, or stone wall. The plants are tolerant of limey or chalky soils.

CULTIVATION: *Grow in well-drained soil in sun. Pinch back after flowering. Remove wayward stems but do not prune hard.*

☼ ◊ ♀ ❀❀   ‡60cm (24in) ↔1m (3ft)

Cistus x *purpureus*

This summer-flowering, rounded, evergreen shrub bears few-flowered clusters of dark pink flowers with maroon blotches at the base of each petal. The dark green leaves are borne on upright, sticky, red-flushed shoots. Good in a large rock garden, on a sunny bank, or in a container. May not survive cold winters.

CULTIVATION: *Grow in well-drained, poor to moderately fertile soil. Choose a sheltered site in full sun. Tolerates lime. Can be trimmed lightly after flowering, but avoid hard pruning.*

☼ ◊ ♀ ❀❀   ‡↔1m (3ft)

Early-flowering Clematis

The early-flowering "species" clematis are valued for their showy displays during spring and early summer. They are generally deciduous climbers with mid- to dark green, divided leaves. The flowers of the earliest species to bloom are usually bell-shaped; those of the later *C. montana* types are either flat or saucer-shaped. Flowers are often followed by decorative seedheads. Many clematis, especially *C. montana* types, are vigorous and will grow very quickly, making them ideal for covering featureless or unattractive walls. Allowed to grow through deciduous shrubs, they may flower before their host comes into leaf.

CULTIVATION: *Grow in well-drained, fertile, humus-rich soil, in full sun or semi-shade. The roots and base of the plant should always be shaded. Tie in young growth carefully and prune out shoots that exceed the allotted space, immediately after flowering.*

☼ ☀ ◊ ❀❀❀

‡2–3m (6–10ft) ↔1.5m (5ft)

‡10m (30ft) ↔4m (12ft)

MORE CHOICES

C. cirrhosa var. *balearica* and *C. cirrhosa* 'Freckles' Cream flowers speckled with pinky-brown.

C. 'Helsingborg' Deep blue and brown.

C. 'White Columbine' White flowers.

‡10m (30ft) ↔2–3m (6–10ft)

‡5m (15ft) ↔2–3m (6–10ft)

1 *Clematis* 'Frances Rivis' ♀ **2** *C.* 'Markham's Pink' ♀ **3** *C. montana* var. *grandiflora* ♀
4 *C. montana* var. *rubens* **5** *C. montana* var. *rubens* 'Tetrarose' ♀

Mid-season Clematis

The mid-season, mainly hybrid clematis are twining and deciduous climbers, bearing an abundance of stunning flowers throughout the summer months. Their flowers are large, saucer-shaped, and outward-facing with a plentiful choice of shapes and colours. Towards the end of the summer, blooms may darken. The leaves are pale to mid-green and divided into several leaflets. Mid-season clematis look very attractive scrambling through other shrubs, especially if they bloom before or after their host. Top growth may be damaged in severe winters, but plants are usually quick to recover.

CULTIVATION: *Grow in well-drained, fertile, humus-rich soil, with the roots in shade and the heads in sun. Pastel flowers may fade in sun; better in semi-shade. Mulch in late winter, avoiding the immediate crown. Tie in young growth carefully, cutting back older stems to strong buds in late winter.*

☼ ☀ ◊ ✿✿✿

1
‡2.5m (8ft) ↔1m (3ft)

2
‡2.5m (8ft) ↔1m (3ft)

3
‡2-3m (6-10ft) ↔1m (3ft)

1 *Clematis* 'Bees' Jubilee' **2** *C.* 'Doctor Ruppel' **3** *C.* 'Elsa Späth'

‡to 3m (10ft) ↔ 1m (3ft)

5

‡2.5m (8ft) ↔ 1m (3ft)

6

‡2.5m (8ft) ↔ 1m (3ft)

7

‡3m (10ft) ↔ 1m (3ft)

8

‡2.5m (8ft) ↔ 1m (3ft)

9

‡3m (10ft) ↔ 1m (3ft)

10

‡2.5m (8ft) ↔ 1m (3ft)

11

‡2-3m (6-10ft) ↔ 1m (3ft)

4 *C.* 'Fireworks' **5** *C.* 'Gillian Blades' ♀ **6** *C.* 'H. F. Young' **7** *C.* 'Henryi'
8 *C.* 'Lasurstern' ♀ **9** *C.* 'Marie Boisselot' ♀ **10** *C.* 'Miss Bateman' **11** *C.* 'Nelly Moser' ♀

‡2–3m (6–10ft) ↔1m (3ft)

‡2–3m (6–10ft) ↔1m (3ft)

‡2m (6ft) ↔1m (3ft)

15

‡2m (6ft) ↔1m (3ft)

MORE CHOICES

'Lord Nevill' Deep blue flowers with purple-red anthers.

'Mrs Cholmondely' Lavender, brown anthers. 'Will Goodwin' Pale blue with yellow anthers.

‡2–3m (6–10ft) ↔1m (3ft)

‡2–3m (6–10ft) ↔1m (3ft)

12 *C.* 'Niobe' ♀ **13** *C.* 'Richard Pennell' ♀ **14** *C.* 'Royalty' **15** *C.* 'Silver Moon'
16 *C.* 'The President' ♀ **17** *C.* 'Vyvyan Pennell'

Late-flowering Clematis

Many large-flowered hybrid clematis flower from mid- to late summer, when the season for the *C. viticella* types, characterized usually by smaller but more profuse flowers, also begins. These are followed by other late-flowering species. They may be deciduous or evergreen, with an enormous variety of flower and leaf shapes and colours. Many, like 'Perle d'Azur', are vigorous and will cover large areas of wall or disguise unsightly buildings. May develop decorative, silvery-grey seedheads which last well into winter. With the exception of 'Bill Mackenzie', most look good when trained up into small trees.

CULTIVATION: *Grow in humus-rich, fertile soil with good drainage, with the base in shade and the upper part in sun or partial shade. Mulch in late winter, avoiding the immediate crown. Cut back hard each year before growth begins, in early spring.*

☼ ☀ ◊ ❊ ❊ ❊

1
↕4m (12ft) ↔1.5m (5ft)

2
↕7m (22ft) ↔2–3m (6–10ft)

3
↕2.5m (8ft) ↔1.5m (5ft)

1 *C.* 'Alba Luxurians' **2** *C.* 'Bill Mackenzie' ♥ **3** *C.* 'Duchess of Albany'

‡2–3m (6–10ft) ↔1m (3ft)

‡3m (10ft) ↔1.5m (5ft)

‡3–5m (10–15ft) ↔1.5m (5ft)

‡3m (10ft) ↔1m (3ft)

‡3m (10ft) ↔1m (3ft)

‡3m (10ft) ↔1m (3ft)

‡3m (10ft) ↔1m (3ft)

‡3m (10ft) ↔1m (3ft)

‡6–7m (20–22ft) ↔2–3m (6–10ft)

4 *C.* 'Comtesse de Bouchaud' ♀ **5** *C.* 'Etoile Violette' ♀ **6** *C.* 'Jackmanii' ♀
7 *C.* 'Madame Julia Correvon' ♀ **8** *C.* 'Minuet' ♀ **9** *C.* 'Perle d'Azur' **10** *C.* 'Venosa Violacea' ♀ **11** *C.* 'Purpurea Plena Elegans' ♀ **12** *C. rehderiana* ♀

Clianthus puniceus

Lobster claw is an evergreen, woody-stemmed, climbing shrub with scrambling shoots. Drooping clusters of claw-like, brilliant red flowers are seen in spring and early summer. (The cultivar 'Albus' has pure white flowers.) The mid-green leaves are divided into many oblong leaflets. Suitable for wall-training; it needs the protection of a cool conservatory in frost-prone climates.

CULTIVATION: *Grow in well-drained, fairly fertile soil, in sun with shelter from wind. Pinch-prune young plants to promote bushiness; otherwise, keep pruning to a minimum.*

☼ ◊ ♀ ❀   ‡4m (12ft) ↔3m (10ft)

Codonopsis convolvulacea

A slender, herbaceous, summer-flowering climber with twining stems that bears delicate, bell- to saucer-shaped, soft blue-violet flowers. The leaves are lance-shaped to oval and bright green. Grow in a herbaceous border or woodland garden, scrambling through other plants.

CULTIVATION: *Grow in moist but well-drained, light, fertile soil, ideally in dappled shade. Provide support or grow through neighbouring shrubs. Cut to the base in spring.*

☼ ◊ ◗ ❀❀❀   ‡to 2m (6ft)

Colchicum speciosum 'Album'

While *Colchicum speciosum*, the autumn crocus, has pink flowers, this cultivar is white. A cormous perennial, it produces thick, weather-resistant, goblet-shaped, snow-white blooms. Its narrow, mid-green leaves appear in winter or spring after the flowers have died down. Grow at the front of a border, at the foot of a bank, or in a rock garden. All parts are highly toxic if ingested.

CULTIVATION: *Grow in moist but well-drained soil, in full sun. Plant bulbs in early autumn, 10cm (4in) below the surface of soil that is deep and fertile.*

☼ ◊ ♀ ❀ ❀ ❀ ‡18cm (7in) ↔10cm (4in)

Convallaria majalis

Lily-of-the-valley is a creeping perennial bearing small, very fragrant white flowers which hang from arching stems in late spring or early summer. The narrowly oval leaves are mid- to dark green. An excellent ground cover plant for woodland gardens and other shady areas; it will spread rapidly under suitable conditions.

CULTIVATION: *Grow in reliably moist, fertile, humus-rich, leafy soil in deep or partial shade. Top-dress with leaf mould in autumn.*

☼ ☼ ◊ ♀ ❀ ❀ ❀
‡23cm (9in) ↔30cm (12in)

Convolvulus cneorum

This compact, rounded, evergreen shrub bears masses of funnel-shaped, shining white flowers with yellow centres, which open from late spring to summer. The narrowly lance-shaped leaves are silvery-green. Excellent as a largish plant in a rock garden, or on a sunny bank. In areas with cold, wet winters, grow in a container and move into a cool greenhouse in winter.

CULTIVATION: *Grow in gritty, very well-drained, poor to moderately fertile soil, in a sunny, sheltered site. Trim back after flowering, if necessary.*

☼ ◊ ♀ ❀❀ ‡60cm (24in) ↔90cm (36in)

Convolvulus sabatius

A small, trailing perennial bearing trumpet-shaped, vibrant blue-purple flowers from summer into early autumn. The slender stems are clothed with small, oval, mid-green leaves. Excellent in crevices between rocks; in cold areas, best grown in containers with shelter under glass during the winter. Sometimes seen as *C. mauritanicus*.

CULTIVATION: *Grow in well-drained, gritty, poor to moderately fertile soil. Provide a sheltered site in full sun.*

☼ ◊ ♀ ❀❀ ‡15cm (6in) ↔50cm (20in)

Coreopsis verticillata 'Moonbeam'

Throughout summer, the light-green, feathery leaves of 'Moonbeam' are smothered by lemon yellow, daisy-shaped flowers, which are attractive to bees. This small, drought-resistant, pretty perennial is suited to edging paths and borders.

CULTIVATION: *Grow in fertile, well-drained soil in full sun or partial shade. Deadhead to prolong flowering.*

☼ ◊ ◑ ❋ ❋ ❋   ‡↔to 50cm (20in)

Cornus alba 'Sibirica'

A deciduous shrub that is usually grown for the winter effect of its bright coral-red, bare young stems. Small clusters of creamy-white flowers appear in late spring and early summer amid oval, dark green leaves which turn red in autumn. Particularly effective in a waterside planting or any situation where the winter stems show up well.

CULTIVATION: *Grow in any moderately fertile soil, in sun. For the best stem effect, cut back hard and feed every spring once established, although this will be at the expense of the flowers.*

☼ ◊ ◑ ♀ ❋ ❋ ❋   ‡↔3m (10ft)

Cornus alba 'Spaethii'

A vigorous and upright, deciduous shrub bearing bright green, elliptic leaves which are margined with yellow. It is usually grown for the effect of its bright red young shoots in winter. In late spring and early summer, small clusters of creamy-white flowers appear amid the foliage. Very effective wherever the stems show up well in winter.

CULTIVATION: *Grow in any moderately fertile soil, preferably in full sun. For the best stem effect, but at the expense of any flowers, cut back hard and feed each year in spring, once established.*

☼ ◊ ◊ ♀ ❀ ❀ ❀   ‡↔3m (10ft)

Cornus florida 'Cherokee Chief'

This small deciduous tree has an attractive rounded crown. It produces tiny flowers with striking ruby-pink bracts in spring, and red fruits and reddish leaves in autumn. Break-resistant branches make it ideal for planting near patios. Alternatively, grow it at the back of a border for spring and autumn colour, or in the lawn as a specimen.

CULTIVATION: *Grow in fertile, humus-rich, well-drained neutral to acid soil. Shade from afternoon sun. Water well during drought.*

☼ ☀ ◊ ❀ ❀ ❀
‡to 6m (20ft) ↔to 8m (25ft)

Cornus kousa var. *chinensis*

This broadly conical, deciduous tree with flaky bark is valued for its dark green, oval leaves which turn an impressive, deep crimson-purple in autumn. The early summer flowers have long white bracts, fading to red-pink. An effective specimen tree, especially in a woodland setting. *C. kousa* 'Satomi' has even deeper autumn colour, and pink flower bracts.

CULTIVATION: *Best in well-drained, fertile, neutral to acid soil that is rich in humus, in full sun or partial shade. Keep pruning to a minimum.*

☼ ☀ ◊ ❀❀❀ ↕7m (22ft) ↔5m (15ft)

Cornus mas

The cornelian cherry is a vigorous and spreading, deciduous shrub or small tree. Clusters of small yellow flowers provide attractive late winter colour on the bare branches. The oval, dark green leaves turn red-purple in autumn, giving a display at the same time as the fruits ripen to red. Particularly fine as a specimen tree for a woodland garden.

CULTIVATION: *Tolerates any well-drained soil, in sun or partial shade. Pruning is best kept to a minimum.*

☼ ☀ ◊ ❀❀❀ ↕↔5m (15ft)

Cornus sericea 'Flaviramea'

This vigorous, deciduous shrub makes a bright display of its bare yellow-green young shoots in winter, before the oval, dark green leaves emerge in spring. Clusters of white flowers appear in late spring and early summer. The leaves redden in autumn. Excellent in a bog garden or in wet soil near water.

CULTIVATION: *Grow in reliably moist soil, in full sun. Restrict spread by cutting out 1 in 4 old stems annually. Prune all stems hard and feed each year in early spring for the best display of winter stems.*

☼ ◑ ♦ ♀ ❀ ❀ ❀   ‡2m (6ft) ↔4m (12ft)

Correa backhouseana

The Australian fuchsia is a dense, spreading, evergreen shrub with small clusters of tubular, pale red-green or cream flowers during late autumn to late spring. The hairy, rust-red stems are clothed with oval, dark green leaves. In cold climates, grow against a warm wall or over-winter in frost-free conditions. *Correa* 'Mannii', with red flowers, is also recommended.

CULTIVATION: *Grow in well drained, fertile, acid to neutral soil, in full sun. Trim back after flowering, if necessary.*

☼ ◊ ♀ ❀
‡1–2m (3–6ft) ↔1.5–2.5m (5–8ft)

Cortaderia selloana 'Aureolineata'

This pampas grass, with rich yellow-margined, arching leaves which age to dark golden-yellow, is a clump-forming, evergreen perennial, also known as 'Gold Band'. Feathery plumes of silvery flowers appear on tall stems in late summer. May not survive in cold areas. The flowerheads can be dried for decoration.

CULTIVATION: *Grow in well-drained, fertile soil, in full sun. In late winter, cut out all dead foliage and remove the previous year's flower stems: wear gloves to protect hands from the sharp foliage.*

☼ ◊ ♀ ❀ ❀
‡to 2.2m (7ft) ↔1.5m (5ft) or more

Cortaderia selloana 'Sunningdale Silver'

This sturdy pampas grass is a clump-forming, evergreen, weather-resistant perennial. In late summer, silky plumes of silvery-cream flowers are borne on strong, upright stems above the narrow, arching, sharp-edged leaves. In cold areas, protect the crown with a dry winter mulch. The flowerheads can be dried.

CULTIVATION: *Grow in well-drained, fertile, not too heavy soil, in full sun. Remove old flower stems and any dead foliage in late winter: wear gloves to protect hands from the sharp foliage.*

☼ ◊ ♀ ❀ ❀
‡3m (10ft) or more ↔to 2.5m (8ft)

Corydalis solida '**George Baker**'

A low, clump-forming, herbaceous perennial bearing upright spires of deep salmon-rose flowers. These appear in spring above the delicate, finely cut, greyish-green leaves. Excellent in a rock garden or in an alpine house.

CULTIVATION: *Grow in sharply drained, moderately fertile soil or compost. Site in full sun, but tolerates some shade.*

☼ ☀ ◊ ♀ ✿✿✿
‡to 25cm (10in) ↔to 20cm (8in)

Corylopsis pauciflora

This deciduous shrub bears hanging, catkin-like clusters of small, fragrant, pale yellow flowers on its bare branches during early to mid-spring. The oval, bright green leaves are bronze when they first emerge. Often naturally well-shaped, it makes a beautiful shrub for sites in dappled shade. The flowers may be damaged by frost.

CULTIVATION: *Grow in moist but well-drained, humus-rich, acid soil, in partial shade with shelter from wind. Allow room for the plant to spread. The natural shape is easily spoilt, so prune only to remove dead wood.*

☼ ◊ ♦ ♀ ✿✿✿ ‡1.5m (5ft) ↔2.5m (8ft)

Corylus avellana 'Contorta'

The corkscrew hazel is a deciduous shrub bearing strongly twisted shoots that are particularly striking in winter; they can also be useful in flower arrangements. Winter interest is enhanced later in the season with the appearance of pale yellow catkins. The mid-green leaves are almost circular and toothed.

CULTIVATION: *Grow in any well-drained, fertile soil, in sun or semi-shade. Once established, the twisted branches tend to become congested and may split, so thin out in late winter.*

☼ ☀ ◊ ♀ ✿✿✿   ↕↔15m (50ft)

Corylus maxima 'Purpurea'

The purple filbert is a vigorous, open, deciduous shrub that, left unpruned, will grow into a small tree. In late winter, purplish catkins appear before the rounded, purple leaves emerge. The edible nuts ripen in autumn. Effective as a specimen tree, in a shrub border, or as part of a woodland planting.

CULTIVATION: *Grow in any well-drained, fertile soil, in sun or partial shade. For the best leaf effect, but at the expense of the nuts, cut back hard in early spring.*

☼ ☀ ◊ ✿✿✿
↕to 6m (20ft) ↔5m (15ft)

Cosmos bipinnatus 'Sonata White'

A branching but compact annual bearing single, saucer-shaped white flowers with yellow centres, from summer to autumn, at the tips of upright stems. The leaves are bright green and feathery. Excellent for exposed gardens. The flowers are good for cutting.

CULTIVATION: *Grow in moist but well-drained, fertile soil, in full sun. Deadhead to prolong flowering.*

☼ ◊ ◑ ❋ ❋ ❋   ‡↔30cm (12in)

Cotinus coggygria 'Royal Purple'

This deciduous shrub is grown for its rounded, red-purple leaves which turn a brilliant scarlet in autumn. Smoke-like plumes of tiny, pink-purple flowers are produced on older wood, but only in areas with long, hot summers. Good in a shrub border or as a specimen tree; where space permits, plant in groups.

CULTIVATION: *Grow in moist but well-drained, fairly fertile soil, in full sun or partial shade. For the best foliage effect, cut back hard to a framework of older wood each spring, before growth begins.*

☼ ☼ ◊ ◑ ♀ ❋ ❋ ❋   ‡↔5m (15ft)

Cotinus 'Grace'

A fast-growing, deciduous shrub or
small tree carrying oval, purple leaves
which turn a brilliant, translucent red
in late autumn. The smoke-like clusters
of tiny, pink-purple flowers only appear
in abundance during hot summers.
Effective on its own or in a border. For
green leaves during the summer but
equally brilliant autumn foliage colour,
look for *Cotinus* 'Flame'.

CULTIVATION: *Grow in moist but well-
drained, reasonably rich soil, in sun or
partial shade. The best foliage is seen
after hard pruning each spring, just
before new growth begins.*

☼ ☀ ◊ ◑ ✿✿✿
‡6m (20ft) ↔ 5m (15ft)

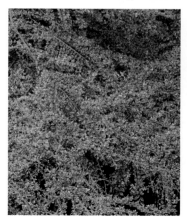

Cotoneaster atropurpureus 'Variegatus'

This compact, low-growing shrub,
sometimes seen as *C. horizontalis*
'Variegatus', has fairly inconspicuous
red flowers in summer, followed in
autumn by a bright display of
orange-red fruits. The small, oval,
white-margined, deciduous leaves
also give autumn colour, turning pink
and red before they drop. Effective as
ground cover or in a rock garden.

CULTIVATION: *Grow in well-drained,
moderately fertile soil, in full sun.
Tolerates dry soil and partial shade.
Pruning is best kept to a minimum.*

☼ ☀ ◊ ♀ ✿✿✿
‡45cm (18in) ↔ 90cm (36in)

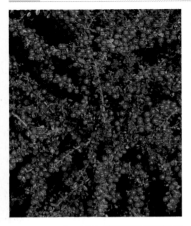

Cotoneaster conspicuus 'Decorus'

A dense, mound-forming, evergreen shrub that is grown for its shiny red berries. These ripen in autumn and will often persist until late winter. Small white flowers appear amid the dark green leaves in summer. Good in a shrub border or under a canopy of deciduous trees.

CULTIVATION: *Grow in well-drained, moderately fertile soil, ideally in full sun, but it tolerates shade. If necessary, trim lightly to shape after flowering.*

☼ ☀ ◊ ♀ ❋ ❋ ❋
‡1.5m (5ft) ↔2–2.5m (6–8ft)

Cotoneaster horizontalis

A deciduous shrub with spreading branches that form a herringbone pattern. The tiny, pinkish-white flowers, which appear in summer, are attractive to bees. Bright red berries ripen in autumn, and the glossy, dark green leaves redden before they fall. Good as groundcover, but most effective when grown flat up against a wall.

CULTIVATION: *Grow in any but water-logged soil. Site in full sun for the best berries, or semi-shade. Keep pruning to a minimum; if wall-trained, shorten outward-facing shoots in late winter.*

☼ ☀ ◊ ◊ ❋ ❋ ❋
‡1m (3ft) ↔1.5m (5ft)

Cotoneaster lacteus

This dense, evergreen shrub has arching branches that bear clusters of small, cup-shaped, milky-white flowers from early to midsummer. These are followed by brilliant red berries in autumn. The oval leaves are dark green and leathery, with grey-woolly undersides. Ideal for a wildlife garden, since it provides food for bees and birds; also makes a good windbreak or informal hedge.

CULTIVATION: *Grow in well-drained, fairly fertile soil, in sun or semi-shade. Trim hedges lightly in summer, if necessary; keep pruning to a minimum.*

☼ ☼ ◊ ♀ ❀ ❀ ❀   ↕↔4m (12ft)

Cotoneaster simonsii

An upright, deciduous or semi-evergreen shrub with small, cup-shaped white flowers in summer. The bright orange-red berries that follow ripen in autumn and persist well into winter. Autumn colour is also seen in the glossy leaves which redden from dark green. Good for hedging, and can also be clipped fairly hard to a semi-formal outline.

CULTIVATION: *Grow in any well-drained soil, in full sun or partial shade. Clip hedges to shape in late winter or early spring, or allow to grow naturally.*

☼ ☼ ◊ ♀ ❀ ❀ ❀   ↕3m (10ft)↔2m (6ft)

Cotoneaster sternianus

This graceful, evergreen or semi-evergreen shrub bears arching branches which produce clusters of pink-tinged white flowers in summer, followed by a profusion of large, orange-red berries in autumn. The grey-green leaves have white under-sides. Good grown as a hedge.

CULTIVATION: *Grow in any well-drained soil, in sun or semi-shade. Trim hedges lightly after flowering, if necessary; pruning is best kept to a minimum.*

☼ ☀ ◊ ♀ ❀ ❀ ❀   ‡↔3m (10ft)

Cotoneaster x *watereri* 'John Waterer'

A fast-growing, evergreen or semi-evergreen shrub or small tree valued for the abundance of red berries that clothe its branches in autumn. In summer, clusters of white flowers are carried among the lance-shaped, dark green leaves. Good on its own or at the back of a shrub border.

CULTIVATION: *Grow in any but water-logged soil, in sun or semi-shade. When young, cut out any badly placed shoots to develop a framework of well-spaced branches. Thereafter, keep pruning to an absolute minimum.*

☼ ☀ ◊ ◊ ❀ ❀ ❀   ‡↔5m (15ft)

Crambe cordifolia

A tall, clump-forming, vigorous perennial grown for its stature and its fragrant, airy, billowing sprays of small white flowers which are very attractive to bees. These are borne on strong stems in summer above the large, elegant, dark green leaves. Magnificent in a mixed border, but allow plenty of space.

CULTIVATION: *Grow in any well-drained, preferably deep, fertile soil, in full sun or partial shade. Provide shelter from strong winds.*

☼ ☀ ◊ ♀ ✽✽✽
‡to 2.5m (8ft) ↔1.5m (5ft)

Crataegus laevigata 'Paul's Scarlet'

This rounded, thorny, deciduous tree is valued for its long season of interest. Abundant clusters of double, dark pink flowers appear from late spring to summer, followed in autumn by small red fruits. The leaves, divided into three or five lobes, are a glossy mid-green. A particularly useful specimen tree for a town, coastal, or exposed garden. The very similar 'Rosea Flore Pleno' makes a good substitute.

CULTIVATION: *Grow in any but water-logged soil, in full sun or partial shade. Pruning is best kept to a minimum.*

☼ ☀ ◊ ◊ ♀ ✽✽✽   ‡↔8m (25ft)

Crataegus x *lavalleei* 'Carrierei'

This vigorous hawthorn is a broadly spreading, semi-evergreen tree with thorny shoots and leathery green leaves which turn red in late autumn and winter. Flattened clusters of white flowers appear in early summer, followed in autumn by round red fruits that persist into winter. Tolerates pollution, so is good for a town garden. Often listed just as *C.* x *lavallei*.

CULTIVATION: *Grow in any but water-logged soil, in full sun or partial shade. Pruning is best kept to a minimum.*

☼ ☀ ◊ ◊ ♀ ❀❀❀
‡7m (22ft) ↔10m (30ft)

Crinodendron hookerianum

The lantern tree is an upright, evergreen shrub, so-called because of its large, scarlet to carmine-red flowers, which hang from the upright shoots during late spring and early summer. The leaves are narrow and glossy dark green. It dislikes alkaline soils, and may not survive in areas with cold winters.

CULTIVATION: *Grow in moist but well-drained, fertile, humus-rich, acid soil, in partial shade with protection from cold winds. Tolerates a sunny site if the roots are kept cool and shaded. Trim lightly after flowering, if necessary.*

☼ ◊ ♀ ❀❀   ‡6m (20ft) ↔5m (15ft)

Crinum x *powellii* 'Album'

A sturdy, bulbous perennial bearing clusters of up to ten large, fragrant, widely flared, pure white flowers on upright stems in late summer and autumn. *C.* x *powellii* is equally striking, with pink flowers. The strap-shaped leaves, to 1.5m (5ft) long, are mid-green and arch over. In cold areas, choose a sheltered site and protect the dormant bulb over winter with a deep, dry mulch.

CULTIVATION: *Grow in deep, moist but well-drained, fertile soil that is rich in humus, in full sun with shelter from frost and cold, drying winds.*

☼ ◊ ❀ ❀   ↕1.5m (5ft) ↔30cm (12in)

Crocosmia x *crocosmiiflora* 'Solfatare'

This clump-forming perennial produces spikes of funnel-shaped, apricot-yellow flowers on arching stems in midsummer. The bronze-green, deciduous leaves are strap-shaped, emerging from swollen corms at the base of the stems. Excellent in a border; the flowers are good for cutting. May be vulnerable to frost in cold areas.

CULTIVATION: *Grow in moist but well-drained, fertile, humus-rich soil, in full sun. Provide a dry mulch over winter.*

☼ ◊ ♀ ❀ ❀
↕60–70cm (24–28in) ↔8cm (3in)

Crocosmia 'Lucifer'

A robust, clump-forming perennial with swollen corms at the base of the stems. These give rise to pleated, bright green leaves and, in summer, arching spikes of upward-facing red flowers. Particularly effective at the edge of a shrub border or by a pool.

CULTIVATION: *Grow in moist but well-drained, moderately fertile, humus-rich soil. Site in full sun or dappled shade.*

☼ ☀ ◊ ◊ ♀ ✿✿✿
‡1–1.2m (3–4ft) ↔8cm (3in)

Crocosmia masoniorum

A robust, late-summer-flowering perennial bearing bright vermilion, upward-facing flowers. These are carried above the dark green foliage on arching stems. The flowers and foliage emerge from a swollen, bulb-like corm. Thrives in coastal gardens. Where frosts are likely, grow in the shelter of a warm wall.

CULTIVATION: *Best in moist but well-drained, fairly fertile, humus-rich soil, in full sun or partial shade. In cold areas, provide a dry winter mulch.*

☼ ☀ ◊ ♀ ✿✿   ‡1.2m (4ft) ↔8cm (3in)

Spring-flowering Crocus

Spring-flowering crocus are indispensable dwarf perennials as they bring a welcome splash of early spring colour into the garden. Some cultivars of *C. sieberi*, such as 'Tricolor' or 'Hubert Edelstein', bloom even earlier, in late winter. The goblet-shaped flowers emerge from swollen, underground corms at the same time as or just before the narrow, almost upright foliage. The leaves are mid-green with silver-green central stripes, and grow markedly as the blooms fade. They are very effective in drifts at the front of a mixed or herbaceous border, or in massed plantings in rock gardens or raised beds.

CULTIVATION: *Grow in gritty, well-drained, poor to moderately fertile soil, in full sun. Water freely during the growing season, and apply a low-nitrogen fertilizer each month. C. corsicus must be kept completely dry over summer. Can be naturalized under the right growing conditions.*

☀ ◊ ✿✿✿

‡8–10cm (3–4in) ↔ 4cm (1½in)

‡7cm (3in) ↔ 5cm (2in)

‡5–8cm (2–3in) ↔ 2.5cm (1in)

‡5–8cm (2–3in) ↔ 2.5cm (1in)

‡5–8cm (2–3in) ↔ 2.5cm (1in)

1 *C. corsicus* ♀ **2** *C. chrysanthus* 'E.A. Bowles' ♀ **3** *C. sieberi* 'Albus'
4 *C. sieberi* 'Tricolor' ♀ **5** *C. sieberi* 'Hubert Edelsten' ♀

Autumn-flowering Crocus

These crocus are invaluable for their late-flowering, goblet-shaped flowers with showy insides. They are dwarf perennials with underground corms which give rise to the foliage and autumn flowers. The leaves are narrow and mid-green with silver-green central stripes, appearing at the same time or just after the flowers. All types are easy to grow in the right conditions, and look excellent when planted in groups in a rock garden. Rapid-spreading crocus, like *C. ochroleucus*, are useful for naturalizing in grass or under deciduous shrubs. *C. banaticus* is effective planted in drifts at the front of a border, but do not allow it to become swamped by larger plants.

CULTIVATION: *Grow in gritty, well-drained, poor to moderately fertile soil, in full sun. Reduce watering during the summer months for all types except* C. banaticus, *which prefers damper soil and will tolerate partial shade.*

☼ ◊ ❀ ❀ ❀

1 ‡10cm (4in) ↔ 5cm (2in)

2 ‡10cm (4in) ↔ 5cm (2in)

3 ‡6–8cm (2½–3in) ↔ 5cm (2in)

4 ‡8cm (3in) ↔ 2.5cm (1in)

5 ‡5cm (2in) ↔ 2.5cm (1in)

6 ‡10–12cm (4–5in) ↔ 4cm (1½in)

1 *C. banaticus* ♀ **2** *C. goulimyi* ♀ **3** *C. kotschyanus* ♀ **4** *C. medius* **5** *C. ochroleucus* **6** *C. pulchellus* ♀

Cryptomeria japonica 'Bandai-sugi'

Japanese cedars are useful in gardens, as an alternative to other conifers. They are reliable, and their leaf colour varies with the seasons. 'Bandai-sugi' is a relatively small variety that would make a good hedge. On its own, it makes a rounded shrub. The mid-green foliage turns bronze in winter.

CULTIVATION: *Grow in moist but well-drained soil, enriched with organic matter, in a sheltered site in sun or partial shade. No formal pruning is necessary.*

☼ ☀ ◊ ◊ ♀ ❀❀❀   ↕↔2m (6ft)

Cryptomeria japonica 'Elegans Compacta'

This small, slow-growing conifer looks good in a heather bed or rock garden with other dwarf conifers. It has feathery sprays of slender, soft green leaves, which turn a rich bronze-purple in winter. Mature trees usually have a neat cone shape.

CULTIVATION: *Grow in any well-drained soil, in full sun or partial shade. Needs no formal pruning, but to renovate untidy trees, cut back to within about 70cm (28in) of ground level, in spring.*

☼ ☀ ◊ ♀ ❀❀❀
↕2–4m (6–12ft) ↔2m (6ft)

x *Cuprocyparis leylandii* 'Gold Rider'

Leyland cypress trees are famous for their vigour, and they will grow very quickly to a great height if left untrimmed. As a result, they must never be planted too close to buildings as they easily get out of control. 'Gold Rider', however, is a restrained form with attractive bright yellow-green, evergreen foliage. It would make a good, thick hedge for screening purposes or as a windbreak.

CULTIVATION: *Grow in any deep, well-drained soil. The foliage colour is best in full sun. Trim regularly.*

☼ ☀ ◊ ◑ ♀ ❀❀❀ ↕↔3m (10ft) or more

x *Cupressocyparis leylandii* 'Haggerston Grey'

This popular cultivar of the Leyland cypress is a fast-growing coniferous tree with a tapering, columnar shape. It has dense, grey-green foliage, and will establish quickly planted as a screen or windbreak. There are no flowers.

CULTIVATION: *Grow in deep, well-drained soil in full sun or partial shade. Needs no formal pruning, unless grown as a hedge, when it should be trimmed 2 or 3 times during each growing season.*

☼ ☀ ◊ ❀❀❀
↕to 35m (120ft) ↔to 5m (15ft)

Cyananthus lobatus

A spreading, mat-forming perennial
grown for its late summer display
of bright blue-purple, broadly
funnel-shaped flowers, on single
stems with hairy brown calyces.
The leaves are fleshy and dull green,
with deeply cut lobes. Perfect in a
rock garden or trough.

CULTIVATION: *Grow in poor to moderately
fertile, moist but well-drained soil,
preferably neutral to slightly acid, and
rich in organic matter. Choose a site in
partial shade.*

☀ ◊ ◊ ♀ ❀❀❀
‡5cm (2in) ↔ to 30cm (12in)

Cyclamen cilicium

A tuberous perennial valued for
its slender, nodding, white or pink
flowers, borne in autumn and often
into winter among patterned, rounded
or heart-shaped, mid-green leaves.
This elegant plant needs a warm and
dry summer, when it is dormant,
and in cool-temperate climates it may
be better grown in a cool greenhouse,
or under trees or shrubs, to avoid
excessive summer moisture.

CULTIVATION: *Grow in moderately fertile,
well-drained soil enriched with organic
matter, in partial shade. Mulch annually
with leafmould after flowering when the
leaves wither.*

☀ ◊ ♀ ❀   ‡5cm (2in) ↔8cm (3in)

Cyclamen coum
Pewter Group

This winter- to spring-flowering, tuberous perennial is excellent for naturalizing beneath trees or shrubs. The compact flowers have upswept petals that vary from white to shades of pink and carmine-red. These emerge from swollen, underground tubers at the same time as the rounded, silver-green leaves. Provide a deep, dry mulch in cold areas.

CULTIVATION: *Grow in gritty, well-drained, fertile soil that dries out in summer, in sun, or light shade. Mulch annually when the leaves wither.*

☀ ☼ ◊ ♀ ❀

‡5–8cm (2–3in) ↔10cm (4in)

Cyclamen hederifolium

An autumn-flowering, tuberous perennial bearing shuttlecock-like flowers which are pale to deep pink and flushed deep maroon at the mouths. The ivy-like leaves, mottled with green and silver, appear after the flowers from a swollen, underground tuber. It self-seeds freely, forming extensive colonies under trees and shrubs, especially where protected from summer rainfall.

CULTIVATION: *Grow in well-drained, fertile soil, in sun or partial shade. Mulch each year after the leaves wither.*

☀ ☼ ◊ ♀ ❀ ❀ ❀

‡10–13cm (4–5in) ↔15cm (6in)

Cynara cardunculus

The cardoon is a clump-forming, statuesque, late-summer-flowering perennial which looks very striking in a border. The large, thistle-like purple flowerheads, which are very attractive to bees, are carried above the deeply divided, silvery leaves on stout, grey-woolly stems. The flowerheads dry well for indoor display and are very attractive to bees; when blanched, the leaf stalks can be eaten.

CULTIVATION: *Grow in any well-drained, fertile soil, in full sun with shelter from cold winds. For the best foliage effect, remove the flower stems as they emerge.*

☼ ◊ ♀ ❀ ❀ ❀   ↕1.5m (5ft) ↔1.2m (4ft)

Cytisus battandieri

Pineapple broom develops a loose and open-branched habit. It is a semi-evergreen shrub bearing dense clusters of pineapple-scented, bright yellow flowers from early to midsummer. The silvery-grey leaves, divided into three, make an attractive backdrop to herbaceous and mixed plantings. Best by a sunny wall in cold areas. The cultivar 'Yellow Tail' is recommended.

CULTIVATION: *Grow in any well-drained, not too rich soil, in full sun. Very little pruning is necessary, but old wood can be cut out after flowering, and will be replaced with strong, young growth. Resents transplanting.*

☼ ◊ ❀ ❀ ❀   ↕↔5m (15ft)

Cytisus x *beanii*

This low-growing and semi-trailing, deciduous, spring-flowering shrub carries an abundance of pea-like, rich yellow flowers on arching stems. The leaves are small and dark green. It is a colourful bush for a rock garden or raised bed, and is also effective if allowed to cascade over a wall. *Cytisus* x *ardoinei* is similar but slightly less spreading where space is limited.

CULTIVATION: *Grow in well-drained, poor to moderately fertile soil, in full sun. Trim lightly after flowering, but avoid cutting into old wood.*

☼ ◊ ♀ ❋❋❋
‡to 60cm (24in) ↔to 1m (3ft)

Cytisus x *praecox* 'Allgold'

This compact, deciduous shrub is smothered by a mass of pea-like, dark yellow flowers from mid- to late spring. The tiny, grey-green leaves are carried on arching stems. Suitable for a sunny shrub border or large rock garden. 'Warminster' is very similar, with paler, cream-yellow flowers.

CULTIVATION: *Grow in well-drained, acid to neutral soil, in sun. Pinch out the growing tips to encourage bushiness, then cut back new growth by up to two-thirds after flowering; avoid cutting into old wood. Replace old, leggy specimens.*

☼ ◊ ♀ ❋❋❋   ‡1.2m (4ft) ↔1.5m (5ft)

Daboecia cantabrica 'Bicolor'

This straggling, heather-like shrub bears slender spikes of urn-shaped flowers from spring to autumn. They are white, pink, or beetroot-red, sometimes striped with two colours. The leaves are small and dark green. Good in a heather bed or among other acid-soil-loving plants. 'Waley's Red', with glowing deep magenta flowers, is also recommended.

CULTIVATION: *Best in sandy, well-drained, acid soil, in full sun; tolerates neutral soil and some shade. Clip over lightly in early spring to remove spent flowers, but do not cut into old wood.*

☼ ☀ ◊ ❀❀❀
‡45cm (18in) ↔ 60cm (24in)

Daboecia cantabrica 'William Buchanan'

This vigorous, compact, heather-like shrub bears slender spikes of bell-shaped, purple-crimson flowers from late spring to mid-autumn. The narrow leaves are dark green above with silver-grey undersides. Good with other acid-soil-loving plants, or among conifers in a rock garden.

CULTIVATION: *Grow in well-drained, sandy, acid to neutral soil, preferably in sun, but it tolerates light shade. Trim in early to mid-spring to remove old flowers, but do not cut into old wood.*

☼ ☀ ◊ ♀ ❀❀❀
‡45cm (18in) ↔ 60cm (24in)

Dahlias

Dahlias are showy, deciduous perennials, grown as annuals in cold climates, with swollen underground tubers which should be stored in frost-free conditions in climates with cold winters. They are valued for their massive variety of brightly coloured flowers, which bloom from midsummer to autumn when many other plants are past their best. The leaves are mid- to dark green and divided. Very effective in massed plantings, wherever space allows: in small gardens, choose a selection of dwarf types to fill gaps in border displays, or grow in containers. The flowers are ideal for cutting.

CULTIVATION: *Best in well-drained soil, in sun. In frost-prone climates, lift tubers and store in cool peat over winter. Plant out tubers once the danger of frost has passed. Feed with high-nitrogen fertilizer every week in early summer. Taller varieties need staking. Deadhead to prolong flowering.*

☼ ◊ ❀ ❀

1 ‡1.1m (3½ft) ↔ 45cm (18in)

2 ‡1.1m (3½ft) ↔ 60cm (24in)

‡60cm (24in) ↔ 45cm (18in)

‡1.2m (4ft) ↔ 60cm (24in)

‡1.1m (3½ft) ↔ 60cm (24in)

1 *D.* 'Bishop of Llandaff' ♀ **2** *D.* 'Clair de Lune' ♀ **3** *D.* 'Fascination' (dwarf) ♀
4 *D.* 'Hamari Accord' ♀ **5** *D.* 'Conway'

‡1.2m (4ft) ↔ 60cm (24in)

‡1.1m (3½ft) ↔ 60cm (24in)

‡1.2m (4ft) ↔ 60cm (24in)

‡1m (3ft) ↔ 45cm (18in)

‡↔ 45–50cm (18–20in)

‡1.1m (3½ft) ↔ 60cm (24in)

‡1.1m (3½ft) ↔ 60cm (24in)

‡60cm (24in) ↔ 45cm (18in)

‡1.2m (4ft) ↔ 60cm (24in)

‡1.2m (4ft) ↔ 60cm (24in)

6 *D.* 'Hamari Gold' ♀ **7** *D.* 'Hillcrest Royal' ♀ **8** *D.* 'Kathryn's Cupid' **9** *D.* 'Rokesley Mini' **10** *D.* 'Sunny Yellow' (dwarf) **11** *D.* 'So Dainty' ♀ **12** *D.* 'Wootton Cupid' ♀ **13** *D.* 'Yellow Hammer' (dwarf) ♀ **14** *D.* 'Zorro' ♀ **15** *D.* 'Wootton Impact' ♀

Daphne bholua 'Gurkha'

An upright, deciduous shrub bearing clusters of strongly fragrant, tubular, white and purplish-pink flowers on its bare stems in late winter. They open from deep pink-purple buds, and are followed by round, black-purple fruits. The lance-shaped leaves are leathery and dark green. A fine plant for a winter garden, but needs protection in cold areas. All parts are highly toxic if ingested.

CULTIVATION: *Grow in well-drained but moist soil, in sun or semi-shade. Mulch to keep the roots cool. Best left unpruned.*

☼ ☀ ◊ ♀ ❀❀
↕2–4m (6–12ft) ↔1.5m (5ft)

Daphne tangutica Retusa Group

These dwarf forms of *D. tangutica*, sometimes listed simply as *D. retusa*, are evergreen shrubs valued for their clusters of very fragrant, white to purple-red flowers which are borne during late spring and early summer. The lance-shaped leaves are glossy and dark green. Useful in a variety of sites, such as a large rock garden, shrub border, or mixed planting. All parts are toxic.

CULTIVATION: *Grow in well-drained, moderately fertile, humus-rich soil that does not dry out, in full sun or dappled shade. Pruning is not necessary.*

☼ ☀ ◊ ♀ ❀❀❀   ↕↔75cm (30in)

Darmera peltata

A handsome, spreading perennial, sometimes included in the genus *Peltiphyllum*, that forms an imposing, umbrella-like clump with large, round, mid-green leaves, to 60cm (24in) across, which turn red in autumn. The foliage appears after the compact clusters of star-shaped, white to bright pink, spring flowers. Ideal for a bog garden or by the edge of a pond or stream.

CULTIVATION: *Grow in reliably moist, moderately fertile soil, in sun or partial shade; tolerates drier soil in shade.*

☼ ☼ ◊ ♀ ❀❀❀ ‡to 2m (6ft) ↔1m (3ft)

Davidia involucrata

The handkerchief tree is named for the showy, large, white flower bracts that adorn the branches in mid- to late spring. For the tree to be seen at its best, it will need space to grow as a free-standing specimen. When not in flower, this deciduous tree is less remarkable, but it has a pleasing conical shape and bright green, heart-shaped leaves. The peeling bark is orange-brown.

CULTIVATION: *Grow in moist but well-drained, fertile soil in sun or partial shade.*

☼ ☼ ◊ ◊ ♀ ❀❀❀
‡15m (50ft) ↔10m (30ft)

Delphiniums

Delphiniums are clump-forming perennials cultivated for their towering spikes of exquisite, shallowly cup-shaped, spurred, single or double flowers. These appear in early to midsummer, and are available in a range of colours from creamy-whites through lilac-pinks and clear sky-blues to deep indigo-blue. The toothed and lobed, mid-green leaves are arranged around the base of the stems. Grow tall delphiniums in a mixed border or island bed with shelter to prevent them being blown over in strong winds; shorter delphiniums are suitable for a rock garden. The flowers are good for cutting.

CULTIVATION: *Grow in well-drained, fertile soil, in full sun. For quality blooms, feed weekly with a balanced fertilizer in spring, and thin out the young shoots when they reach 7cm (3in) tall. Most cultivars need staking. Remove spent flower spikes, and cut back all growth in autumn.*

☼ ◊ ❀ ❀ ❀

1 ‡1.7m (5½ft) ↔ 60–90cm (24–36in)
2 ‡2m (6ft) ↔ 60–90cm (24–36in)
3 ‡1.7m (5½ft) ↔ 60–90cm (24–36in)

1 *Delphinium* 'Blue Nile' ♀ **2** *D.* 'Bruce' ♀ **3** *D.* 'Cassius'

4 ‡to 2m (6ft) ↔ 60–90cm (24–36in)

5 ‡to 1.5m (5ft) ↔ 60–90cm (24–36in)

6 ‡1.7m (5½ft) ↔ 60–90cm (24–36in)

7 ‡2m (6ft) ↔ 60–90cm (24–36in)

8 ‡1.7m (5½ft) ↔ 60–90cm (24–36in)

4 *D.* 'Claire' **5** *D.* 'Conspicuous' ♀ **6** *D.* 'Emily Hawkins' ♀ **7** *D.* 'Fanfare' **8** *D.* 'Giotto'

MORE CHOICES

'Blue Dawn' Pinky-blue flowers with brown eyes.

'Faust' Cornflower blue with indigo eyes.

'Fenella' Deep blue with black eyes.

'Foxhill Nina' Pale pink.

'Gillian Dallas' White.

'Loch Leven' Mid-blue with white eyes.

'Margaret' Mid-blue.

'Michael Ayres' Violet.

'Min' Mauve veined with deep purple, brown eyes.

'Oliver' Light blue/mauve with dark eyes.

'Spindrift' Cobalt blue with cream eyes.

‡ to 2m (6ft) ↔ 60–90cm (24–36in)   ‡ to 1.5m (5ft) ↔ 60–90cm (24–36in)   ‡ to 1.5m (5ft) ↔ 60–90cm (24–36in)

9 *D.* 'Kathleen Cooke' **10** *D.* 'Langdon's Royal Flush' **11** *D.* 'Lord Butler' ♀

‡2m (6ft) ↔75cm (30in)

‡1.7m (5½ft) ↔60–90cm (24–36in)

‡to 1.5m (5ft) ↔60–90cm (24–36in)

‡to 1.5m (5ft) ↔75cm (30in)

‡to 1.5m (5ft) ↔60–90cm (24–36in)

‡1.2m (4ft) ↔60–90cm (24–36in)

‡1.7m (5½ft) ↔60–90cm (24–36in)

12 *D.* 'Mighty Atom' **13** *D.* 'Our Deb' ♀ **14** *D.* 'Rosemary Brock' ♀ **15** *D.* 'Sandpiper'
16 *D.* 'Sungleam' ♀ **17** *D.* 'Thamesmead' **18** *D.* 'Walton Gemstone' ♀

Deschampsia cespitosa 'Goldtau'

Tufted hair grass is a compact, semi-evergreen, clump-forming grass that bears clusters of airy reddish-brown spikelets above the foliage in summer. Both spikelets and the linear, dark green leaves turn gold in autumn. Suitable for mass plantings, edging, or as ground cover.

CULTIVATION: *Grow in dry to damp, neutral to acid soil in sun or partial shade. Remove flowerheads before new growth begins in early spring.*

☼ ☀ ◊ ❀ ❀ ❀
‡70cm (28in) ↔ 50cm (20in)

Deutzia x *elegantissima* 'Rosealind'

This compact, rounded, deciduous shrub bears profuse clusters of small, deep carmine-pink flowers. These are carried from late spring to early summer amid the oval, dull green leaves. Very suitable for a mixed border.

CULTIVATION: *Grow in any well-drained, fertile soil that does not dry out, in full sun or partial shade. Tip-prune when young to encourage bushiness; after flowering, thin out by cutting some older stems right back to the ground.*

☼ ☀ ◊ ♀ ❀ ❀ ❀
‡1.2m (4ft) ↔ 1.5m (5ft)

Border Carnations (*Dianthus*)

This group of *Dianthus* are annuals or evergreen perennials of medium height, suitable for mixed or herbaceous borders. They are grown for their midsummer flowers, which are good for cutting. Each flower stem bears five or more double flowers to 8cm (3in) across, with no fewer than 25 petals. These may be of one colour only, as in 'Golden Cross'; gently flecked, like 'Grey Dove'; or often with white petals margined and striped in strong colours. Some have clove-scented flowers. The linear leaves of all carnations are blue-grey or grey-green with a waxy bloom.

CULTIVATION: *Ideal in well-drained, neutral to alkaline soil enriched with well-rotted manure or garden compost; apply a balanced fertilizer in spring. Plant in full sun. Provide support in late spring using thin canes or twigs, or wire rings. Deadhead to prolong flowering and maintain a compact habit.*

☼ ◊ ❀❀❀

‡45–60cm (18–24in) ↔ 45cm (18in)　　‡45–60cm (18–24in) ↔ 45cm (18in)　　‡45–60cm (18–24in) ↔ 45cm (18in)

‡45–60cm (18–24in) ↔ 45cm (18in)　　‡45–60cm (18–24in) ↔ 45cm (18in)　　‡45–60cm (18–24in) ↔ 45cm (18in)

1 *Dianthus* 'David Russell' ♀ **2** *D.* 'Devon Carla' **3** *D.* 'Golden Cross' ♀
4 *D.* 'Grey Dove' ♀ **5** *D.* 'Ruth White' **6** *D.* 'Spinfield Wizard'

Pinks (*Dianthus*)

Summer-flowering pinks belong, like carnations, to the genus *Dianthus* and are widely grown for their charming, often clove-scented flowers and attractive, narrow, blue-grey leaves. They flower profusely over long periods; when cut, the stiff-stemmed blooms last exceptionally well. Thousands are available, usually in shades of pink, white, carmine, salmon, or mauve with double or single flowers; they may be plain, marked with a contrasting colour, or laced around the margins. Most are excellent border plants; tiny alpine pinks like 'Pike's Pink' and magenta-flowered 'Joan's Blood' are well suited to a rock garden or raised bed.

CULTIVATION: *Best in well-drained, neutral to alkaline soil, in an open, sunny site. Alpine pinks need very sharp drainage. Feed with a balanced fertilizer in spring. Deadhead all types to prolong flowering and to maintain a compact habit.*

☼ ◊ ✼ ✼ ✼

‡25–45cm (10–18in) ↔ 40cm (16in)   ‡25–45cm (10–18in) ↔ 40cm (16in)   ‡25–45cm (10–18in) ↔ 40cm (16in)

‡25–45cm (10–18in) ↔ 40cm (16in)   ‡25–45cm (10–18in) ↔ 40cm (16in)   ‡25–45cm (10–18in) ↔ 40cm (16in)

1 *Dianthus* 'Becky Robinson' ♀ **2** *D.* 'Devon Pride' **3** *D.* 'Doris' ♀
4 *D.* 'Gran's Favourite' ♀ **5** *D.* 'Haytor White' ♀ **6** *D.* 'Houndspool Ruby' ♀

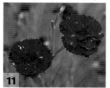

‡25–45cm (10–18in) ↔ 40cm (16in)

‡8–10cm (3–4in) ↔ 20cm (8in)

‡25–45cm (10–18in) ↔ 40cm (16in)

‡25–45cm (10–18in) ↔ 40cm (16in)

‡25–45cm (10–18in) ↔ 40cm (16in)

‡25–45cm (10–18in) ↔ 40cm (16in)

MORE CHOICES

D. alpinus The original alpine pink, with pale-flecked pink flowers.

D. alpinus 'Joan's Blood' Dark-centred deep pink flowers.

'Bovey Belle' Clove-scented, fuchsia-pink.

'Coronation Ruby' Clove-scented, warm pink flowers with ruby markings.

'Inschriach Dazzler' Alpine pink with fringed carmine flowers.

'Valda Wyatt' Clove-scented, double lavender flowers.

‡8–10cm (3–4in) ↔ 20cm (8in)

‡25–45cm (10–18in) ↔ 40cm (16in)

‡25–45cm (10–18in) ↔ 40cm (16in)

‡25–45cm (10–18in) ↔ 40cm (16in)

7 *D.* 'Kathleen Hitchcock' ♀ **8** *D.* 'La Bourboule' ♀ **9** *D.* 'Monica Wyatt' ♀
10 *D.* 'Natalie Saunders' ♀ **11** *D.* 'Oakwood Romance' **12** *D.* 'Oakwood Splendour'
13 *D.* 'Pike's Pink' ♀ **14** *D.* 'Suffolk Pride' **15** *D.* 'Trisha's Choice' **16** *D.* 'White Joy' ♀

Diascia barberae 'Blackthorn Apricot'

A mat-forming perennial bearing loose spikes of apricot flowers. These are produced over a long period from summer to autumn, above the narrowly heart-shaped, mid-green leaves. Good in a rock garden, at the front of a mixed border, or on a sunny bank. *D. barberae* 'Fisher's Flora' and 'Ruby Field' are also recommended.

CULTIVATION: *Grow in moist but well-drained, fertile soil, in full sun. In cold climates, shelter from heavy frost and overwinter young plants under glass.*

☼ ◊ ♀ ❀ ❀
‡25cm (10in) ↔ to 50cm (20in)

Diascia rigescens

This trailing perennial is valued for its tall spires of salmon-pink flowers. These appear above the mid-green, heart-shaped leaves during summer. Excellent in a rock garden or raised bed, or at the front of a border. In cold areas, grow at the base of a warm, sunny wall.

CULTIVATION: *Grow in moist but well-drained, fertile soil that is rich in humus. Site in full sun. Overwinter young plants under glass.*

☼ ◊ ♀ ❀ ❀
‡30cm (12in) ↔ to 50cm (20in)

Dicentra spectabilis

Bleeding heart is an elegant, hummock-forming perennial. In late spring and early summer, it bears rows of hanging, red-pink, heart-shaped flowers on arching stems. The leaves are deeply cut and mid-green. A beautiful plant for a shady border or woodland garden.

CULTIVATION: *Grow in reliably moist, fertile, humus-rich, neutral or slightly alkaline soil. Prefers a site in partial shade, but tolerates full sun.*

☼ ☀ ◊ ❀ ❀ ❀
‡to 1.2m (4ft) ↔45cm (18in)

Dicentra spectabilis **'Alba'**

This white-flowered form of bleeding heart is otherwise very similar to the species (above), so to avoid possible confusion, purchase plants during the flowering period. The leaves are deeply cut and light green, and the heart-shaped flowers are borne along arching stems in late spring and early summer.

CULTIVATION: *Grow in reliably moist but well-drained soil that is enriched with well-rotted organic matter. Prefers partial shade, but tolerates full sun.*

☼ ☀ ◊ ❀ ❀ ❀
‡to 1.2m (4ft) ↔45cm (18in)

Dicentra 'Stuart Boothman'

A spreading perennial that bears heart-shaped pink flowers along the tips of arching stems during late spring and summer. The very finely cut foliage is fern-like and grey-green. Very attractive for a shady border or woodland planting.

CULTIVATION: *Best in moist but well drained, humus-rich, neutral to slightly alkaline soil, in partial shade. Divide and replant clumps in early spring, after the leaves have died down.*

☀ ◊ ◑ ♀ ❀ ❀ ❀
‡30cm (12in) ↔40cm (16in)

Dicksonia antarctica

Several species of tree fern are available to gardeners, but this is perhaps the hardiest. Tree ferns have fibrous trunks topped by a tuft of large, feathery, bright green fronds, up to 3m (10ft) in length. Dicksonias do not flower and growth is slow. They make fine specimens for sheltered sites in damp shade, or in a cool conservatory.

CULTIVATION: *Grow in humus-rich, moist, neutral to acid soil in full or partial shade. Outdoors, wrap the trees up in winter to protect them from frost.*

☀ ☀ ◑ ♀ ❀ ❀
‡to 4m (12ft) or more ↔4m (12ft)

Dictamnus albus var. *purpureus*

This purple-flowered perennial is otherwise identical to the species. Upright spikes of flowers are borne in early summer, above the highly aromatic, light green leaves. Striking in a mixed or herbaceous border.

CULTIVATION: *Grow in any dry, well-drained, moderately fertile soil, in full sun or partial shade.*

☼ ☀ ◊ ♀ ❀❀❀
‡40–90cm (16–36in) ↔60cm (24in)

Digitalis x *mertonensis*

An evergreen, clump-forming, short-lived perennial grown for its spires of flared, strawberry-pink flowers which are borne from late spring to early summer. The leaves are dark green and lance-shaped. A beautiful plant for mixed or herbaceous borders or woodland plantings. The dusky pink cultivar 'Raspberry' is also recommended.

CULTIVATION: *Best in moist but well-drained soil, in partial shade, but tolerates full sun and dry soil.*

☼ ☀ ◊ ♀ ❀❀❀
‡to 90cm (36in) ↔30cm (12in)

Digitalis purpurea f. *albiflora*

This ghostly form of the common foxglove is a tall biennial producing robust, stately spires of tubular, pure white flowers during summer. The coarse, lance-shaped leaves are bright green. A lovely addition to a woodland garden or mixed border. May be sold as *D. purpurea* 'Alba'.

CULTIVATION: *Grow in moist but well-drained soil, in partial shade, but also tolerates dry soil in full sun.*

☼ ☼ ◊ ◊ ♀ ❀❀❀
↕1–2m (3–6ft) ↔ to 60cm (24in)

Disanthus cercidifolius

It is quite unusual to see this medium-sized, rounded shrub in gardens, but it is a fine plant for autumn colour, performing best on acid soil. The heart-shaped leaves turn various shades of orange, red, purple, and yellow in autumn before they fall. Lightly fragrant red flowers appear in autumn. Protect from late spring frosts.

CULTIVATION: *Grow in moist but well-drained, neutral to acid soil in partial sun or light shade. Choose a sheltered site; hard frosts in late spring may kill the plant if they occur after bud burst.*

☼ ◊ ◊ ♀ ❀❀❀
↕↔3m (10ft)

Dodecatheon meadia f. *album*

This herbaceous perennial is valued for its open clusters of creamy-white flowers with strongly reflexed petals. These are borne on arching stems during mid- to late spring, above the rosettes of oval, pale green, toothed leaves. Suits a woodland or shady rock garden; flowering is followed by a period of dormancy. *D. dentatum* looks very similar, but is only half the height, ideal for a small rock garden.

CULTIVATION: *Grow in moist, humus-rich soil, in partial shade. May be prone to slug and snail damage in spring.*

☼ ◊ ♀ ❀ ❀ ❀
↕40cm (16in) ↔25cm (10in)

Dryopteris filix-mas

The male fern is a deciduous foliage perennial forming large clumps of lance-shaped, mid-green fronds which emerge from a stout, scaly crown in spring. No flowers are produced. Ideal for a shady border or corner, by the side of a stream or pool, or in woodland. 'Cristata', with crested fronds, is a handsome cultivar of this fern.

CULTIVATION: *Grow in reliably moist, humus-rich soil. Site in partial shade with shelter from cold, drying winds.*

☼ ◊ ♀ ❀ ❀ ❀   ↕↔1m (3ft)

Dryopteris wallichiana

Wallich's wood fern is a deciduous foliage perennial with a strongly upright, shuttlecock-like habit. The fronds are yellow-green when they emerge in spring, becoming dark green in summer. A fine architectural specimen for a moist, shady site.

CULTIVATION: *Best in damp soil that is rich in humus. Choose a sheltered position in partial shade.*

☀ ◐ ◊ ♀ ❀ ❀ ❀
‡1–2m (3–6ft) ↔75cm (30in)

Eccremocarpus scaber

The Chilean glory flower is a fast-growing, scrambling climber with clusters of brilliant orange-red, tubular flowers in summer. The leaves are divided and mid-green. Grow as a short-lived perennial to clothe an arch or pergola, or up into a large shrub. In frost-prone areas, use as a trailing annual.

CULTIVATION: *Grow in free-draining, fertile soil, in a sheltered, sunny site. Cut back to within 30–60cm (12–24in) of the base in spring.*

☀ ◊ ❀ ❀
‡2–3m (6–10ft) or more

Echinacea purpurea 'Magnus'

Coneflowers are increasingly popular late summer perennials, and there are now many colours to choose from. The daisy-like flowers have a raised, textured centre or "cone". They come into bloom at a time when many other summer perennials are flagging, extending the summer season. 'Magnus' has large, purple-pink flowers with orange cones.

CULTIVATION: *Grow in any deep, well-drained soil in sun. Enrich the soil with well-rotted organic matter before planting. Divide plants in spring, if necessary.*

☼ ◊ ◖ ❋❋❋   ↕↔60cm (24in)

Echinops ritro

This globe thistle is a compact perennial forming clumps of eye-catching flower-heads, metallic blue at first, turning a brighter blue as the flowers open. The leathery green leaves have white-downy undersides. Excellent for a wild garden. The flowers dry well if cut before fully open. *E. bannaticus* 'Taplow Blue' makes a good substitute if this plant cannot be found; it may grow a little taller.

CULTIVATION: *Grow in well-drained, poor to moderately fertile soil. Site in full sun, but tolerates partial shade. Deadhead to prevent self-seeding.*

☼ ◑ ◊ ♀ ❋❋❋
↕60cm (24in) ↔45cm (18in)

Elaeagnus x *ebbingei* 'Gilt Edge'

A large, dense, evergreen shrub grown for its oval, dark green leaves which are edged with rich golden-yellow. Inconspicuous yet highly scented, creamy-white flowers are produced in autumn. Makes an excellent, fast-growing, well-shaped specimen shrub; can also be planted as an informal hedge.

CULTIVATION: *Grow in well-drained, fertile soil, in full sun. Dislikes very chalky soil. Trim to shape in late spring; completely remove any shoots with plain green leaves as soon as seen.*

☼ ◊ ♀ ✸✸✸   ↕↔4m (12ft)

Elaeagnus pungens 'Maculata'

This large, evergreen foliage shrub bears oval, dark green leaves that are generously splashed in the centre with dark yellow. Small but very fragrant flowers are produced from mid-autumn, followed by red fruits. Tolerates coastal conditions.

CULTIVATION: *Best in well-drained, fairly fertile soil, in sun. Dislikes very chalky soil. Trim lightly in spring, as needed. Remove completely any shoots with plain green leaves as soon as seen.*

☼ ◊ ✸✸✸   ↕4m (12ft) ↔5m (15ft)

Elaeagnus 'Quicksilver'

A fast-growing shrub with an open habit bearing small yellow flowers in summer which are followed by yellow fruits. The silver, deciduous leaves are lance-shaped, and are carried on silvery shoots. Makes an excellent specimen shrub, or can be planted to great effect with other silver-leaved plants. May be known as *E. angustifolia* var. *caspica*.

CULTIVATION: *Grow in any but chalky soil that is fertile and well-drained, in full sun. Tolerates dry soil and coastal winds. Keep pruning to a minimum.*

☼ ◊ ❀❀❀   ‡↔4m (12ft)

Enkianthus campanulatus

A spreading, deciduous shrub grown for its dense, hanging clusters of bell-shaped, creamy-yellow flowers in late spring. The dull green leaves give a fine autumn display when they change to orange-red. Suits an open site in a woodland garden; can become tree-like with age. *E. cernuus* var. *rubens*, with deep pink flowers, is more suited to a small garden as it grows to only 2.5m (8ft) tall.

CULTIVATION: *Grow in reliably moist but well-drained, peaty or humus-rich, acid soil. Best in full sun, but tolerates some shade. Keep pruning to a minimum.*

☼ ☼ ◊ ◊ ♀ ❀❀❀   ‡↔4–5m (12–15ft)

Epimedium x *perralchicum*

A robust, clump-forming, evergreen perennial that produces spikes of delicate, bright yellow flowers above the glossy dark green foliage in mid- and late spring. The leaves are tinged bronze when young. Very useful as ground cover under trees or shrubs.

CULTIVATION: *Grow in moist but well-drained, moderately fertile soil that is enriched with humus, in partial shade. Provide shelter from cold, drying winds.*

☀ ◊ ◊ ❀❀❀
‡40cm (16in) ↔ 60cm (24in)

Epimedium x *rubrum*

A compact perennial bearing loose clusters of pretty crimson flowers with yellow spurs, in spring. The divided leaves are tinted bronze-red when young, ageing to mid-green, then reddening in autumn. Clump together to form drifts in a damp, shady border or woodland garden. *E. grandiflorum* 'Rose Queen' is very similar.

CULTIVATION: *Grow in moist but well-drained, moderately fertile, humus-rich soil, in partial shade. Cut back old, tatty foliage in late winter so that the flowers can be seen in spring.*

☀ ◊ ◊ ♀ ❀❀❀   ‡↔30cm (12in)

Epimedium x *youngianum* 'Niveum'

A clump-forming perennial that bears dainty clusters of small white flowers in late spring. The bright green foliage is bronze-tinted when young and, despite being deciduous, persists well into winter. Makes good ground cover in damp, shady borders and woodland areas.

CULTIVATION: *Grow in moist but well-drained, fertile, humus-rich soil, in partial shade. In late winter, cut back old, tatty foliage so that the new flowers can be seen in spring.*

☀ ◊ ◑ ♀ ✿✿✿
‡20–30cm (8–12in) ↔30cm (12in)

Eranthis hyemalis

Winter aconite is one of the earliest spring-flowering bulbs, bearing buttercup-like, bright yellow flowers. These sit on a ruff of light green leaves, covering the ground from late winter until early spring. Ideal for naturalizing beneath deciduous trees and large shrubs, as is the cultivated variety, 'Guinea Gold', with similar flowers and bronze-green leaves.

CULTIVATION: *Grow in moist but well-drained, fertile soil that does not dry out in summer. Best in dappled shade.*

☀ ◊ ◑ ♀ ✿✿✿
‡5–8cm (2–3in) ↔5cm (2in)

Erica arborea var. *alpina*

This tree heath is an upright shrub, much larger than other heathers, densely clothed with clusters of small, honey-scented white flowers from late winter to late spring. The evergreen leaves are needle-like and dark green. A fine centrepiece for a heather garden.

CULTIVATION: *Grow in well-drained, ideally sandy, acid soil, in an open, sunny site. Tolerates alkaline conditions. Cut back young plants in early spring by about two-thirds to promote bushy growth; in later years, pruning is unnecessary. Tolerates hard pruning to renovate.*

☼ ◊ ♀ ❀ ❀ ❀   ↕2m (6ft) ↔85cm (34in)

Erica x *veitchii* 'Exeter'

An upright, open, evergreen shrub that bears masses of highly scented, tubular to bell-shaped white flowers from midwinter to spring. The needle-like leaves are bright green. Good in a large rock garden with conifers or as a focal point among low-growing heathers, but may not survive winter in cold climates.

CULTIVATION: *Grow in sandy soil that is well-drained, in sun. Best in acid soil, but tolerates slightly alkaline conditions. When young, cut back by two-thirds in spring to encourage a good shape; reduce pruning as the plant gets older.*

☼ ◊ ♀ ❀ ❀   ↕2m (6ft) ↔65cm (26in)

Early-flowering Ericas

The low-growing, early flowering heaths are evergreen shrubs valued for their urn-shaped flowers in winter and spring. The flowers come in white and a wide range of pinks, bringing invaluable early interest during the winter months. This effect can be underlined by choosing cultivars with colourful foliage; *E. erigena* 'Golden Lady', for example, has bright golden-yellow leaves, and *E. carnea* 'Foxhollow' carries yellow, bronze-tipped foliage which turns a deep orange in cold weather. Excellent as ground cover, either in groups of the same cultivar, or with other heathers and dwarf conifers.

CULTIVATION: *Grow in open, well-drained, preferably acid soil, but they tolerate alkaline conditions. Choose a site in full sun. Cut back flowered stems in spring to remove most of the previous year's growth; cultivars of* E. erigena *may be scorched by frosts; remove any affected growth in spring.*

☼ ◊ ❀ ❀ ❀

‡15cm (6in) ↔ 25cm (10in)

‡15cm (6in) ↔ 40cm (16in)

‡15cm (6in) ↔ 45cm (18in)

‡15cm (6in) ↔ 35cm (14in)

‡30cm (12in) ↔ 60cm (24in)

‡30cm (12in) ↔ 40cm (16in)

1 *E. carnea* 'Ann Sparkes' ♀ **2** *E. carnea* 'Foxhollow' ♀ **3** *E. carnea* 'Springwood White' ♀
4 *E. carnea* 'Vivellii' ♀ **5** *E.* x *darleyensis* 'Jenny Porter' ♀ **6** *E. erigena* 'Golden Lady'

Late-flowering Ericas

Mostly low and spreading in form, the summer-flowering heaths are fully hardy, evergreen shrubs. They look attractive on their own or mixed with other heathers and dwarf conifers. Flowers are borne over a very long period; *E. x stuartii* 'Irish Lemon' starts in late spring, and cultivars of *E. ciliaris* and *E. vagans* bloom well into autumn. Their season of interest can be further extended by choosing types with colourful foliage; the young shoots of *E. williamsii* 'P.D. Williams'

are tipped with yellow in spring, and the golden foliage of *E. cinerea* 'Windlebrooke' turns a rich deep red in winter.

CULTIVATION: *Grow in well-drained, acid soil, although* E. vagans *and* E. williamsii *tolerate alkaline conditions. Choose an open site in full sun. Prune or trim lightly in early spring, cutting back to strong shoots below the flower clusters, or trim over with shears en masse.*

☼ ◊ ❀ ❀ ❀

‡22cm (9in) ↔ 35cm (14in)

‡to 40cm (16in) ↔ 45cm (18in)

‡20cm (8in) ↔ 50cm (20in)

‡25cm (10in) ↔ 50cm (20in)

‡15cm (6in) ↔ 45cm (18in)

1 *E. ciliaris* 'Corfe Castle' **2** *E. ciliaris* 'David McClintock' **3** *E. cinerea* 'Eden Valley'
4 *E. cinerea* 'C.D. Eason' ♀ **5** *E. cinerea* 'Windlebrooke' **6** *E. cinerea* 'Fiddler's Gold'

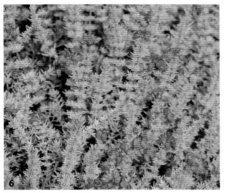

‡25cm (10in) ↔ 50cm (20in)

‡25cm (10in) ↔ 45cm (18in)

‡20cm (8in) ↔ 30cm (12in)

‡30cm (12in) ↔ 50cm (20in)

‡25cm (10in) ↔ 50cm (20in)

‡30cm (12in) ↔ 45cm (18in)

‡15cm (6in) ↔ 30cm (12in)

‡20cm (8in) ↔ to 85cm (34in)

‡30cm (12in) ↔ 45cm (18in)

7 *E.* x *stuartii* 'Irish Lemon' ♀ **8** *E. tetralix* 'Alba Mollis' ♀ **9** *E. vagans* 'Birch Glow' ♀ **10** *E. vagans* 'Lyonesse' ♀ **11** *E. vagans* 'Mrs. D.F. Maxwell' ♀ **12** *E. vagans* 'Valerie Proudley' ♀ **13** *E.* x *watsonii* 'Dawn' **14** *E.* x *williamsii* 'P.D. Williams'

Erigeron karvinskianus

This carpeting, evergreen perennial with grey-green foliage produces an abundance of yellow-centred, daisy-like flowerheads in summer. The outer petals are initially white, maturing to pink and purple. Ideal for wall crevices, or cracks in paving, but may not survive through winter in frost-prone climates. Sometimes sold as *E. mucronatus*.

CULTIVATION: *Grow in well-drained, fertile soil. Choose a site in full sun, ideally with some shade at midday.*

☀ ◊ ♀ ❀❀
‡15–30cm (6–12in) ↔1m (3ft) or more

Erinus alpinus

The fairy foxglove is a tiny, short-lived, evergreen perennial producing short spikes of pink, purple, or white, daisy-like flowers in late spring and summer. The lance- to wedge-shaped leaves are soft, sticky and mid-green. Ideal for a rock garden or in crevices in old walls. 'Mrs Charles Boyle' is a recommended cultivar.

CULTIVATION: *Grow in light, moderately fertile soil that is well-drained. Tolerates semi-shade, but best in full sun.*

☀ ☀ ◊ ♀ ❀❀❀
‡8cm (3in) ↔10cm (4in)

Eryngium alpinum

This spiky, upright, thistle-like perennial bears cone-shaped, purple-blue flowerheads in summer; these are surrounded by prominent, feathery bracts. The deeply toothed, mid-green leaves are arranged around the base of the stems. An excellent textural plant for a sunny garden. The flower heads can be cut, and dry well for arrangements.

CULTIVATION: *Grow in free-draining but not too dry, poor to moderately fertile soil, in full sun. Choose a site not prone to excessive winter wet.*

☼ ◊ ❀❀❀
‡70cm (28in) ↔ 45cm (18in)

Eryngium x *oliverianum*

An upright, herbaceous perennial bearing cone-shaped, bright silver-blue flowerheads with a flat ring of silvery, dagger-like bracts around the base, produced from midsummer to early autumn, above spiny-toothed, dark green leaves. Longer-lived than its parent *E. giganteum*, and an essential architectural addition to a dry, sunny border with a theme of silver- or grey-leaved plants.

CULTIVATION: *Grow in free-draining, fairly fertile soil, in full sun. Choose a site not prone to excessive winter wet.*

☼ ◊ ♀ ❀❀❀
‡90cm (36in) ↔ 45cm (18in)

Eryngium x *tripartitum*

This delicate but spiky, upright perennial bears cone-like heads of tiny, metallic blue flowers in late summer and autumn. The flowerheads sit on a ring of pointed bracts, and are carried above rosettes of grey-green foliage on the tips of wiry, blue-tinted stems. The flowers dry well if cut before fully open.

CULTIVATION: *Grow in free-draining, moderately fertile soil. Choose a position not prone to excessive winter wet, in full sun. Trim lightly after flowering to prevent legginess.*

☼ ◊ ♀ ✿✿✿
‡60–90cm (24–36in) ↔50cm (20in)

Erysimum 'Bowles' Mauve'

This vigorous, shrubby wallflower is one of the longest-flowering of all perennials. It forms a rounded, evergreen bush of narrow, grey-green leaves and produces dense spikes of small, four-petalled, rich mauve flowers all year, most freely in spring and summer. An excellent border plant that benefits from the shelter of a warm wall in frost-prone climates.

CULTIVATION: *Grow in any well-drained, preferably alkaline soil, in full sun. Trim lightly after flowering to keep compact. Often short-lived, but easily propagated by cuttings in summer.*

☼ ◊ ♀ ✿✿
‡to 75cm (30in) ↔60cm (24in)

Erysimum 'Wenlock Beauty'

A bushy, evergreen perennial that produces clusters of bluish-pink and salmon, bronze-shaded flowers from early to late spring. The lance-shaped leaves are softly hairy and mid-green. Good for early-season colour in a sunny rock garden or dry wall; in areas with cold winters, choose a warm, sheltered spot.

CULTIVATION: *Grow in poor or fairly fertile, ideally alkaline soil that has good drainage, in full sun. Trim lightly after flowering to keep compact.*

☼ ◊ ❀❀   ↕↔45cm (18in)

Erythronium dens-canis

The dog's tooth violet is a woodland or mountain meadow plant in the wild and thrives in damp, part-shady conditions in the garden. A clump-forming perennial, it has paired, mid-green leaves with variable bronze marking. From spring to early summer, slender stems rise from the foliage with delicate white, pink, or lilac pendent flowers with reflexed tepals. Effective in drifts.

CULTIVATION: *Plant bulbs in autumn in humus-rich soil in dappled shade. Keep bulbs slightly damp in storage and plant at least 10cm (4in) deep.*

☼ ◐ ♀ ❀❀❀
↕10–15cm (4–6in) ↔10cm (4in)

Erythronium 'Pagoda'

This very vigorous, clump-forming, bulbous perennial is related to the dog's tooth violet, *E. dens-canis* and, like it, looks good planted in groups under deciduous trees and shrubs. Clusters of pale sulphur-yellow flowers droop from slender stems in spring, above the large, oval, bronze-mottled, glossy dark green leaves.

CULTIVATION: *Grow in moist but well-drained soil that is rich in humus. Choose a position in partial shade.*

☀ ◊ ◊ ♀ ❋ ❋ ❋
‡15–35cm (6–14in) ↔10cm (4in)

Escallonia 'Apple Blossom'

A compact, evergreen shrub carrying dense, glossy dark foliage and, from early to midsummer, a profusion of small, pink-flushed white flowers. Valuable in a shrub border, it can also be grown as a hedge, barrier, or windbreak. Very useful in coastal areas, if prevailing winds are not harshly cold. *E.* 'Donard Seedling', also usually readily available, looks similar and is a little hardier.

CULTIVATION: *Grow in any well-drained, fertile soil, in full sun. In especially cold areas, shelter from wind. Cut out old or damaged growth after flowering.*

☀ ◊ ♀ ❋ ❋ ❋   ‡↔2.5m (8ft)

Escallonia 'Iveyi'

A vigorous, upright, evergreen shrub bearing large clusters of fragrant, pure white flowers from mid- to late summer. The rounded leaves are glossy dark green. Grow in a shrub border, or use as a hedge where winters are reliably mild; the foliage often takes on bronze tints in cold weather, but choose a sheltered position in frost-prone areas.

CULTIVATION: *Grow in any fertile soil with good drainage, in full sun with shelter from cold, drying winds. Remove damaged growth in autumn, or in spring if flowering finishes late.*

☼ ◊ ♀ ❀❀　　　　‡↔3m (10ft)

Escallonia 'Langleyensis'

This graceful, semi-evergreen shrub produces abundant clusters of small, rose-pink flowers. These are borne from early to midsummer above the oval, glossy bright green leaves. Thrives in relatively mild coastal gardens as an informal hedge or in a shrub border.

CULTIVATION: *Best in well-drained, fertile soil, in a sunny site. Protect from cold, drying winds in frost-prone areas. Cut out dead or damaged growth after flowering; old plants can be renovated by hard pruning in spring.*

☼ ◊ ♀ ❀❀❀　　‡2m (6ft) ↔3m (10ft)

Eschscholzia caespitosa

A tufted annual bearing a profusion of scented, bright yellow flowers in summer. The blue-green leaves are finely divided and almost thread-like. Suitable for a sunny border, rock garden, or gravel patch. The flowers close up in dull weather.

CULTIVATION: *Grow in well-drained, poor soil. Choose a site in full sun. For early flowers the following year, sow seed directly outdoors in autumn.*

☼ ◊ ❋❋❋   ‡↔to 15cm (6in)

Eschscholzia californica

The California poppy is a mat-forming annual with cup-shaped flowers borne on slender stems throughout summer. Colours are mixed, including white, red, or yellow, but most are usually orange. The cultivar 'Dali' flowers in scarlet only. The leaves are finely cut and greyish-green. Grow in a sunny border or rock garden; the flowers last well when cut.

CULTIVATION: *Best in light, poor soil with good drainage, in full sun. For early flowers the following year, sow seed directly outdoors in autumn.*

☼ ◊ ❋❋❋
‡to 3cm (12in) ↔to 15cm (6in)

Eucalyptus gunnii

The cider gum is a vigorous, evergreen tree useful as a fast-growing feature in a new garden. The new yellow- to greyish-green bark is revealed in late summer as the old, whitish-green layer is shed. Young plants have rounded, grey-blue leaves; on adult growth they are lance-shaped. Protect in cold areas with a thick, dry, winter mulch, especially when young.

CULTIVATION: *Grow in well-drained, fertile soil, in sun. To keep compact, and for the best display of young foliage, cut back hard each spring.*

☼ ◊ ♀ ✿✿
‡10–25m (30–80ft) ↔6–15m (20–50ft)

Eucalyptus pauciflora subsp. *niphophila*

The snow gum is a handsome, silvery, evergreen tree with open, spreading branches and attractively peeling, white and grey bark. Young leaves are oval and dull blue-green; on mature stems they are lance-shaped and deep blue-green. The flowers are less significant. Popularly grown as a shrub, pruned hard back at regular intervals.

CULTIVATION: *Grow in well-drained, fertile soil, in full sun. For the best display of young foliage, cut back hard each year in spring.*

☼ ◊ ♀ ✿✿✿   ‡↔to 6m (20ft)

Eucryphia x *nymansensis* 'Nymansay'

A column-shaped, evergreen tree that bears clusters of large, fragrant, glistening white flowers with yellow stamens. These are borne in late summer to early autumn amid the oval, glossy dark green leaves. Makes a magnificent flowering specimen tree; best in mild and damp, frost-free climates.

CULTIVATION: *Grow in well-drained, reliably moist soil, preferably in full sun with shade at the roots, but will tolerate semi-shade. Shelter from cold winds. Remove damaged growth in spring.*

☼ ◊ ◑ ♀ ❁ ❁ ‡15m (50ft) ↔5m (15ft)

Euonymus alatus

The winged spindle is a dense, deciduous shrub with winged stems, much-valued for its spectacular autumn display; small, purple and red fruits split to reveal orange seeds as the oval, deep green foliage turns to scarlet. The flowers are much less significant. Excellent in a shrub border or light woodland. The fruits are poisonous. 'Compactus' is a dwarf version of this shrub, only half its height.

CULTIVATION: *Grow in any well-drained, fertile soil. Tolerates light shade, but fruiting and autumn colour are best in full sun. Keep pruning to a minimum.*

☼ ☀ ◊ ❁ ❁ ❁ ‡2m (6ft) ↔3m (10ft)

Euonymus europaeus 'Red Cascade'

A tree-like, deciduous shrub that produces colourful autumn foliage. Inconspicuous flowers in early summer are followed by rosy-red fruits that split to reveal orange seeds. The oval, mid-green leaves turn scarlet-red at the end of the growing season. The fruits are toxic.

CULTIVATION: *Grow in any fertile soil with good drainage, but thrives on chalky soil. Tolerates light shade, but fruiting and autumn colour are best in full sun. Two or more specimens are required to guarantee a good crop of fruits. Very little pruning is necessary.*

☼ ☀ ◊ ♀ ❀❀❀ ‡3m (10ft) ↔2.5m (8ft)

Euonymus fortunei 'Emerald 'n' Gold'

A small and scrambling, evergreen shrub that will climb if supported. The bright green, oval leaves have broad, bright golden-yellow margins, and take on a pink tinge in cold weather. The spring flowers are insignificant. Use to fill gaps in a shrub border, or wall-train.

CULTIVATION: *Grow in any but waterlogged soil. The leaves colour best in full sun, but tolerates light shade. Trim over in midspring. Trained up a wall, it may reach a height of up to 5m (15ft).*

☼ ☀ ◊ ◊ ♀ ❀❀❀ ‡60cm (24in) or more ↔90cm (36in)

Euonymus fortunei 'Silver Queen'

A compact, upright or scrambling, evergreen shrub that looks most effective when grown as a climber against a wall or up into a tree. The dark green leaves have broad white edges that become pink-tinged in prolonged frost. The greenish-white flowers in spring are insignificant.

CULTIVATION: *Grow in any but water-logged soil, in sun or light shade. Leaf colour is best in full sun. Trim shrubs in mid-spring. Allowed to climb, it can grow up to 6m (20ft) tall.*

☼ ☀ ◊ ◊ ❀ ❀ ❀
‡2.5m (8ft) ↔1.5m (5ft)

Eupatorium maculatum Atropurpureum Group

Joe-Pye weed is a stately perennial that bears clusters of dusky pink flowers on tall red stems, from late summer to early autumn; these attract butterflies and bees. Handsome seed heads persist into winter. Plant in groups, at the back of the border, or interspersed with ornamental grasses. 'Gateway' is a popular compact form.

CULTIVATION: *Grow in moist, fertile, well-drained soil, in full sun or partial shade. Divide when necessary.*

☼ ☀ ◊ ❀ ❀ ❀
‡to 2.2m (7ft) ↔to 1m (3ft)

Euphorbia amygdaloides var. *robbiae*

Mrs. Robb's bonnet, also known simply as *E. robbiae*, is a spreading, evergreen perennial bearing open heads of yellowish-green flowers in spring. The long, dark green leaves are arranged in rosettes at the base of the stems. Particularly useful in shady areas. Can be invasive.

CULTIVATION: *Grow in well-drained but moist soil, in full sun or partial shade. Tolerates poor, dry soil. Dig up invasive roots to contain spread. The milky sap can irritate skin.*

☼ ◑ ◊ ◔ ❀❀❀
‡75–80cm (30–32in) ↔30cm (12in)

Euphorbia characias subsp. *wulfenii* 'John Tomlinson'

This billowing, shrubby perennial bears larger flowerheads than the species. The flowers are bright yellow-green, without dark eyes, and are borne in rounded heads above the narrow, grey-green leaves. In frost-prone climates, best grown at the base of a warm, sunny wall.

CULTIVATION: *Grow in light soil that has good drainage, in full sun. Shelter from cold winds. Cut the flowered stems back to the base in autumn; wear gloves, since the milky sap can irritate skin.*

☼ ◊ ♀ ❀❀
‡↔1.2m (4ft)

Euphorbia x *martini*

This upright, clump-forming, evergreen subshrub bears spikes of yellow-green flowers with very distinctive, dark red nectar glands. These are carried on red-tinged shoots from spring to midsummer, above lance-shaped, mid-green leaves which are often tinged purple when young. A choice plant with an architectural look for a hot, dry site. May suffer in cold winters.

CULTIVATION: *Grow in well-drained soil, in a sheltered, sunny site. Deadhead after flowering, wearing gloves to protect hands from the milky sap which may irritate skin.*

☼ ◊ ❀❀   ↕↔1m (3ft)

Euphorbia myrsinites

A small, evergreen perennial bearing sprawling stems which are densely clothed with a spiral arrangement of fleshy, elliptic, blue-green leaves. Clusters of yellow-green flowers brighten the tips of the stems in spring. Excellent in a dry, sunny rock garden, or trailing over the edge of a raised bed.

CULTIVATION: *Grow in well-drained, light soil, in full sun. Deadhead after flowering as it self-seeds freely. Wear gloves to avoid contact with the milky sap, which is a potential skin irritant.*

☼ ◊ ♀ ❀❀
↕10cm (4in) ↔ to 30cm (12in)

Euphorbia polychroma

This evergreen perennial forms a neat, rounded clump of softly hairy, mid-green foliage. It is covered with clusters of brilliant greenish-yellow flowers over long periods in spring. Excellent with spring bulbs, in a border or light woodland. Tolerates a wide range of soil types. Its cultivar 'Major' grows a little taller.

CULTIVATION: *Grow in either well-drained, light soil, in full sun, or moist, humus-rich soil, in light dappled shade. Deadhead after flowering. The milky sap can irritate skin.*

☼ ☀ ◊ ◑ ❀ ❀ ❀
‡40cm (16in) ↔60cm (24in)

Euphorbia schillingii

This vigorous, clump-forming perennial which, unlike many euphorbias, dies back in winter, bears clusters of long-lasting, yellowish-green flowers from midsummer to mid-autumn. The lance-shaped leaves are dark green and have pale green or white central veins. Ideal for lighting up a woodland planting or shady wild garden.

CULTIVATION: *Grow in reliably moist, humus-rich soil, in light dappled shade. Deadhead after flowering. The milky sap can irritate skin.*

☼ ◑ ♉ ❀ ❀ ❀   ‡1m (3ft) ↔30cm (12in)

Exochorda x *macrantha* 'The Bride'

A dense, spreading, deciduous shrub grown for its gently arching habit and abundant clusters of fragrant, pure white flowers. These are borne in late spring and early summer amid the oval, fresh green leaves. An elegant foil to other plants in a mixed border.

CULTIVATION: *Grow in any well-drained soil, in full sun or light dappled shade. Does not like shallow, chalky soil. Very little pruning is necessary.*

☼ ☀ ◊ ♀ ❋ ❋ ❋   ↕2m (6ft) ↔3m (10ft)

Fagus sylvatica 'Dawyck Gold'

The golden beech is more upright and conical than the common beech, and bears vivid golden leaves in spring, gradually mellowing to pale green as the season progresses. In autumn, the leaves begin to yellow before they fall. The dried, coppery leaves persist through winter, which makes it better for screening than most other deciduous hedges.

CULTIVATION: *Grow in any well-drained soil in sun or partial shade. The golden leaf colour is best in partial shade. Remove diseased or crossing branches in winter.*

☼ ☀ ◊ ◊ ♀ ❋ ❋ ❋
↕18m (60ft) ↔7m (22ft)

Fargesia nitida

Fountain bamboo is a slow-growing perennial which forms a dense clump of upright, dark purple-green canes. In the second year after planting, cascades of narrow, dark green leaves are produced from the top of the clump. Handsome in a wild garden; to restrict spread, grow in a large container. May be sold as *Sinarundinaria nitida*. *F. murielae* is equally attractive, similar but with yellow stems and brighter leaves.

CULTIVATION: *Grow in reliably moist, fertile soil, in light dappled shade. Shelter from cold, drying winds.*

☀ ◐ ❀❀❀
↕to 5m (15ft) ↔1.5m (15ft) or more

x *Fatshedera lizei*

The tree ivy is a mound-forming, evergreen shrub cultivated for its large, handsome, glossy dark green, ivy-like leaves. Small white flowers appear in autumn. Excellent in shady areas; in cold climates, grow in a sunny, sheltered site. Given support, it can be trained against a wall; it also makes a fine house- or conservatory plant. For leaves with cream edges, look for 'Variegata'.

CULTIVATION: *Best in moist but well-drained, fertile soil, in sun or partial shade. No regular pruning is required. Tie in to grow against a support.*

☀ ☀ ◐ ◐ ♔ ❀❀
↕1.2–2m (4–6ft) or more ↔3m (10ft)

Fatsia japonica

The Japanese aralia is a spreading, evergreen shrub grown for its large, palm-shaped, glossy green leaves. Broad, upright clusters of rounded, creamy-white flowerheads appear in autumn. An excellent architectural plant for a shady border. Tolerates atmospheric pollution and thrives in sheltered city gardens. 'Variegata' has cream-edged leaves.

CULTIVATION: *Grow in any well-drained soil, in sun or shade. Provide shelter from cold winds and hard frosts, especially when young. Little pruning is necessary, except to cut out wayward shoots and damaged growth in spring.*

☼ ☀ ◊ ♀ ❀❀❀   ↕↔1.5–4m (5–12ft)

Festuca glauca 'Blaufuchs'

This bright blue fescue is a densely tufted, evergreen, perennial grass. It is excellent in a border or rock garden, as a foil to other plants. Spikes of not particularly striking, violet-flushed, blue-green flowers are borne in early summer, above the foliage. The narrow, bright blue leaves are its chief attraction.

CULTIVATION: *Grow in dry, well-drained, poor to moderately fertile soil, in sun. For the best foliage colour, divide and replant clumps every 2 or 3 years.*

☼ ◊ ❀❀❀
↕to 30cm (12in) ↔25cm (10in)

Ficus carica 'Brown Turkey'

Grown as a specimen tree or trained against a sunny wall, the common fig is remarkable for its large, lobed, bright green leaves and its heavy crop of edible fruit. These only ripen fully in long, hot summers however. The leaves drop in autumn, leaving unripened figs to wither on the branches. Wall-trained figs are more easily managed than specimen trees.

CULTIVATION: *Grow in moist but well-drained soil in sun, and confine the roots to encourage better fruiting. Protect emerging figs in spring from late frosts.*

☼ ◊ ◊ ❀ ❀ ❀
‡3m (10ft) or more ↔ 4m (12ft)

Filipendula purpurea

An upright, clump-forming perennial that looks well planted in groups to form drifts of elegant, dark green foliage. Feathery clusters of red-purple flowers are carried above the leaves on purple-tinged stems in summer. Suitable for a waterside planting or bog garden; can also be naturalized in damp woodland.

CULTIVATION: *Grow in reliably moist, moderately fertile, humus-rich soil, in partial shade. Can be planted in full sun where the soil does not dry out.*

☼ ☀ ◊ ◊ ❀ ❀ ❀
‡1.2m (4ft) ↔ 60cm (24in)

Filipendula rubra 'Venusta'

A vigorous, upright perennial that produces feathery plumes of tiny, soft pink flowers on tall branching stems in midsummer. The large, dark green leaves are jaggedly cut into several lobes. Excellent in a bog garden, in moist soil by the side of water, or in a damp wild garden.

CULTIVATION: *Grow in reliably moist, moderately fertile soil, in partial shade. Thrives in wet or boggy soil, where it will tolerate full sun.*

☀ ◊ ◆ ♀ ❋ ❋ ❋
‡2–2.5m (6–8ft) ↔1.2m (4ft)

Forsythia x *intermedia* 'Lynwood Variety'

A vigorous, deciduous shrub with upright stems that arch slightly at the tips. Its golden-yellow flowers appear in profusion on bare branches in early spring. Mid-green, oval leaves emerge after flowering. A reliable shrub for a mixed border or as an informal hedge: particularly effective with spring bulbs. The flowering stems are good for cutting.

CULTIVATION: *Grow in well-drained, fertile soil, in full sun. Tolerates partial shade, although flowering will be less profuse. On established plants, cut out some old stems after flowering.*

☀ ☀ ◊ ♀ ❋ ❋ ❋   ‡↔3m (10ft)

Forsythia suspensa

This upright, deciduous shrub, sometimes called golden bell, is less bushy than *F. x intermedia* (see facing page, below), with more strongly arched stems. It is valued for the nodding, bright yellow flowers which open on its bare branches from early to mid-spring. The leaves are oval and mid-green. Good planted on its own, as an informal hedge, or in a shrub border.

CULTIVATION: *Best in well-drained, fertile soil, ideally in full sun; flowering is less spectacular in shade. Cut out some of the older stems on established plants after flowering.*

☼ ☀ ◊ ❀ ❀ ❀   ↕↔3m (10ft)

Fothergilla major

This slow-growing, upright shrub bears spikes of bottlebrush-like, fragrant white flowers in late spring. It is also valued for its blaze of autumn foliage. The deciduous leaves are glossy dark green in summer, then turn orange, yellow, and red before they fall. An attractive addition to a shrub border or in light woodland.

CULTIVATION: *Best in moist but well-drained, acid soil that is rich in humus. Choose a position in full sun for the best flowers and autumn colour. Requires very little pruning.*

☼ ◊ ◗ ♀ ❀ ❀ ❀   ↕2.5m (8ft) ↔2m (6ft)

Fremontodendron 'California Glory'

A vigorous, upright, semi-evergreen shrub producing large, cup-shaped, golden-yellow flowers from late spring to mid-autumn. The rounded leaves are lobed and dark green. Excellent for wall-training; in frost-prone areas, grow against walls that receive plenty of sun.

CULTIVATION: *Best in well-drained, poor to moderately fertile, neutral to alkaline soil, in sun. Shelter from cold winds. Best with no pruning, but wall-trained plants can be trimmed in spring.*

☼ ◊ ♀ ❋❋ ‡6m (20ft) ↔4m (12ft)

Fritillaria imperialis 'Rubra'

This red-flowered variety of the crown imperial bears tall upright stems in early summer, topped by clusters of hanging bell flowers above whorls of fresh green foliage. A bunch of narrow leaves forms a tuft above the flowers. This is an impressively tall perennial for a large border. Plant bulbs in late summer. The species form has orange flowers.

CULTIVATION: *Choose a sunny border in fertile, well-drained soil. Best left undisturbed.*

☼ ◊ ◊ ❋❋❋ ‡1.5m (5ft) ↔to 1m (3ft) or more

Fritillaria meleagris

The snake's head fritillary is a bulbous perennial, which naturalizes well in grass where summers are cool and damp. The drooping, bell-shaped flowers are pink, pinkish-purple, or white, and are strongly chequered. They are carried singly or in pairs during spring. The narrow leaves are grey-green. There is a white-flowered form, f. *alba*, which is also recommended.

CULTIVATION: *Grow in any moist but well-drained soil that is rich in humus, in full sun or light shade. Divide and replant bulbs in late summer.*

☀ ☼ ◊ ◑ ♀ ✿✿✿
‡to 30cm (12in) ↔5–8cm (2–3in)

Fritillaria pallidiflora

In late spring, this robust, bulbous perennial bears bell-shaped, foul-smelling flowers above the grey-green, lance-shaped leaves. They are creamy-yellow with green bases, chequered brown-red inside. Suits a rock garden or border in areas with cool, damp summers. Naturalizes easily in damp meadows.

CULTIVATION: *Grow in moist but well-drained, moderately fertile soil, in sun or partial shade. Divide and replant bulbs in late summer.*

☀ ☼ ◊ ◑ ♀ ✿✿✿
‡to 40cm (16in) ↔5–8cm (2–3in)

Hardy Fuchsias

The hardy fuchsias are wonderfully versatile shrubs: as well as being useful in mixed borders and as flowering hedges, they can be trained as espaliers or fans against warm walls, or grown as free-standing standards or pillars. Hanging flowers, varying in form from single to double, appear throughout summer and into autumn. During cold winters, even hardy fuchsias may lose some of their upper growth after severe frosts. They are fast to recover, however, and most will retain their leaves if overwintered in a cool greenhouse, or temperatures stay above 4°C (39°F). The hardiest cultivar is 'Genii'.

CULTIVATION: *Grow in well-drained but moist, fertile soil. Choose a position in sun or semi-shade with shelter from cold winds. Provide a deep winter mulch. Pinch-prune young plants to encourage a bushy habit. In early spring, remove frost damaged stems; cut back healthy growth to the lowest buds.*

☼ ☀ ◊ ◊ ◐

1 ↕↔75-90cm (30-36in)

2 ↕15-30cm (6-12in) ↔45cm (18in)

3 ↕↔1-1.1m (3-3¹/₂ft)

4 ↕2-3m (6-10ft) ↔1-2m (3-6ft)

5 ↕↔15-30cm (6-12in)

6 ↕to 3m (10ft) ↔2-3m (6-10ft)

1 *F.* 'Genii' ♀ ❋❋❋ **2** *F.* 'Lady Thumb' ♀ ❋❋ **3** *F.* 'Mrs. Popple' ♀ ❋❋
4 *F.* 'Riccartonii' ♀ ❋❋ **5** *F.* 'Tom Thumb' ♀ ❋❋❋ **6** *F. magellanica* 'Versicolor' ❋❋

Half-hardy and Tender Fuchsias

The half-hardy and tender fuchsias are flowering shrubs that require at least some winter protection in cold climates. The compensation for this extra care is an increased range of beautiful summer flowers for the garden, greenhouse, and conservatory. All can be grown out of doors in the summer months and make superb patio plants when grown in containers, whether pinch-pruned into dense bushes or trained as columns or standards. In frost-prone areas, shelter half-hardy fuchsias in a greenhouse or cool conservatory over the winter months; the tender species fuchsias, including 'Thalia', need a minimum temperature of 10°C (50°F) at all times.

CULTIVATION: *Grow in moist but well-drained, fertile soil or compost, in sun or partial shade. Pinch-prune young plants to promote bushiness, and trim after flowering to remove spent blooms. Prune back to an established framework in early spring.*

☼ ☀ ◊ ◐

↕↔ 30–60cm (12–24in)

↕ to 60cm (24in) ↔ to 45cm (18in)

↕ to 4m (12ft) ↔ 1–1.2m (3–4ft)

1 *Fuchsia* 'Annabel' ♀ ❀❀ **2** *F.* 'Billy Green' ♀❀ **3** *F. boliviana* var. *alba* ❀

‡↔45–75cm (18–30in)　　‡to 90cm (36in) ↔ to 75cm (30in)　　‡to 90cm (36in) ↔ to 60cm (24in)

‡1.5m (5ft) ↔ to 80cm (32in)　　‡to 45cm (18in) ↔ to 60cm (24in)　　‡to 75cm (18in) ↔ to 60cm (24in)

4 *Fuchsia* 'Celia Smedley' ♀ ❀❀❀　**5** *F.* 'Checkerboard' ♀ ❀❀❀　**6** *F.* 'Coralle' ♀ ❀
7 *F. fulgens* ♀❦　**8** *F.* 'Joy Patmore' ❀❀　**9** *F.* 'Leonora' ❀❀

↕↔30–60cm (12–24in)

↕to 45cm (18in) ↔45cm (18in)

↕to 75cm (30in) ↔to 60cm (24in)

↕to 60cm (24in) ↔75cm (30in)

↕↔45–90cm (18–36in)

MORE CHOICES

'Brookwood Belle' ❀❀
Cerise and white flowers.

'Pacquesa' ❀❀ Red and
red-veined white.

'Winston Churchill' ❀❀
Lavender and pink.

↕60cm (24in) ↔45cm (18in)

10 *F.* 'Mary' ♀ ❀ **11** *F.* 'Nellie Nuttall' ♀ ❀❀ **12** *F.* 'Royal Velvet' ♀ ❀❀
13 *F.* 'Snowcap' ♀ ❀❀ **14** *F.* 'Swingtime' ♀ ❀❀ **15** *F.* 'Thalia' ♀❀

Trailing Fuchsias

Fuchsia cultivars with a trailing or spreading habit are the perfect plants to have trailing over the edge of a tall container, window box, or hanging basket, where their pendulous flowers will be shown to great effect. They bloom continuously throughout summer and into early autumn, and can be left in place undisturbed right up until the end of the season. After this, they are best discarded and new plants bought the following year. An alternative, longer-lived planting is to train trailing fuchsias into attractive weeping standards for container plantings, but they must be kept frost-free in winter.

CULTIVATION: *Grow in fertile, moist but well-drained soil or compost, in full sun or partial shade. Shelter from cold, drying winds. Little pruning is necessary, except to remove wayward growth. Regularly pinch out the tips of young plants to encourage bushiness and a well-balanced shape.*

☼ ☀ ◊ ◑ ❄ ❄

1 ‡15–30cm (6–12in) ↔ 45cm (18in)

2 ‡15–30cm (6–12in) ↔ 45cm (18in)

‡45cm (18in) ↔ 60cm (24in)

‡ to 60cm (24in) ↔ 75cm (30in)

1 *F.* 'La Campanella' ♀ **2** *F.* 'Golden Marinka' ♀ **3** *F.* 'Jack Shahan' **4** *F.* 'Lena' ♀

Gaillardia 'Dazzler'

A bushy, short-lived perennial that
bears large, daisy-like flowers over a
long period in summer. These have
yellow-tipped, bright orange petals
surrounding an orange-red centre.
The leaves are soft, lance-shaped
and mid-green. Effective in a sunny,
mixed, or herbaceous border; the
flowers are good for cutting. May
not survive winter in cold climates.

CULTIVATION: *Grow in well-drained, not
too fertile soil, in full sun. May need
staking. Often short-lived, but can be
reinvigorated by division in winter.*

☀ ◊ ✿✿
‡60–85cm (24–34in) ↔45cm (18in)

Galanthus elwesii

This robust snowdrop is a bulbous
perennial that produces slender,
honey-scented, pure white flowers
in late winter, above the bluish-green
foliage. The inner petals have green
markings. Good for borders and
rock gardens; naturalizes easily
in light woodland.

CULTIVATION: *Grow in moist but well-
drained, humus-rich soil that does not
dry out in summer. Choose a position in
partial shade.*

☀ ◊ ◊ ♀ ✿✿✿
‡10–15cm (4–6in) ↔8cm (3in)

Galanthus 'Magnet'

This tall snowdrop is a vigorous bulbous perennial bearing drooping, pear-shaped, pure white flowers during late winter and early spring; the inner petals have a deep green, V-shaped mark at the tips. The strap-shaped, grey-green leaves are arranged around the base of the plant. Good for naturalizing in grass or in a woodland garden.

CULTIVATION: *Grow in moist but well-drained, fertile soil that does not dry out in summer. Choose a position in partial shade.*

☀ ◊ ◊ ♀ ❀❀❀
‡20cm (8in) ↔8cm (3in)

Galanthus nivalis 'Flore Pleno'

This double-flowered form of the common snowdrop, *G. nivalis*, is a robust, bulbous perennial. Drooping, pear-shaped, pure white flowers appear from late winter to early spring, with green markings on the tips of the inner petals. The narrow leaves are grey-green. Good for naturalizing under deciduous trees or shrubs.

CULTIVATION: *Grow in reliably moist but well-drained, fertile soil, in light shade. Divide and replant every few years after flowering to maintain vigour.*

☀ ◊ ◊ ♀ ❀❀❀   ‡↔10cm (4in)

Galanthus 'S. Arnott'

This honey-scented snowdrop, which
has even larger flowers than 'Magnet'
(facing page, above), is a fast-growing,
bulbous perennial. Nodding,
pear-shaped, pure white flowers
appear in late winter to early spring,
and have a green, V-shaped mark at
the tip of each inner petal. The
narrow leaves are grey-green. Suits
a rock garden or raised bed. 'Atkinsii'
is another recommended large
snowdrop, similar to this one.

CULTIVATION: *Grow in moist but well-
drained, fertile soil, in dappled shade.
Keep reliably moist in summer. Divide
and replant clumps after flowering.*

☼ ◊ ◊ ♥ ❀ ❀ ❀ ‡20cm (8in) ↔8cm (3in)

Garrya elliptica 'James Roof'

This silk-tassel bush is an upright,
evergreen shrub, becoming tree-like
with age. It is grown for its long,
silver-grey catkins, which dangle
from the branches in winter and
early spring. The leaves are dark
sea-green and have wavy margins.
Excellent in a shrub border, against
a shady wall or as hedging; tolerates
coastal conditions.

CULTIVATION: *Grow in well-drained,
moderately fertile soil, in full sun or
partial shade. Tolerates poor, dry soil.
Trim after flowering, as necessary.*

☼ ☼ ◊ ♥ ❀ ❀ ❀ ‡↔4m (12ft)

Gaultheria mucronata 'Mulberry Wine' (female)

This evergreen, spreading shrub, sometimes included in *Pernettya*, is much-valued for its autumn display of large, rounded, magenta to purple berries which show off well against the toothed, glossy dark green leaves. Small white flowers are borne throughout summer.

CULTIVATION: *Grow in reliably moist, peaty, acid to neutral soil, in partial shade or full sun. Plant close to male varieties to ensure a reliable crop of berries. Chop away spreading roots with a spade to restrict the overall size.*

☼ ☀ ◑ ♀ ✿ ✿ ✿   ↕↔1.2m (4ft)

Gaultheria mucronata 'Wintertime' (female)

An evergreen, spreading shrub, sometimes included in the genus *Pernettya*, bearing large, showy white berries which persist well into winter. Small white flowers are borne from late spring into early summer. The glossy dark green leaves are elliptic to oblong and toothed.

CULTIVATION: *Grow in reliably moist, acid to neutral, peaty soil. Best in light shade, but tolerates sun. Plant close to male varieties to ensure a good crop of berries. Dig out spreading roots as necessary to restrict the overall size.*

☼ ☀ ◑ ♀ ✿ ✿ ✿   ↕↔1.2m (4ft)

Gaultheria procumbens

Checkerberry is a creeping shrub which bears drooping, urn-shaped, white or pink flowers in summer, followed by aromatic scarlet fruits. As these usually persist until spring, they give winter colour. The glossy, dark green leaves have a strong fragrance when crushed, giving the plant its other common name, winter-green. Good ground cover in shade.

CULTIVATION: *Grow in acid to neutral, peaty, moist soil in partial shade; full sun is tolerated only where the soil is always moist. Trim after flowering.*

☼ ◑ ♀ ❀❀❀
‡15cm (6in) ↔ to 1m (3ft) or more

Gaura lindheimeri

A tall, clump-forming perennial with basal leaves and slender stems. From late spring to autumn, it bears loose spires of pinkish-white buds, which open in the morning to display white flowers. A graceful plant for a mixed flower border, tolerating both heat and drought.

CULTIVATION: *Grow in fertile, moist but well-drained soil, ideally in full sun, but some partial shade is tolerated. If necessary, divide clumps in spring.*

☼ ◐ ◑ ♀ ❀❀
‡to 1.5m (5ft) ↔ 90cm (36in)

Gazanias

These useful summer bedding plants are vigorous, spreading perennials, usually grown as annuals. They are cultivated for their long display of large and very colourful, daisy-like flowers, being quite similar to sunflowers, although they close in dull or cool weather. Flowers may be orange, white, golden yellow, beige, bronze, or bright pink, often with striking contrasting central zones. Hybrid selections with variously coloured flowerheads are also popular. The lance-shaped leaves are dark green with white-silky undersides. Gazanias grow well in containers and tolerate coastal conditions.

CULTIVATION: *Grow in light, sandy, well-drained soil in full sun. Remove old or faded flowerheads to prolong flowering. Water freely during the growing season. The plants will die back on arrival of the first frosts.*

☼ ◊ ❀

MORE CHOICES

G. Chansonette Series Mixed colours, zoned in a contrasting shade.

G. Mini-star Series Mixed or single zoned colours.

G. Talent Series Mixed or single colours.

‡20cm (8in) ↔25cm (10in)

‡20cm (8in) ↔25cm (10in)

‡20cm (8in) ↔25cm (10in)

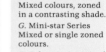

‡20cm (8in) ↔25cm (10in)

‡20cm (8in) ↔25cm (10in)

1 *Gazania* 'Aztec' **2** *G.* 'Cookei' **3** *G.* 'Daybreak Garden Sun' **4** *G.* 'Michael'
5 *G. rigens* var. *uniflora*

Genista aetnensis

The Mount Etna broom is an upright, almost leafless, deciduous shrub which is excellent on its own or at the back of a border in hot, dry situations. Masses of fragrant, pea-like, golden-yellow flowers cover the weeping, mid-green stems during midsummer.

CULTIVATION: *Grow in well-drained, light, poor to moderately fertile soil, in full sun. Keep pruning to a minimum; old, straggly plants are best replaced.*

☼ ◊ ♥ ❀❀❀   ↕↔8m (25ft)

Genista lydia

This low, deciduous shrub forms a mound of arching, prickle-tipped branches which are covered with yellow, pea-like flowers in early summer. The small, narrow leaves are blue-green. Ideal for hot, dry sites such as a rock garden or raised bed.

CULTIVATION: *Grow in well-drained, light, poor to moderately fertile soil, in full sun. Keep pruning to a minimum; old, straggly plants are best replaced.*

☼ ◊ ♥ ❀❀❀
↕to 60cm (24in) ↔to 1m (3ft)

Genista 'Porlock'

The Porlock broom is a semi-evergreen, medium-sized shrub that bears heavy clusters of fragrant, bright yellow, pea-like flowers in spring. The mid-green leaves are divided into three leaflets. Clip back after flowering to keep the plant in shape, but do not prune hard. This is a good plant choice for coastal sites.

CULTIVATION: *Grow in well-drained, fertile soil in sun, sheltered from hard frosts.*

☼ ◊ ♀ ❀ ❀ ❀   ↕↔2.5m (8ft)

Gentiana acaulis

The trumpet gentian is a mat-forming, evergreen perennial which bears large, trumpet-shaped, vivid deep blue flowers in spring. The leaves are oval and glossy dark green. Good in a rock garden, raised bed, or trough; thrives in areas with cool, damp summers.

CULTIVATION: *Grow in reliably moist but well-drained, humus-rich soil, in full sun or partial shade. Protect from hot sun in areas with warm, dry summers.*

☼ ☀ ◊ ◊ ♀ ❀ ❀ ❀
↕8cm (3in) ↔to 30cm (12in)

Gentiana asclepiadea

The willow gentian is an arching, clump-forming perennial producing trumpet-shaped, dark blue flowers in late summer and autumn; these are often spotted or striped with purple on the insides. The leaves are lance-shaped and fresh green. Suitable for a border or large rock garden.

CULTIVATION: *Grow in moist, fertile, humus-rich soil. Best in light shade, but tolerates sun if the soil is reliably moist.*

☀ ◐ ◊ ♀ ❀❀❀
‡60–90cm (24–36in) ↔45cm (18in)

Gentiana septemfida

This late-summer-flowering gentian is a spreading to upright, clump-forming herbaceous perennial. Clusters of narrowly bell-shaped, bright blue flowers with white throats are borne amid the oval, mid-green leaves. Good for a rock garden or raised bed; thrives in cool, damp summers. The variety var. *lagodechiana*, also recommended, has just one flower per stem.

CULTIVATION: *Grow in moist but well-drained, humus-rich soil, in full sun or partial shade. Protect from hot sun in areas with warm, dry summers.*

☀ ☀ ◊ ◊ ♀ ❀❀❀
‡to 15–20cm (6–8in) ↔to 30cm (12in)

Small Hardy Geraniums

The low-growing members of the genus *Geranium* (often incorrectly referred to as pelargoniums, see pages 372–377) are versatile plants, useful not only at the front of borders but also in rock gardens or as ground cover. These fully hardy, evergreen perennials are long-lived and undemanding, tolerating a wide range of sites and soil types. The lobed and toothed leaves are often variegated or aromatic. In summer, they bear typically saucer-shaped flowers ranging in colour from white through soft blues such as 'Johnson's Blue' to the intense pink of *G. cinereum* var. *caulescens*, often with contrasting veins, eyes, or other markings.

CULTIVATION: *Grow in sharply drained, humus-rich soil, in full sun. Feed with a balanced fertilizer every month during the growing season, but water sparingly in winter. Remove withered flower stems and old leaves to encourage fresh growth.*

☼ ◊ ❀ ❀ ❀

‡to 45cm (18in) ↔1m (3ft) or more   ‡to 15cm (6in) ↔to 30cm (12in)

1 *G.* 'Ann Folkard' ♥ **2** *G.* 'Ballerina' ♥

‡to 15cm (6in) ↔ to 30cm (12in)

‡to 45cm (18in) ↔ indefinite

‡to 15cm (6in) ↔ 50cm (20in)

‡30cm (12in) ↔ 60cm (24in)

‡to 45cm (18in) ↔ 75cm (30in)

‡to 30cm (12in) ↔ to 1m (3ft)

‡30cm (12in) ↔ 1.2m (4ft)

3 *G. subcaulescens* ♀ **4** *G. clarkei* 'Kashmir White' **5** *G. dalmaticum* ♀
6 *G. himalayense* 'Gravetye' **7** *G.* 'Johnson's Blue' **8** *G.* x *riversleaianum*
'Russell Prichard' ♀ **9** *G. wallichianum* 'Buxton's Variety'

Large Hardy Geraniums

The taller, clump-forming types of hardy perennial geranium – not to be confused with pelargoniums (see pages 372–377) – make effective, long-lived border plants or infill among shrubs, requiring a minimum of attention. They are especially suited to cottage garden plantings and between roses. Their lobed, evergreen leaves may be coloured or aromatic, and provide a long season of interest. Throughout summer, this is heightened by an abundance of saucer-shaped flowers, in white and shades of blue, pink, and purple. Markings on flowers vary, from the dramatic, contrasting dark eyes of *G. psilostemon* to the delicate venation of *G. sanguineum* var. *striatum*.

CULTIVATION: *Best in well-drained, fairly fertile soil, in full sun or partial shade, but tolerant of any soil that is not waterlogged. Remove old leaves and withered flower stems to encourage new growth.*

☼ ☀ ◊ ◑ ❋ ❋ ❋

‡45cm (18in) ↔ 60cm (24in)

‡↔ 60cm (24in)

‡↔ 50cm (20in)

1 *G. endressii* ♀ **2** *G.* × *magnificum* ♀ **3** *G.* ROZANNE ('Gerwat') ♀

‡60–90cm (24–36in) ↔ 60cm (24in)

‡60–120cm (24–48in) ↔ 60cm (24in)

‡↔ 30cm (12in)

‡10cm (4in) ↔ 30cm (12in)

‡60cm (24in) ↔ 90cm (36in)

‡60cm (24in) ↔ 90cm (36in)

4 *G. pratense* 'Mrs. Kendall Clark' ♀ **5** *G. psilostemon* ♀ **6** *G. renardii* ♀ **7** *G. sanguineum* var. *striatum* ♀ **8** *G. sylvaticum* 'Mayflower' ♀ **9** *G.* x *oxonianum* 'Wargrave Pink'

Geum 'Lady Stratheden'

A clump-forming perennial bearing double, bright yellow flowers on arching stems over a long period in summer. The mid-green leaves are large and lobed. An easy, long-flowering plant for brightening up a mixed or herbaceous border. The taller 'Fire Opal' is closely related, with reddish-orange flowers on purple stems.

CULTIVATION: *Grow in moist but well-drained, fertile soil, in full sun. Avoid sites that become waterlogged in winter.*

☼ ◊ ◑ ◊ ♀ ✽ ✽ ✽
‡40–60cm (16–24in) ↔ 60cm (24in)

Geum montanum

A small, clump-forming perennial grown for its solitary, cup-shaped, deep golden-yellow flowers. These are produced in spring and early summer above large, lobed, dark green leaves. Excellent for a rock garden, raised bed, or trough.

CULTIVATION: *Grow in well-drained, preferably gritty, fertile soil, in full sun. Will not tolerate waterlogging in winter.*

☼ ◊ ◊ ♀ ✽ ✽ ✽
‡15cm (6in) ↔ to 30cm (12in)

Gillenia trifoliata

An upright, graceful perennial that forms clumps of olive green leaves which turn red in autumn. Delicate white flowers with slender petals are borne on wiry red stems in summer. Effective in a shady border or light woodland; the cut flowers last well.

CULTIVATION: *Grow in moist but well drained, fertile soil. Best in partial shade, but tolerates some sun if shaded during the hottest part of the day.*

☼ ◊ ◊ ♈ ❈ ❈ ❈
‡to 1m (3ft) ↔60cm (24in)

Gladiolus communis subsp. *byzantinus*

In late spring, this upright, cormous perennial produces a blaze of magenta flowers with purple-marked lips. These are arranged in spikes above fans of narrow, mid-green leaves. An elegant subject for a mixed or herbaceous border; the flowers are good for cutting.

CULTIVATION: *Grow in fertile soil, in full sun. Plant corms on a bed of sharp sand to improve drainage. No need to lift corms in winter, but benefits from a winter mulch in frost-prone climates.*

☼ ◊ ♈ ❈ ❈ ❈   ‡to 1m (3ft) ↔8cm (3in)

Gladiolus murielae

An upright cormous perennial that bears strongly scented white flowers with purple-red throats. These hang from elegant stems during summer, above fans of narrow, mid-green leaves. Ideal for mixed borders; the flowers are suitable for cutting. Will not survive cold winters.

CULTIVATION: *Grow in fertile soil, in full sun. Plant on a layer of sand to improve drainage. In frost-prone climates lift the corms when the leaves turn yellow-brown, snap off the leaves, and store the corms in frost-free conditions.*

☀ ◊ ♀ ❀
↕70–100cm (28–39in) ↔5cm (2in)

Gleditsia triacanthos **'Sunburst'**

This fast-growing honey locust is a broadly conical, deciduous tree valued for its beautiful foliage and light canopy. The finely divided leaves are bright gold-yellow when they emerge in spring, maturing to dark green, then yellowing again before they fall. A useful, pollution-tolerant tree for a small garden.

CULTIVATION: *Grow in any well-drained, fertile soil, in full sun. Prune only to remove dead, damaged, or diseased wood, from late summer to midwinter.*

☀ ◊ ❀❀❀   ↕12m (40ft) ↔10m (30ft)

Griselinia littoralis 'Variegata'

A dense and attractive, evergreen shrub that carries glossy leaves variegated with irregular, creamy white margins and greyish streaks. It makes a good evergreen hedge or specimen shrub, particularly for coastal sites, and it establishes quickly. The shrub responds well to trimming in spring and can be cut back hard, if necessary. The flowers are insignificant.

CULTIVATION: *Grow in well-drained soil in sun.*

☼ ◊ ♀ ❋ ❋ ❋   ‡8m (25ft) ↔5m (15ft)

Gunnera manicata

A massive, clump-forming perennial that produces the largest leaves of any hardy garden plant, to 2m (6ft) long. These are rounded, lobed, sharply toothed, and dull green, with stout, prickly stalks. Tall spikes of tiny, greenish-red flowers appear in summer. An imposing plant by water or in a bog garden. Needs protection to survive winter in cold climates.

CULTIVATION: *Best in permanently moist, fertile soil, in full sun or partial shade. Shelter from cold winds. Protect from frost by folding the dead leaves over the dormant crown before winter.*

☼ ☀ ◊ ♦ ♀ ❋ ❋   ‡2.5m (8ft) ↔3–4m (10–12ft) or more

Gymnocarpium dryopteris 'Plumosum'

The oak fern is a delicate woodland plant useful as deciduous ground cover for shady spots with reasonably moist soil. Its characteristic triangular fronds arise from long, creeping rhizomes in spring, and darken with age. The rhizomes spread widely but are not invasive. Propagate from rhizomes in spring.

CULTIVATION: *Grow in moist soil in dappled shade.*

☀ ◊ ♀ ❋❋❋   ↕20cm (8in) ↔indefinite

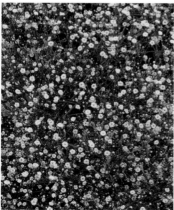

Gypsophila paniculata 'Bristol Fairy'

This herbaceous perennial forms a mound of slightly fleshy, lance-shaped, blue-green leaves on very slender, wiry stems. The profusion of tiny, double white flowers in summer, forms a cloud-like display. Very effective cascading over a low wall or as a foil to more upright, sharply defined flowers.

CULTIVATION: *Grow in well-drained, deep, moderately fertile, preferably alkaline soil, in full sun. Resents being disturbed after planting.*

☀ ◊ ❋❋❋   ↕↔to 1.2m (4ft)

Gypsophila 'Rosenschleier'

In summer, tiny, double white flowers which age to pale pink are carried in airy sprays, forming a dense cloud of blooms. The slightly fleshy leaves are lance-shaped and blue-green. The flowers dry well for decoration.

CULTIVATION: *Grow in well-drained, deep, moderately fertile, preferably alkaline soil. Choose a position in full sun. Resents root disturbance.*

☼ ◊ ♀ ❀ ❀ ❀
↕40–50cm (16–20in) ↔1m (3ft)

Hakonechloa macra 'Aureola'

This colourful grass is a deciduous perennial forming a clump of narrow, arching, bright yellow leaves with cream and green stripes. They flush red in autumn, and persist well into winter. Reddish-brown flower spikes appear in late summer. A versatile plant that can be used in a border, rock garden, or containers.

CULTIVATION: *Grow in moist but well-drained, fertile, humus-rich soil. Leaf colour is best in partial shade, but tolerates full sun.*

☼ ☼ ◊ ◊ ♀ ❀ ❀ ❀
↕35cm (14in) ↔40cm (16in)

x *Halimiocistus sahucii*

A compact shrub that forms mounds of linear, dark green leaves with downy undersides. Masses of saucer-shaped white flowers are produced throughout summer. Good in a border, at the base of a warm wall, or in a rock garden.

CULTIVATION: *Best in freely draining, poor to moderately fertile, light, gritty soil, in a sunny site. Shelter from excessive winter wet.*

☼ ◊ ♀ ✿✿✿
‡45cm (18in) ↔ 90cm (36in)

x *Halimiocistus wintonensis* 'Merrist Wood Cream'

A spreading, evergreen shrub bearing creamy-yellow flowers with red bands and yellow centres in late spring and early summer. The lance-shaped leaves are grey-green. Good at the front of a mixed border or at the foot of a warm wall. Also suits a raised bed or rock garden. May not survive cold winters.

CULTIVATION: *Grow in freely draining, poor to moderately fertile soil, in full sun. Choose a position protected from excessive winter wet.*

☼ ◊ ♀ ✿✿
‡60cm (24in) ↔ 90cm (36in)

Halimium lasianthum

A spreading bush with clusters of saucer-shaped, golden-yellow flowers in late spring and early summer. Each petal normally has a brownish-red mark at the base. The foliage is grey-green. Halimiums flower best in regions with long, hot summers, and this one is suited to a coastal garden.

CULTIVATION: *Best in well-drained, moderately fertile sandy soil in full sun, with shelter from cold, drying winds. Established plants dislike being moved. After flowering, trim to maintain symmetry.*

☼ ◊ ❀❀　　　　↕1m (3ft) ↔1.5m (5ft)

Halimium 'Susan'

A small, spreading, evergreen shrub valued for its single or semi-double summer flowers, bright yellow with deep purple markings. The leaves are oval and grey-green. Good for rock gardens in warm, coastal areas. Provide shelter at the foot of a warm wall where winters are cold. Flowers are best during long, hot summers. *H. ocymoides*, similar but more upright in habit, is also recommended.

CULTIVATION: *Grow in freely draining, fairly fertile, light, sandy soil, in full sun. Provide shelter from cold winds. Trim lightly in spring, as necessary.*

☼ ◊ ♀ ❀❀
↕45cm (18in) ↔60cm (24in)

Witch Hazels (*Hamamelis*)

These spreading, deciduous shrubs are grown for their large clusters of usually yellow spidery flowers, which appear on the branches when the shrubs are bare in winter; those of *H. vernalis* may coincide with the unfurling of the leaves. Each flower has four narrow petals and an enchanting fragrance. Most garden species also display attractive autumn foliage; the broad, bright green leaves turn red and yellow before they fall. Witch hazels bring colour and scent to the garden in winter; they are good as specimen plants or grouped in a shrub border or woodland garden.

CULTIVATION: *Grow in moderately fertile, moist but well-drained, acid to neutral soil in full sun or partial shade, in an open but not exposed site. Witch hazels also tolerate deep, humus-rich soil over chalk. Remove wayward or crossing shoots when dormant in late winter or early spring to maintain a healthy, permanent framework.*

☀ ☼ ◊ ❀ ❀ ❀

‡↔4m (12ft)   ‡↔4m (12ft)

1 *Hamamelis* × *intermedia* 'Arnold Promise' ♀ **2** *H.* × *intermedia* 'Barmstedt Gold' ♀

3 ‡↔4m (12ft)

4 ‡↔4m (12ft)

5 ‡↔4m (12ft)

6 ‡↔4m (12ft)

7 ‡↔5m (15ft)

3 *H.* × *intermedia* 'Diane' ♀ **4** *H.* × *intermedia* 'Jelena' ♀ **5** *H.* × *intermedia* 'Pallida' ♀
6 *H.* × *intermedia* 'Aphrodite' ♀ **7** *H. vernalis* 'Sandra'

Hebe albicans

A neat, mound-forming shrub with tightly packed, glaucous grey-green foliage. It bears short, tight clusters of white flowers at the ends of the branches in the first half of summer. A useful evergreen hedging plant in mild coastal areas, or grow in a shrub border.

CULTIVATION: *Grow in poor to moderately fertile, moist but well-drained, neutral to slightly alkaline soil in full sun or partial shade. Shelter from cold, drying winds. It will need little or no pruning.*

☼ ◐ ◊ ◊ ♥ ❀ ❀ ❀
‡60cm (24in) ↔ 90cm (36in)

Hebe cupressoides 'Boughton Dome'

This dwarf, evergreen shrub is grown for its neat shape and dense foliage which forms a pale green dome. Flowers are infrequent. The congested, slender, greyish-green branches carry scale-like, pale green leaves. Excellent in a rock garden; gives a topiary effect without any clipping. Thrives in coastal gardens.

CULTIVATION: *Grow in moist but well-drained, poor to moderately fertile soil, in full sun or partial shade. No regular pruning is necessary.*

☼ ◐ ◊ ◊ ❀ ❀
‡30cm (12in) ↔ 60cm (24in)

Hebe 'Great Orme'

An open, rounded, evergreen shrub that carries slender spikes of small, deep pink flowers which fade to white. These are borne from midsummer to mid-autumn amid the lance-shaped, glossy dark green leaves. Good in a mixed or shrub border; shelter at the base of a warm wall in cold climates.

CULTIVATION: *Grow in moist but well-drained, poor to moderately fertile soil, in sun or light shade. Shelter from cold winds. Pruning is unnecessary, but leggy plants can be cut back in spring.*

☼ ☼ ◊ ◊ ♀ ❈ ❈   ‡↔1.2m (4ft)

Hebe macrantha

This upright, spreading shrub has leathery green leaves. It bears relatively large flowers for a hebe, produced in clusters of three in early summer. A useful evergreen edging plant for seaside gardens in mild areas.

CULTIVATION: *Grow in moist but well-drained, reasonably fertile, neutral to slightly alkaline soil in full sun or partial shade. Shelter from cold, drying winds. It needs little or no pruning.*

☼ ☼ ◊ ◊ ♀ ❈ ❈
‡60cm (24in) ↔ 90cm (36in)

Hebe ochracea 'James Stirling'

A compact shrub that bears its medium-sized white flowers in clusters from late spring to early summer. Like other whipcord hebes, it has small, scale-like leaves that lie flat against the stems to give the appearance of a dwarf conifer. An excellent evergreen for a rock garden, the rich ochre-yellow foliage looking very attractive in winter.

CULTIVATION: *Best in moist but well-drained, neutral to slightly alkaline soil. Site in full sun or partial shade, and protect from cold, drying winds. Do not prune unless absolutely necessary.*

☼ ☀ ◊ ◐ ♀ ❀ ❀ ❀
‡45cm (18in) ↔ 60cm (24in)

Hebe pinguifolia 'Pagei'

A low-growing, evergreen shrub bearing purple stems with four ranks of leathery, oval, blue-green leaves. Abundant clusters of white flowers appear at the tips of the shoots in late spring and early summer. Plant in groups as ground cover, or in a rock garden.

CULTIVATION: *Grow in moist but well-drained, poor to moderately fertile soil, in sun or partial shade. Best with some shelter from cold, drying winds. Trim to neaten in early spring, if necessary.*

☼ ☀ ◊ ◐ ♀ ❀ ❀ ❀
‡30cm (12in) ↔ 90cm (36in)

Hebe rakaiensis

A rounded, evergreen shrub bearing spikes of white flowers from early to midsummer. The leaves are elliptic and glossy bright green. Ideal either as a small, spreading specimen shrub or as a focal point in a large rock garden.

CULTIVATION: *Grow in moist but well-drained, poor to moderately fertile soil, in sun or partial shade. Best with some shelter from cold, drying winds. Trim to shape in early spring, if necessary.*

☀ ☀ ◊ ◑ ♀ ❀ ❀ ❀ ‡1m (3ft) ↔1.2m (4ft)

Hebe '**Silver Queen**'

A dense, rounded, evergreen shrub bearing colourful, oval leaves; these are mid-green with creamy-white margins. Purple flowers that contrast well with the foliage are carried in dense spikes during summer and autumn. A fine, pollution-tolerant plant for a mixed border or rock garden. In cold areas, shelter at the foot of a warm wall. 'Blue Gem' is a similar plant with plain leaves.

CULTIVATION: *Grow in moist but well-drained, poor to moderately fertile soil, in sun or light shade. Shelter from cold, drying winds. No pruning is necessary.*

☀ ☀ ◊ ◑ ♀ ❀ ‡↔60–120cm (24–48in)

Hedera algeriensis 'Ravensholst'

This vigorous cultivar of Canary Island ivy is a self-clinging, evergreen climber useful as ground cover or to mask a bare wall. The leaves are shallowly lobed, glossy, and dark green. There are no flowers. It may be damaged in areas with severe winters, but it usually grows back quickly.

CULTIVATION: *Best in fertile, moist but well-drained soil that is rich in organic matter. It tolerates shade, and can be pruned or trimmed at any time of year.*

☀ ☼ ◊ ◊ ♀ ❀ ❀   ‡5m (15ft)

Hedera colchica 'Dentata'

This Persian ivy is a very vigorous, evergreen, self-clinging climber carrying large, heart-shaped, drooping, glossy green leaves. The stems and leaf-stalks are flushed purple. A handsome plant for covering an unattractive wall in shade; also effective as ground cover. 'Dentata Variegata' has mottled grey-green leaves edged with cream.

CULTIVATION: *Best in moist but well-drained, fertile, ideally alkaline soil, in partial to deep shade. Prune at any time of the year to restrict size.*

☀ ☼ ◊ ◊ ♀ ❀ ❀ ❀   ‡10m (30ft)

Hedera colchica 'Sulphur Heart'

This coloured-leaf Persian ivy is a very vigorous, self-clinging evergreen climber which can also be grown as ground cover. The large, heart-shaped leaves are dark green suffused with creamy-yellow; as they mature, the colour becomes more even. Will quickly cover a wall in shade.

CULTIVATION: *Grow in moist but well-drained, fertile, preferably alkaline soil. Tolerates partial shade, but leaf colour is more intense in sun. Prune at any time of the year to restrict size.*

☼ ☀ ◊ ◊ ♥ ❀ ❀ ❀   ‡5m (15ft)

Hedera hibernica

Irish ivy is a vigorous, evergreen, self-clinging climber valued for its broadly oval, dark green leaves which have grey-green veins and five triangular lobes. Useful for a wall or against a robust tree, or as fast-growing ground cover under trees or shrubs.

CULTIVATION: *Best in moist but well-drained, fertile, ideally alkaline soil, in partial to full shade. Prune at any time of the year to restrict spread.*

☼ ☀ ◊ ◊ ❀ ❀ ❀   ‡to 10m (30ft)

English Ivies (*Hedera helix*)

Hedera helix, the common or English ivy, is an evergreen, woody-stemmed, self-clinging climber and the parent of an enormous selection of cultivars. Leaf forms vary from heart-shaped to deeply lobed, ranging in colour from the bright gold 'Buttercup' to the deep purple 'Atropurpurea'. They make excellent ground cover, tolerating even dry shade, and will quickly cover featureless walls; they can damage paintwork or invade gutters if not kept in check. Variegated cultivars are especially useful for enlivening dark corners and shaded walls. Small ivies make good houseplants and can be trained over topiary frames.

CULTIVATION: *Best in moist but well-drained, humus-rich, alkaline soil. Choose a position in full sun or shade; ivies with variegated leaves may lose their colour in shade. Trim as necessary to keep under control. 'Goldchild' and 'Little Diamond' may suffer in areas with heavy frosts.*

☼ ☀ ◐ ◊ ◊

‡8m (25ft)   ‡2m (6ft)   ‡2m (6ft)

‡1m (3ft)   ‡1m (3ft)   ‡30cm (12in)

1 *H. helix* 'Atropurpurea' ❀❀❀ **2** *H. helix* 'Buttercup' ♀ ❀❀❀ **3** *H. helix* 'Glacier' ♀ ❀❀❀
4 *H. helix* 'Goldchild' ♀ ❀❀ **5** *H. helix* 'Ivalace' ❀❀❀ **6** *H. helix* 'Little Diamond' ❀❀

Helenium 'Moerheim Beauty'

This rewarding summer perennial is a firm garden favourite for its lovely, rich coppery red, daisy-like flowers that appear towards the end of summer when many other flowers are fading. Combined with ornamental grasses and coneflowers (*Echinacea* and *Rudbeckia*), it can bring a late summer garden to life. Deadhead regularly and divide large clumps to maintain vigour.

CULTIVATION: *Grow in moist but well-drained soil in sun.*

☼ ◊ ◗ ♀ ❀❀❀
‡90cm (36in) ↔60cm (24in)

Helenium 'Sahin's Early Flowerer'

This helenium was spotted as an accidental seedling in the 1990s and noted for its long flowering season. The reddish orange to yellow daisy-like flowers appear earlier than most heleniums, sometimes before midsummer, and the display can continue until autumn's end. The colour combines well with orange or red dahlias and crocosmias, and most daylilies (*Hemerocallis*).

CULTIVATION: *Grow in moist but well-drained soil in sun.*

☼ ◊ ◗ ♀ ❀❀❀   ‡↔60cm (24in)

Helianthemum 'Henfield Brilliant'

This rock rose is a small, spreading, evergreen shrub bearing saucer-shaped, brick-red flowers in late spring and summer. The leaves are narrow and grey-green. Effective in groups on a sunny bank; also good in a rock garden or raised bed, or at the front of a border.

CULTIVATION: *Grow in moderately fertile, well-drained, neutral to alkaline soil, in sun. Trim after flowering to keep bushy. Often short-lived, but easily propagated by softwood cuttings in late spring.*

☼ ◊ ♀ ❁❁❁

‡20–30cm (8–12in) ↔30cm (12in) or more

Helianthemum 'Rhodanthe Carneum'

This long-flowering rock rose, also sold as 'Wisley Pink', is a low and spreading, evergreen shrub. Pale pink, saucer-shaped flowers with yellow-flushed centres appear from late spring to summer amid narrow, grey-green leaves. Good in a rock garden, raised bed, or mixed border.

CULTIVATION: *Best in well-drained, moderately fertile, neutral to alkaline soil, in full sun. Trim over after flowering to encourage further blooms.*

☼ ◊ ♀ ❁❁❁

‡to 30cm (12in) ↔to 45cm (18in) or more

Helianthemum 'Wisley Primrose'

This primrose-yellow rock rose is a fast-growing, spreading, evergreen shrub. It bears a profusion of saucer-shaped flowers with golden centres over long periods in late spring and summer. The leaves are narrowly oblong and grey-green. Group in a rock garden, raised bed, or sunny bank. For paler, creamy flowers, look for 'Wisley White'.

CULTIVATION: *Grow in well-drained, moderately fertile, preferably neutral to alkaline soil, in full sun. Trim after flowering to encourage further blooms.*

☼ ◊ ♀ ❀ ❀ ❀
‡to 30cm (12in) ↔to 45cm (18in) or more

Helianthus 'Loddon Gold'

This double-flowered sunflower is a tall, spreading perennial with coarse, oval, mid-green leaves which are arranged along the upright stems. Grown for its large, bright yellow flowers, which open during late summer and last into early autumn. Use to extend the season of interest in herbaceous and mixed borders.

CULTIVATION: *Grow in moist to well-drained, moderately fertile soil that is rich in humus. Choose a sheltered site in full sun. Flowers are best during long, hot summers. Stake flower stems.*

☼ ◊ ◑ ♀ ❀ ❀ ❀
‡to 1.5m (5ft) ↔90cm (36in)

Helianthus 'Monarch'

This semi-double sunflower is a tall, spreading perennial with sturdy, upright stems bearing oval and toothed, mid-green leaves. Large, star-like, bright golden-yellow flowers with yellow-brown centres are produced from late summer to autumn. A statuesque plant for late summer interest in a herbaceous or mixed border. 'Capenoch Star' is another recommended perennial sunflower, lemon-yellow with golden centres, to 1.5m (5ft) tall.

CULTIVATION: *Grow in any well-drained, moderately fertile soil. Choose a sunny, sheltered site. The stems need support.*

☼ ◊ ♀ ❋ ❋ ❋   ‡to 2m (6ft) ↔1.2m (4ft)

Helichrysum splendidum

A compact, white-woolly, evergreen perennial bearing linear, aromatic, silver-grey foliage. Small, bright yellow flowerheads open at the tips of the upright stems from midsummer to autumn, and last into winter. Suitable for a mixed border or rock garden; the flowers can be dried for winter decoration. May not survive outside during cold winters.

CULTIVATION: *Grow in well-drained, poor to moderately fertile, neutral to alkaline soil, in full sun. Remove dead or damaged growth in spring, cutting back leggy shoots into old wood.*

☼ ◊ ♀ ❋ ❋   ‡↔1.2m (4ft)

Helictotrichon sempervirens

Blue oat grass is a useful plant in a mixed garden scheme for its hummocks of greyish blue, evergreen leaves. From early summer, tall, oat-like flower spikes emerge above the foliage; these are very graceful with nodding tips and last all summer. Mix in with other blue- or silver-foliaged plants, or grow in clumps or as a specimen plant. Tidy plants and divide clumps in autumn or spring.

CULTIVATION: *Grow in well-drained, ideally alkaline soil in full sun or light shade.*

☼ ◊ ♀ ❀❀❀
‡to 1.4m (4¹/₂ft) ↔ 60cm (24in)

Heliopsis helianthoides var. *scabra* 'Light of Loddon'

The vivid golden yellow, semi-double flowers of this vigorous perennial appear from midsummer to early autumn. Unobtrusive support, put into place in spring so that the plant can grow through it, and regular deadheading, improves and lengthens the display. The flowers are good for cutting. Divide large clumps in spring to maintain vigour.

CULTIVATION: *Grow in fertile, moist but well-drained soil in sun.*

☼ ◊ ◊ ♀ ❀❀❀
‡to 1.1m (3¹/₂ft) ↔ 60cm (24in)

Helleborus argutifolius

The large Corsican hellebore, sometimes known as *H. corsicus*, is an early-flowering, clump-forming, evergreen perennial bearing large clusters of nodding, pale green flowers. These appear in winter and early spring above the handsome dark green leaves which are divided into three sharply toothed leaflets. Excellent for early interest in a woodland garden or mixed border.

CULTIVATION: *Grow in moist, fertile, preferably neutral to alkaline soil, in full sun or partial shade. Often short-lived, but self-seeds readily.*

☼ ☀ ◊ ♦ ♀ ❀ ❀ ❀
↕ to 1.2m (4ft) ↔ 90cm (36in)

Helleborus x *hybridus*

The perennial Lenten rose blooms in winter and early spring. Single or double, pendent, saucer-shaped flowers last for a few months and open in range of whites, yellows, pinks, and purples, often spotted or with darker edges. Most are evergreen with large, toothed, divided leaves. Use as ground cover or at the base of taller plants.

CULTIVATION: *Grows best in moist, neutral to alkaline soil in dappled shade, but tolerant of all but very poorly drained or dry soils. Hellebores need protection from strong, winter winds. They will naturally hybridize and self-seed.*

☀ ◊ ❀ ❀ ❀   ↕↔ to 45cm (18in)

Helleborus niger

A clump-forming, usually evergreen
perennial valued for its nodding
clusters of cup-shaped white flowers
in winter and early spring. The dark
green leaves are divided into several
leaflets. Effective with snowdrops
beneath winter-flowering shrubs,
but can be difficult to naturalize.
'Potter's Wheel' is a pretty cultivar
of this hellebore, with large flowers.

CULTIVATION: *Grow in deep, fertile, neutral
to alkaline soil that is reliably moist. Site
in dappled shade with shelter from cold,
drying winds.*

☀ ◊ ❉ ❉ ❉
‡to 30cm (12in) ↔ 45cm (18in)

Helleborus x *nigercors*

This hybrid hellebore combines
vigour with a long-lasting flower
display from late winter into early
spring. Its large, green-tinged, white
flowers are a welcome arrival in
winter. The thick green leaves are
quite profuse and will need to be
trimmed back if they are not to
obscure the flowers. Remove old
flowering stems in late spring.

CULTIVATION: *Grow in moist but well
drained soil in sun or shade.*

☼ ☀ ☀ ◊ ❉ ❉ ❉
‡to 30cm (12in) ↔ 90cm (36in)

Daylilies (*Hemerocallis*)

Daylilies are clump-forming, herbaceous perennials, so-called because each of their showy flowers only lasts for about a day; in nocturnal daylilies, such as 'Green Flutter', the flowers open in late afternoon and last through the night. The blooms are abundant and rapidly replaced, some appearing in late spring while other cultivars flower into late summer. Flower shapes vary from circular to spider-shaped, in shades of yellow and orange. The leaves are strap-like and evergreen. Tall daylilies make a dramatic contribution to a mixed or herbaceous border; dwarf types, such as 'Stella de Oro', are useful for small gardens or in containers.

CULTIVATION: *Grow in well-drained but moist, fertile soil, in sun; 'Corky' and 'Green Flutter' tolerate semi-shade. Mulch in spring, and feed with a balanced fertilizer every two weeks until buds form. Divide and replant every few years, in spring or autumn.*

☼ ☀ ◊ ◑ ❀ ❀ ❀

‡60cm (24in) ↔ 1m (3ft)

‡90cm (36in) ↔ 45cm (18in)

‡50cm (20in) ↔ 1m (3ft)

‡↔ 1m (3ft)

‡60cm (24in) ↔ 1m (3ft)

‡30cm (12in) ↔ 45cm (18in)

1 *H.* 'Buzz Bomb' **2** *H.* 'Golden Chimes' **3** *H.* 'Green Flutter' **4** *H. lilioasphodelus*
5 *H.* 'Oriental Ruby' **6** *H.* 'Stella de Oro'

Hepatica nobilis

A small, slow-growing, anemone-like,
semi-evergreen perennial bearing
saucer-shaped, purple, white, or
pink flowers. These appear in early
spring, usually before the foliage has
emerged. The mid-green, sometimes
mottled leaves are leathery and
divided into three lobes. Suits a shady
rock garden. *H.* x *media* 'Ballardii' is
a very similar plant, with reliably
deep blue flowers.

CULTIVATION: *Grow in moist but well-
drained, fertile, neutral to alkaline soil,
in partial shade. Provide a mulch of leaf
mould in autumn or spring.*

☼ ◊ ◑ ♀ ✿✿✿✿
‡10cm (4in) ↔15cm (6in)

Heuchera micrantha var. *diversifolia* 'Palace Purple'

A clump-forming perennial valued
for its glistening, dark purple-red,
almost metallic foliage, which is
topped by airy sprays of white
flowers in summer. The leaves have
five pointed lobes. Plant in groups
as ground cover for a shady site, but
can be slow to spread.

CULTIVATION: *Grow in moist but well-
drained, fertile soil, in sun or partial
shade. Tolerates full shade where the
ground is reliably moist. Lift and divide
clumps every few years, after flowering.*

☼ ☀ ◊ ◑ ✿✿✿
‡↔45–60cm (18–24in)

Heuchera 'Red Spangles'

This clump-forming, evergreen perennial is valued for its sprays of small, bell-shaped, crimson-scarlet flowers. These are borne in early summer, with a repeat bloom in late summer, on dark red stems above the lobed, heart-shaped, purplish-green leaves. Effective as groundcover when grouped together.

CULTIVATION: *Grow in moist but well-drained, fertile soil, in sun or partial shade. Tolerates full shade where the ground is reliably moist. Lift and divide clumps every three years, after flowering.*

☼ ☀ ◊ ◊ ❀ ❀ ❀
‡50cm (20in) ↔25cm (10in)

Hibiscus syriacus 'Oiseau Bleu'

Also known as 'Blue Bird', this vigorous, upright, deciduous shrub bears large, mallow-like, lilac-blue flowers with red centres. These are borne from mid- to late summer amid deep green leaves. Ideal for a mixed or shrub border. Flowers are best in hot summers; plant against a warm wall in cold areas.

CULTIVATION: *Grow in moist but well-drained, fertile, neutral to slightly alkaline soil, in full sun. Prune young plants hard in late spring to encourage branching at the base; keep pruning to a minimum once established.*

☼ ◊ ◊ ♥ ❀ ❀ ❀   ‡3m (10ft) ↔2m (6ft)

Hibiscus syriacus **'Woodbridge'**

A fast-growing, upright, deciduous shrub producing large, deep rose-pink flowers with maroon blotches around the centres. These are borne from late summer to mid-autumn amid the lobed, dark green leaves. Valuable for its late season of interest. In cold areas, plant against a warm wall for the best flowers.

CULTIVATION: *Grow in moist but well-drained, fertile, slightly alkaline soil, in full sun. Prune young plants hard to encourage branching; keep pruning to a minimum once established.*

☼ ◊ ◑ ♚ ❀ ❀ ❀   ‡3m (10ft) ↔2m (6ft)

Hippophae rhamnoides

Sea buckthorn is a spiny, deciduous shrub with attractive fruits and foliage. Small yellow flowers in spring are followed by orange berries on female plants, which persist well into winter. The silvery-grey leaves are narrow and claw-like. Good for hedging, especially in coastal areas.

CULTIVATION: *Best in sandy, moist but well-drained soil, in full sun. For good fruiting, plants of both sexes must grow together. Little pruning is required; trim hedges in late summer, as necessary.*

☼ ◊ ◑ ❀ ❀ ❀   ‡↔6m (20ft)

Hoheria glabrata

This deciduous, spreading tree is grown for its graceful habit and clusters of cup-shaped, fragrant white flowers in midsummer, which are attractive to butterflies. The broad, dark green leaves turn yellow in autumn before they fall. Best in a shrub border in a maritime climate, against a sunny wall in cold areas.

CULTIVATION: *Grow in moderately fertile, well-drained, neutral to alkaline soil in full sun or partial shade, sheltered from cold, drying winds. Pruning is seldom necessary, but if branches are damaged by frost, cut them back in spring.*

☼ ☀ ◊ ❀❀❀   ‡↔7m (22ft)

Hoheria sexstylosa

Ribbonwood is an evergreen tree or shrub with narrow, glossy, mid-green leaves with toothed margins. It is valued for its graceful shape and abundant clusters of white flowers; attractive to butterflies, they appear in late summer. Grow against a warm, sunny wall in areas with cold winters. The cultivar 'Stardust' is recommended.

CULTIVATION: *Best in moderately fertile, well-drained, neutral to alkaline soil in full sun or partial shade, sheltered from cold, drying winds. Protect the roots with a thick winter mulch. Pruning is seldom necessary.*

☼ ☀ ◊ ❀❀   ‡8m (25ft) ↔6m (20ft)

Hostas

Hostas are deciduous perennials grown principally for their dense mounds of large, overlapping, lance- to heart-shaped leaves. A wide choice of foliage colour is available, from the cloudy blue-green of 'Halcyon' to the bright yellow-green 'Golden Tiara'. Many have leaves marked with yellow or white around the edges; *H. fortunei* var. *albopicta* has bold, central splashes of creamy-yellow. Upright clusters of funnel-shaped flowers, varying from white through lavender-blue to purple, are borne on tall stems in summer. Hostas are effective at the front of a mixed border or as ground cover under deciduous trees, and are also suitable for containers.

CULTIVATION: *Grow in well-drained but reliably moist, fertile soil, in full sun or partial shade with shelter from cold winds. Yellow-leaved hostas colour best in full sun with shade at midday. Mulch in spring to conserve moisture throughout the summer.*

☼ ☀ ◗ ✿✿✿

1 ‡50cm (20in) ↔1m (3ft)

2 ‡55cm (22in) ↔1m (3ft)

3 ‡55cm (22in) ↔1m (3ft)

4 ‡55cm (22in) ↔1m (3ft)

5 ‡60cm (24in) ↔1m (3ft)

6 ‡30cm (12in) ↔50cm (20in)

1 *Hosta crispula* **2** *H. fortunei* var. *albopicta* **3** *H. fortunei* var. *aureomarginata* ♀
4 *H.* 'Francee' ♀ **5** *H.* 'Frances Williams' ♀ **6** *H.* 'Golden Tiara' ♀

‡1m (3ft) ↔75cm (30in)

‡45cm (18in) ↔75cm (30in)

‡45cm (18in) ↔1m (3ft)

‡60cm (24in) ↔1.2m (4ft)

‡45cm (18in) ↔75cm (30in)

‡1m (3ft) ↔1.2m (4ft)

7 *Hosta* 'Honeybells' **8** *H. lancifolia* **9** *H.* 'Love Pat' ♀ **10** *H.* 'Royal Standard' ♀
11 *H.* 'Shade Fanfare' **12** *H. sieboldiana* var. *elegans* ♀

13
‡75cm (30in) ↔ 1.2m (4ft)

‡35–40cm (14–16in) ↔ 70cm (28in)

15
‡1m (3ft) ↔ 45cm (18in)

16
‡45cm (18in) ↔ 70cm (28in)

17
‡50cm (20in) ↔ 1m (3ft)

18
‡5cm (2in) ↔ 25cm (10in)

19
‡75cm (30in) ↔ 1m (3ft)

13 *H.* 'Sum and Substance' ♀ **14** *H.* Tardiana Group 'Halcyon' ♀ **15** *H. undulata* var. *undulata* ♀ **16** *H. undulata* var. *univittata* ♀ **17** *H. ventricosa* ♀ **18** *H. venusta* ♀ **19** *H.* 'Wide Brim' ♀

Humulus lupulus 'Aureus'

The golden hop is a twining, perennial climber grown for its attractively lobed, bright golden-yellow foliage. Hanging clusters of papery, cone-like, greenish-yellow flowers appear in autumn. Train over a fence or trellis, or up into a small tree. The flowers dry well for garlands and swags.

CULTIVATION: *Grow in moist but well drained, moderately fertile, humus-rich soil. Tolerates partial shade, but leaf colour is best in full sun. Give the twining stems support. Cut back any dead growth to ground level in early spring.*

☼ ☀ ◊ ◑ ♀ ❀ ❀ ❀   ‡6m (20ft)

Hyacinthoides italica

The Italian bluebell is a smaller plant than the similar-looking English bluebell. The pretty blue and purple, starry flowers appear in spring on upright stems above the narrow leaves. They are carried in clusters of up to 30 flowers. This is an attractive plant for a shady or woodland border. Plant bulbs in autumn; lift and divide clumps after flowering.

CULTIVATION: *Grow in moist but well-drained soil in light shade.*

☀ ◊ ◑ ♀ ❀ ❀ ❀
‡10–20cm (4–8in) ↔5cm (2in)

Hyacinthus orientalis 'Blue Jacket'

This navy blue hyacinth is a bulbous perennial bearing dense, upright spikes of fragrant, bell-shaped flowers with purple veins in early spring. Good for spring bedding; specially prepared bulbs can be planted in pots during autumn for an indoor display of early flowers. With the paler 'Delft Blue', one of the best true blue hyacinths.

CULTIVATION: *Grow in any well-drained, moderately fertile soil or compost, in sun or partial shade. Protect container-grown bulbs from winter wet.*

☼ ☼ ◊ ♀ ❀❀❀
‡20–30cm (8–12in) ↔8cm (3in)

Hyacinthus orientalis 'City of Haarlem'

This primrose-yellow hyacinth is a spring-flowering, bulbous perennial bearing upright spikes of fragrant, bell-shaped flowers. The lance-shaped leaves are bright green and emerge from the base of the plant. Good for spring bedding or in containers; specially prepared bulbs can be planted in autumn for early flowers indoors.

CULTIVATION: *Grow in well-drained, fairly fertile soil or compost, in sun or partial shade. Protect container-grown plants from excessive winter wet.*

☼ ☼ ◊ ♀ ❀❀❀
‡20–30cm (8–12in) ↔8cm (3in)

Hyacinthus orientalis 'Pink Pearl'

This deep pink hyacinth, bearing dense, upright spikes of fragrant, bell-shaped flowers with paler edges, is a spring-flowering, bulbous perennial. The leaves are narrow and bright green. Excellent in a mixed or herbaceous border; specially prepared bulbs can be planted in pots during autumn for an indoor display of early flowers.

CULTIVATION: *Grow in any well-drained, moderately fertile soil or compost, in sun or partial shade. Protect container-grown bulbs from winter wet.*

☼ ☀ ◊ ❀ ❀ ❀
‡20–30cm (8–12in) ↔8cm (3in)

Hydrangea anomala subsp. *petiolaris*

The climbing hydrangea, often sold simply as *H. petiolaris*, is a woody-stemmed, deciduous, self-clinging climber, usually grown on shady walls for its large, lacecap-like heads of creamy-white, summer flowers. The mid-green leaves are oval and coarsely toothed. Often slow to establish, and may not survive in areas with cold, frost-prone winters.

CULTIVATION: *Grow in any reliably moist, fertile soil in sun or deep shade. Little pruning is required, but as the allotted space is filled, cut back overlong shoots after flowering.*

☼ ☀ ◐ ◊ ♀ ❀ ❀
‡15m (50ft)

Hydrangea arborescens 'Annabelle'

An upright, deciduous shrub bearing large, rounded heads of densely-packed, creamy-white flowers from midsummer to early autumn. The leaves are broadly oval and pointed. Good on its own or in a shrub border; the flowerheads can be dried for winter decoration. 'Grandiflora' has even larger flowerheads.

CULTIVATION: *Grow in moist but well-drained, moderately fertile, humus-rich soil, in sun or partial shade. Keep pruning to a minimum, or cut back hard each spring to a low framework.*

☼ ☀ ◊ ◑ ♀ ✿✿✿
↕1.5m (5ft) ↔2.5m (8ft)

Hydrangea aspera Villosa Group

A group of spreading to upright, deciduous shrubs that can become tree-like with age. In late summer, they produce flattened, lacecap-like heads of small, blue-purple or rich blue flowers, surrounded by larger, lilac-white or rose-lilac flowers. The leaves are lance-shaped and dark green. Excellent in a woodland or wild garden; in cold areas they are best trained against a warm wall.

CULTIVATION: *Grow in moist but well-drained, moderately fertile soil that is rich in humus. Site in full sun or semi-shade. Little pruning is necessary.*

☼ ☀ ◊ ◑ ✿✿   ↕1–4m (3–12ft)

Hydrangea macrophylla

Cultivars of the common hydrangea, *H. macrophylla*, are rounded shrubs with oval, mid- to dark green, deciduous leaves. Their large, showy flowerheads, borne from mid- to late summer, are available in two distinct forms: lacecaps, such as 'Veitchii', have flat flowerheads, and mop-head hydrangeas (Hortensias), such as 'Altona', have round flowerheads. Except in white-flowered cultivars, flower colour is directly influenced by soil pH; acid soils produce blue flowers, and alkaline soils give rise to pink flowers. All types of hydrangea are useful for a range of garden sites, and the flowerheads dry well for indoor arrangements.

CULTIVATION: *Grow in moist but well-drained, fertile soil, in sun or partial shade with shelter from cold winds. Prune hard in spring to enhance flowering, cutting stems right back to strong pairs of buds. Flower colour can be influenced on neutral soils.*

☼ ☀ ◊ ◐ ✽✽

1 ↕1m (3ft) ↔1.5m (5ft)

2 ↕2m (6ft) ↔2.5m (8ft)

3 ↕↔1.5m (5ft)

4 ↕2m (6ft) ↔2.5m (8ft)

5 ↕2m (6ft) ↔2.5m (8ft)

1 *H. macrophylla* 'Altona' (Mop-head) ♀ **2** *H. macrophylla* 'Mariesii Perfecta' (Lacecap) **3** *H. macrophylla* 'Lanarth White' (Lacecap) ♀ **4** *H. macrophylla* 'Générale Vicomtesse de Vibraye' (Mop-head) ♀ **5** *H. macrophylla* 'Veitchii' (Lacecap) ♀

Hydrangea paniculata

Cultivars of *H. paniculata* are fast-growing, upright, deciduous shrubs, with oval, mid- to dark green leaves. They are cultivated for their tall clusters of lacy flowers which usually appear during late summer and early autumn; some cultivars, such as 'Praecox', bloom earlier in the summer. Flowers are mostly creamy-white, with some forms, such as 'Floribunda', becoming pink-tinged as they age. These versatile shrubs are suitable for many different garden uses; as specimen plants, in groups, or in containers. The flower-heads look very attractive when dried for indoor decoration.

CULTIVATION: *Grow in moist but well-drained, fertile soil. Site in sun or partial shade with shelter from cold, drying winds. Pruning is not essential, but plants flower much better if pruned back annually, in early spring, to the lowest pair of healthy buds on a permanent, woody framework.*

☼ ☀ ◊ ◑ ❄ ❄ ❄

‡3–7m (10–22ft) ↔2.5m (8ft)

‡3–7m (10–22ft) ↔2.5m (8ft)

‡3–7m (10–22ft) ↔2.5m (8ft)

1 *H. paniculata* 'Floribunda' **2** *H. paniculata* 'Grandiflora' **3** *H. paniculata* 'Praecox'

Hydrangea quercifolia

The oak-leaved hydrangea is a mound-forming, deciduous shrub bearing conical heads of white flowers that fade to pink, from midsummer to autumn. The deeply lobed, mid-green leaves turn bronze-purple in autumn. Useful in a range of garden sites.

CULTIVATION: *Prefers well-drained but moist, moderately fertile soil, in sun or partial shade. Leaves may become yellow in shallow, chalky soil. Keep pruning to a minimum in spring.*

☼ ☀ ◊ ❀ ❀   ↕2m (6ft) ↔2.5m (8ft)

Hydrangea serrata 'Bluebird'

A compact, upright, long-flowering, deciduous shrub bearing flattened heads of tiny, rich blue flowers surrounded by larger, pale blue flowers from summer to autumn. The narrowly oval, pointed, mid-green leaves turn red in autumn. The flowerheads can be dried for indoor arrangements. 'Grayswood' is a similar shrub with mauve flowers.

CULTIVATION: *Grow in moist but well-drained, moderately fertile, humus-rich soil, in sun or partial shade. Flowers may turn pink in alkaline soils. Cut back weak, thin shoots in mid-spring.*

☼ ☀ ◊ ♀ ❀ ❀   ↕↔1.2m (4ft)

Hydrangea serrata 'Rosalba'

An upright, compact, deciduous shrub valued for its flat flowerheads which appear from summer to autumn; these are made up of tiny pink flowers in the centre, surrounded by larger white flowers that become red-marked as they age. The leaves are oval, mid-green, and pointed. Ideal as a specimen plant or in a shrub border.

CULTIVATION: *Grow in well-drained but moist, moderately fertile, humus-rich soil, in full sun or partial shade. Flowers may turn blue on acid soils. Very little pruning is needed.*

☼ ☀ ◑ ♀ ❀❀   ‡↔1.2m (4ft)

Hypericum 'Hidcote'

This dense, evergreen or semi-evergreen shrub produces abundant clusters of large, cupped, golden-yellow flowers which open from midsummer to early autumn. The leaves are dark green and lance-shaped. Suitable for a shrub border; for a taller but narrower shrub, to 2m (6ft) high but otherwise very similar, look for 'Rowallane'.

CULTIVATION: *Grow in well-drained but moist, moderately fertile soil, in sun or partial shade. Deadhead regularly, and trim annually in spring to increase the flowering potential.*

☼ ☀ ◊ ◑ ♀ ❀❀❀
‡1.2m (4ft) ↔1.5m (5ft)

Hypericum kouytchense

Sometimes known as *H.* 'Sungold', this species of St. John's wort is a rounded, semi-evergreen bush with arching shoots. It has dark blue-green leaves, but its biggest asset is the large clusters of golden-yellow star-shaped flowers borne in profusion during summer and autumn, followed by bright bronze-red fruits. Grow in a shrub or mixed border; a useful rabbit-proof shrub.

CULTIVATION: *Grow in moderately fertile, moist but well-drained soil in full sun or partial shade. Prune or trim after flowering, if necessary.*

☼ ☀ ◊ ◗ ♀ ❀ ❀ ❀
‡1m (3ft) ↔1.5m (5ft)

Iberis sempervirens

A spreading, evergreen subshrub bearing dense, rounded heads of small, unevenly shaped white flowers which are often flushed with pink or lilac. These appear in late spring and early summer, covering the spoon-shaped, dark green leaves. Best in a rock garden or large wall pocket. The recommended cultivar 'Schneeflocke' can be even more floriferous.

CULTIVATION: *Grow in well-drained, poor to moderately fertile, neutral to alkaline soil, in full sun. Trim lightly after flowering for neatness.*

☼ ◊ ❀ ❀ ❀
‡to 30cm (12in) ↔to 40cm (16in)

Ilex x *altaclerensis* 'Golden King' (female)

A compact, evergreen shrub with glossy, dark green leaves edged in gold. The leaf margins may be smooth or toothed. The flowers are insignificant, but develop into red berries in autumn. Tolerant of pollution and coastal exposure; a good tall windbreak or hedge where winters are not too severe.

CULTIVATION: *Grow in moist but well-drained, moderately fertile soil rich in organic matter. For berries, a male holly must grow nearby. A position in full sun is ideal. Trim or prune in early spring, if necessary.*

☼ ◊ ◑ ♀ ❁❁❁
‡30m (20ft) ↔ 4m (12ft)

Ilex x *altaclerensis* 'Lawsoniana' (female)

This dense and bushy holly forms a compact, evergreen tree or shrub. It bears large, usually spineless, oval, bright green leaves, which are splashed with gold and paler green in the centres. Red-brown berries, ripening to red, develop in autumn.

CULTIVATION: *Grow in moist but well-drained soil, in sun for best leaf colour. Grow a male holly nearby to ensure a display of berries. Free-standing plants may need some shaping when young. Remove any all-green shoots as seen.*

☼ ◊ ♀ ❁❁❁
‡to 6m (20ft) ↔ 5m (15ft)

English Hollies (*Ilex aquifolium*)

Ilex aquifolium, the English holly, has many different cultivars, all upright, evergreen trees or large shrubs, which are usually grown on their own or as spiny hedges. They have purple stems, grey bark and dense, glossy foliage. Most cultivars have multi-coloured, spiny leaves, although those of 'J.C. van Tol' are spineless and dark green. 'Ferox Argentea' has extra-spiny leaves. Male and female flowers are borne on separate plants, so female hollies, like 'Madame Briot', must be near males, such as 'Golden Milkboy', if they are to bear a good crop of berries. Tall specimens make effective windbreaks.

CULTIVATION: *Grow in moist, well-drained, fertile, humus-rich soil. Choose a site in full sun for good leaf variegation, but tolerates partial shade. Remove any damaged wood and shape young trees in spring; hedging should be trimmed in late summer. Over-enthusiastic pruning will spoil their form.*

☼ ☀ ◊ ❀❀❀

‡to 25m (80ft) ↔8m (25ft)   ‡to 6m (20ft) ↔2.5m (8ft)   ‡to 15m (50ft) ↔4m (12ft)

1 *Ilex aquifolium* ♀ **2** 'Amber' (F) ♀ **3** 'Argentea Marginata' (F) ♀

‡to 8m (25ft) ↔ 4m (12ft)

‡6m (20ft) ↔ 4m (12ft)

‡8m (25ft) ↔ 5m (15ft)

‡6m (20ft) ↔ 4m (12ft)

‡6m (20ft) ↔ 5m (15ft)

‡6m (20ft) ↔ 5m (15ft)

‡6m (20ft) ↔ 4m (12ft)

‡6m (20ft) ↔ 4m (12ft)

‡10m (30ft) ↔ 4m (12ft)

4 'Ferox Argentea' (M) ♀ **5** 'Golden Milkboy' (M) ♀ **6** 'Handsworth New Silver' (F) ♀
7 'J. C. van Tol' (F) ♀ **8** 'Madame Briot' (F) ♀ **9** 'Pyramidalis' (F) ♀ **10** 'Pyramidalis Fructo Luteo' (F) ♀ **11** 'Silver Milkmaid' (F) ♀ **12** 'Silver Queen' (M) ♀

Ilex crenata 'Convexa' (female)

This bushy form of the box-leaved holly is a dense, evergreen shrub with purple-green stems and spineless, oval to elliptic, glossy, mid- to dark green leaves. It bears an abundance of small, black berries in autumn. Lends itself for use as hedging or topiary.

CULTIVATION: *Needs moist but well-drained, humus rich soil, in full sun or partial shade. Grow near a male holly for a good crop of berries. Cut out badly placed growth in early spring, and trim shaped plants in summer.*

☼ ☀ ◊ ♀ ❀ ❀ ❀
‡to 2.5m (8ft) ↔2m (6ft)

Ilex x meserveae 'Blue Princess' (female)

This blue holly is a vigorous, dense, evergreen shrub with oval, softly spiny, very glossy, greenish-blue leaves. White to pinkish-white, late spring flowers are followed by a profusion of glossy red berries in autumn. The dark purplish-green young stems show well when hedging plants are regularly clipped. Dislikes coastal conditions.

CULTIVATION: *Grow in moist but well-drained, moderately fertile soil, in full sun or semi-shade. For berries, a male holly will need to be nearby. Prune in late summer to maintain shape.*

☼ ☀ ◊ ♀ ❀ ❀ ❀
‡↔3m (10ft)

Indigofera amblyantha

This spreading, deciduous shrub with arching stems is grown for its pretty, pea-like flowers and grey-green foliage. It carries dense, more-or-less upright clusters of small pink flowers from early summer to autumn. In cold climates, it is best trained against a warm, sheltered wall. Cut all stems to just above ground level in early spring.

CULTIVATION: *Grow in fertile, moist but well-drained soil in sun.*

☼ ◊ ◊ ❀ ❀ ❀
↕2m (6ft) ↔ to 2.5m (8ft)

Indigofera heterantha

A medium-sized, spreading shrub grown for its pea-like flowers and elegant foliage. The arching stems carry grey-green leaves made up of many oval to oblong leaflets. Dense, upright clusters of small, purple-pink flowers appear from early summer to autumn. Train against a warm wall in cold climates. *I. amblyantha* is a very similar shrub, also recommended.

CULTIVATION: *Grow in well-drained but moist, moderately fertile soil, in full sun. Prune in early spring, cutting back to just above ground level.*

☼ ◊ ♀ ❀ ❀   ↕↔2–3m (6–10ft)

Ipheion uniflorum 'Froyle Mill'

This variety of ipheion is very similar to 'Wisley Blue' but with dusky violet flowers in spring. Grass-like leaves appear in autumn well before the flowers. Plant the bulbs 8cm (3in) deep, 5cm (2in) apart in autumn in a sunny position. Provide a protective mulch in winter in areas where temperatures regularly fall below -10°C (14°F).

CULTIVATION: *Grow in well-drained soil in full sun.*

☼ ◊ ♀ ❀ ❀

↕10–15cm (4–6in) ↔5–8cm (2–3in)

Ipheion uniflorum 'Wisley Blue'

A vigorous, clump-forming, mainly spring-flowering, bulbous perennial bearing scented, star-shaped, lilac-blue flowers; each petal has a pale base and a dark midrib. Narrow, strap-like, light blue-green leaves are produced in autumn. Useful in a rock garden or for underplanting herbaceous plants.

CULTIVATION: *Grow in moist but well-drained, moderately fertile, humus-rich soil, in full sun. In colder areas, provide a mulch in winter.*

☼ ◊ ♀ ❀ ❀

↕10–15cm (4–6in) ↔5–8cm (2–3in)

Ipomoea 'Heavenly Blue'

This summer-flowering, twining, fast-growing form of morning glory is grown as a climbing annual. The large, funnel-shaped flowers, azure-blue with pure white throats, appear singly or in clusters of two or three. The heart-shaped, light to mid-green leaves have slender tips. Suitable for a summer border, scrambling among other plants. Seeds are highly toxic if ingested.

CULTIVATION: *Grow in well-drained, moderately fertile soil, in sun with shelter from cold, drying winds. Plant out after danger of frost has passed.*

☼ ◊ ♀ ❀❀ ‡to 3–4m (10–12ft)

Ipomoea indica

The blue dawn flower is a vigorous, evergreen climber, perennial in frost-free conditions. Abundant, rich purple-blue, funnel-shaped flowers that often fade to red are borne in clusters of three to five from late spring to autumn. The mid-green leaves are heart-shaped or three-lobed. In mild areas, grow as annuals in a warm conservatory or summer border. The seeds are toxic.

CULTIVATION: *Grow in well-drained, faily fertile soil, in sun with shelter from cold, drying winds. Plant out after all danger of frost has passed. Minimum temperature 7°C (45°F).*

☼ ◊ ♀❀ ‡to 6m (20ft)

Irises for Water Gardens

Irises that flourish in reliably moist or wet soils produce swollen, horizontal creeping stems known as rhizomes that lie just below the ground. These produce several new offsets each year, so give plants plenty or room, or divide them regularly. They have strap-shaped leaves and flower in blues, mauves, white, or yellow in spring and early summer. The true water irises will grow not only in damp ground but also in shallow water; these include *I. laevigata* and

I. pseudacorus, very vigorous plants that will soon overwhelm a small pond. Where space is limited, try *I. ensata* or, in moist or even in well-drained soil around the water, plant *I. sibirica* or one of its many attractive cultivars.

CULTIVATION: *Grow in deep, acid soil enriched with well-rotted organic matter, in sun or light shade. Best in areas with hot summers. Plant rhizomes in early autumn; divide plants after flowering.*

☼ ◐ ◗ ♦ ❀ ❀ ❀

1 ↕90cm (36in) ↔indefinite **2** ↕90cm (36in) ↔indefinite **3** ↕80cm (32in) ↔indefinite

4 ↕80cm (32in) ↔indefinite **5** ↕0.9–1.5m (3–5ft) ↔indefinite **6** ↕0.9–1.5m (3–5ft) ↔indefinite

1 *Iris ensata* 'Flying Tiger' ♀ **2** *I. ensata* 'Variegata' ♀ **3** *I. laevigata* **4** *I. laevigata* 'Variegata' ♀ **5** *I. pseudacorus* **6** *I. pseudacorus* 'Variegata' ♀ **7** *I.sibirica* 'Annemarie Troeger' **8** *I.versicolor* **9** *I. sibirica* 'Crème Chantilly'

‡1m (3ft) ↔ indefinite ‡20–80cm (8–32in) ↔ indefinite ‡1m (3ft) ↔ indefinite

‡80cm (32in) ↔ indefinite ‡80cm (32in) ↔ indefinite ‡1m (3ft) ↔ indefinite

‡1m (3ft) ↔ indefinite ‡80cm (32in) ↔ indefinite ‡1m (3ft) ↔ indefinite

‡80cm (32in) ↔ indefinite ‡to 1m (3ft) ↔ indefinite ‡to 1m (3ft) ↔ indefinite

10 *I.sibirica* 'Dreaming Yellow' **11** *I.sibirica* 'Harpswell Happiness' ♀ **12** *I.sibirica* 'Mikiko' **13** *I.sibirica* 'Oban' **14** *I.sibirica* 'Perfect Vision' ♀ **15** *I.sibirica* 'Roisin' **16** *I.sibirica* 'Smudger's Gift' **17** *I.sibirica* 'Uber den Wolken' **18** *I.sibirica* 'Zakopane'

Iris bucharica

A fast-growing, spring-flowering, bulbous perennial that carries up to six golden-yellow to white flowers on each stem. The glossy, strap-like leaves die back after flowering. The most commonly available form of this iris has yellow and white flowers.

CULTIVATION: *Grow in rich but well-drained, neutral to slightly alkaline soil, in full sun. Water moderately when in growth; after flowering, maintain a period of dry dormancy.*

☼ ◊ ❀❀

↕20–40cm (8–16in) ↔12cm (5in)

Iris confusa

This freely spreading, rhizomatous perennial with bamboo-like foliage produces a succession of up to 30 short-lived flowers on each stem during spring. They are white with yellow crests surrounded by purple or yellow spots. The leaves are arranged in fans at the base of the plant. Suitable for a sheltered, mixed or herbaceous border, but may not survive in areas with cold winters.

CULTIVATION: *Grow in moist but well-drained, rich soil, in sun or semi-shade. Water moderately when in growth. Keep tidy-looking by removing flowered stems.*

☼ ◐ ◊ ♀ ❀❀

↕1m (3ft) or more ↔indefinite

Iris delavayi

This rhizomatous, deciduous perennial bears three-branched flower stems in summer, each one topped by two light to dark purple-blue flowers. The rounded fall petals have white and yellow flecks. The foliage is grey-green. A handsome perennial with its tall stems, it is easily grown in moist ground.

CULTIVATION: *Grow in moist soil in full sun or partial shade. Lift and divide congested clumps after flowering.*

☼ ☀ ◑ ◊ ♀ ❀ ❀ ❀
‡to 1.5m (5ft) ↔ indefinite

Iris douglasiana

A robust, rhizomatous perennial with branched flower stems that each bear two or three white, cream, blue, lavender-blue, or red-purple flowers in late spring and early summer. The stiff, glossy, dark green leaves are often red at the bases. A good display plant for a raised bed or trough.

CULTIVATION: *Grow in well-drained, neutral to slightly acid loam. Site in full sun for the best flowers, or light shade. Does not transplant well, so do not lift and divide unnecessarily.*

☼ ☀ ◊ ❀ ❀ ❀
‡15–70cm (6–28in) ↔ indefinite

Iris foetidissima 'Variegata'

The stinking iris is not as unpleasant as it sounds, although the evergreen, silvery leaves, with white stripes in this cultivar, do have a dreadful scent if crushed. A vigorous rhizomatous perennial, it bears yellow-tinged, dull purple flowers in early summer, followed by seed capsules which split in autumn to display decorative scarlet, yellow or, rarely, white seeds. A useful plant for dry shade.

CULTIVATION: *Prefers well-drained, neutral to slightly acid loam in shade. Divide congested clumps in autumn.*

☼ ☀ ◊ ♀ ❀❀❀
‡30cm–90cm (12–36in) ↔ indefinite

Iris forrestii

An elegant, early summer-flowering rhizomatous perennial with slender flower stems that each carry one or two scented pale yellow flowers with brown markings. The very narrow glossy leaves are mid-green above and grey-green below. Easy to grow in an open border.

CULTIVATION: *Grow in moist but well-drained, neutral to slightly acid loam. Position in full sun or partial shade.*

☼ ☀ ◊ ◊ ♀ ❀❀❀
‡35–40cm (14–16in) ↔ indefinite

Iris graminea

A deciduous, rhizomatous perennial bearing bright green, strap-like leaves. From late spring, rich purple-violet flowers, with fall petals tipped white and violet-veined, are borne either singly or in pairs; they are often hidden among the leaves. The flowers have a fruity fragrance.

CULTIVATION: *Grow in moist but well-drained, neutral to slightly acid loam. Choose a site in full sun or semi-shade. Does not respond well to transplanting.*

☼ ☀ ◊ ◊ ♀ ❀❀❀
‡20–40cm (8–16in) ↔indefinite

Iris 'Katharine Hodgkin'

This very vigorous, tiny but robust, deciduous, bulbous perennial bears delicately patterned, pale blue and yellow flowers, with darker blue and gold markings, in late winter and early spring. The pale to mid-green leaves grow after the flowers have faded. Excellent in a rock garden or at the front of a border, where it will spread slowly to form a group.

CULTIVATION: *Grow in well-drained, neutral to slightly alkaline soil, in an open site in full sun.*

☼ ◊ ♀ ❀❀❀
‡12cm (5in) when flowering
↔5–8cm (2–3in)

Bearded Irises (*Iris*)

These upright, rhizomatous perennials send up fans of sword-shaped, usually broad leaves and simple or branched stems. The flowers are produced in a wide range of colours, with well-developed, often frilly fall and standard petals, and a "beard" of white or coloured hairs in the centre of each fall petal. These are the most widely cultivated group of irises for garden display, usually producing several flowers per stem from spring into early summer, sometimes again later in the season. Taller irises suit a mixed border, and smaller ones may be grown in a rock garden, raised bed, or trough.

CULTIVATION: *Grow in well-drained, moderately fertile, neutral to slightly acid soil in full sun. Plant rhizomes in late summer or early autumn, thinly covered with soil. They must not be shaded by other plants. Do not mulch. Divide large or congested clumps in early autumn.*

☼ ◊ ❀ ❀ ❀

‡70cm (28in) or more ↔ 60cm (24in)

‡70cm (28in) or more ↔ 60cm (24in)

3
‡55m (22in) ↔ 45–60cm (18–24in)

‡70cm (28in) or more ↔ 60cm (24in)

‡↔ 30cm (12in)

1 *Iris* 'Apricorange' **2** *I.* 'Breakers' ♀ **3** *I.* 'Brown Lasso' ♀ **4** *I.* 'Early Light' ♀
5 *I.* 'Eyebright'

‡to 70cm (28in) ↔ to 60cm (24in)

‡20–40cm (8–16in) ↔ 30cm (12in)

‡to 70cm (28in) ↔ to 60cm (24in)

‡to 70cm (28in) ↔ to 60cm (24in)

‡70cm (28in) or more ↔ 60cm (24in)

‡to 70cm (28in) ↔ to 60cm (24in)

‡70cm (28in) or more ↔ 60cm (24in)

‡to 70cm (28in) ↔ to 60cm (24in)

‡70cm (28in) or more ↔ 60cm (24in)

‡85cm (33in) ↔ 60cm (24in)

6 *I.* 'Happy Mood' **7** *I.* 'Honington' **8** *I.* 'Katie-Koo' ♀ **9** *I.* 'Maui Moonlight' ♀
10 *I.* 'Meg's Mantle' **11** *I.* 'Miss Carla' **12** *I.* 'Nicola Jane' **13** *I.* 'Orinoco Flow' ♀
14 *I.* 'Paradise' **15** *I.* 'Paradise Bird'

‡70cm (28in) or more ↔ 60cm (24in)

‡40–70cm (16–28in) ↔ 45–60cm (18–24in)

‡70cm (28in) or more ↔ 60cm (24in)

‡to 70cm (28in) ↔ to 60cm (24in)

16 *I.* 'Phil Keen' ♀ **17** *I.* 'Pink Parchment' **18** *I.* 'Precious Heather' **19** *I.* 'Quark'

20

‡25cm (10in) ↔ 30cm (12in)

21

‡to 70cm (28in) ↔ to 60cm (24in)

24

‡70cm (28in) or more ↔ 60cm (24in)

22

‡to 70cm (28in) ↔ to 60cm (24in)

23

‡to 70cm (28in) ↔ to 60cm (24in)

MORE CHOICES

'Arctic Fancy' White,
50cm (20in) tall.

'Blue-Eyed Brunette'
Brown flowers with
a lilac blaze, to 90cm
(36in) tall.

'Bromyard' Blue-grey
and ochre flowers,
28cm (11in) tall.

'Stepping Out' White
with blue petal edges,
1m (3ft) tall.

25

‡to 70cm (28in) ↔ to 60cm (24in)

26

‡90cm (36in) ↔ 60cm (24in)

20 *I.* 'Rain Dance' ♀ **21** *I.* 'Sherbet Lemon' ♀ **22** *I.* 'Sparkling Lemonade' **23** *I.* 'Sunny Dawn' **24** *I.* 'Sun Miracle' **25** *I.* 'Templecloud' ♀ **26** *I.* 'Vanity'

Iris lacustris

This dwarf, deciduous, rhizomatous perennial bears small flowers in late spring. These are purple-blue to sky-blue with gold crests and a white patch on each of the fall petals; they arise from basal fans of narrow leaves. Suitable for growing in a rock garden or trough.

CULTIVATION: *Grow in reliably moist, lime-free soil that is rich in humus, in sun or partial shade. Water moderately when in growth.*

☼ ◐ ◊ ❀ ❀ ❀
↕10cm (4in) ↔indefinite

Iris pallida 'Variegata'

This semi-evergreen, rhizomatous perennial is probably the most versatile and attractive variegated iris. The strap-like, bright green leaves are clearly striped with golden-yellow. (For silver-striped leaves, look for 'Argentea Variegata'.) The large, scented, soft blue flowers with yellow beards are borne in clusters of two to six on branched stems in late spring and early summer. Grow in a mixed or herbaceous border.

CULTIVATION: *Best in well-drained, fertile, slightly alkaline soil, in sun. Water moderately when in growth.*

☼ ◊ ♀ ❀ ❀ ❀
↕to 1.2m (4ft) ↔indefinite

Iris setosa

This rhizomatous perennial flowers
in late spring and early summer.
Each flowering stem bears several
beautiful, blue or blue-purple flowers
above the narrow, mid-green leaves.
Easily grown in moist soil.

CULTIVATION: *Grow in moist, neutral to
slightly acid soil in full sun or partial
shade. Lift and divide congested clumps
after flowering.*

☼ ☀ ◑ ♀ ✿ ✿ ✿
‡15–90cm (6–36in) ↔ indefinite

Iris unguicularis

A fast-growing, evergreen, rhizomatous
perennial, sometimes called *I. stylosa*,
with short flower stems bearing large,
fragrant blooms from late winter
(sometimes even earlier) to early
spring. The pale lavender to deep violet
petals have contrasting veins and a
band of yellow on each fall petal. The
leaves are grass-like and mid-green.
Ideal for the base of a sunny wall.

CULTIVATION: *Grow in sharply drained,
neutral to alkaline soil. Choose a warm,
sheltered site in full sun. Does not like to
be disturbed. Keep tidy by removing dead
leaves in late summer and spring.*

☼ ◊ ✿ ✿ ✿
‡30cm (12in) ↔ indefinite

Iris variegata

This slender and robust, deciduous, rhizomatous perennial bears three to six flowers on each branched stem from midsummer. The striking flowers are pale yellow with brown or violet veins on the fall petals; there are many other colour combinations available. The deep green leaves are strongly ribbed.

CULTIVATION: *Grow in well-drained, neutral to alkaline soil, in sun or light shade. Avoid mulching with organic matter, which may encourage rot.*

☼ ☀ ◊ ♀ ❀❀❀
‡20–45cm (8–18in) ↔indefinite

Itea ilicifolia

An evergreen shrub bearing upright at first, then spreading, arching shoots. The oval, holly-like leaves are sharply toothed. Tiny, greenish-white flowers are borne in long, catkin-like clusters from midsummer to early autumn. Needs a sheltered position in cold areas.

CULTIVATION: *Grow in well-drained but moist, fertile soil, preferably against a warm wall in full sun. Protect with a winter mulch when young.*

☼ ◊ ♀ ❀❀
‡3–5m (10–15ft) ↔3m (10ft)

Jasminum mesnyi

The primrose jasmine is a half hardy, scrambling, evergreen shrub with large, usually semi-double, bright yellow flowers. These appear singly or in small clusters during spring and summer, amid the glossy, dark green leaves which are divided into three oblong to lance-shaped leaflets. Will climb if tied to a support, but may not survive in frosty areas.

CULTIVATION: *Grow in any well-drained, fertile soil, in full sun or partial shade. Cut back flowered shoots in summer to encourage strong growth from the base.*

☼ ☀ ◊ ♀ ❋
‡to 3m (10ft) ↔1–2m (3–6ft)

Jasminum nudiflorum

Winter jasmine is a lax, mound-forming, deciduous shrub with slender, arching stems. Small, tubular yellow flowers are borne singly on the leafless, green shoots in late winter. The dark green leaves, which develop after the flowers, are divided into three leaflets. Tie in a framework of stems against a wall, or let it sprawl unsupported.

CULTIVATION: *Grow in well-drained, fertile soil. Tolerates semi-shade, but flowers best in sun. Encourage strong growth by cutting back flowered shoots.*

☼ ☀ ◊ ♀ ❋ ❋ ❋   ‡↔to 3m (10ft)

Jasminum officinale '**Argenteovariegatum**'

This variegated form of the common jasmine, *J. officinale*, is a vigorous, deciduous or semi-deciduous, woody climber. The grey-green, cream-edged leaves are made up of 5–9 sharply pointed leaflets. Clusters of fragrant white flowers open from summer to early autumn. If tied in initially, it will twine over supports, such as a trellis or an arch. May not survive in cold, exposed areas.

CULTIVATION: *Grow in well-drained, fertile soil. Tolerates shade, but flowers best in full sun. Thin out crowded growth after flowering.*

☼ ☀ ◊ ♀ ❀ ❀ ❀   ‡to 12m (40ft)

Juglans nigra

The black walnut is an impressive deciduous tree with furrowed bark and large glossy leaves divided into many leaflets. It needs space to grow as it will make a big tree. Catkins in late spring are followed by edible walnuts. Good fruiting varieties include 'Broadview' and 'Buccaneer'.

CULTIVATION: *Grow in deep, well-drained soil in full sun or part shade.*

☼ ☀ ◊ ♀ ❀ ❀ ❀
‡30m (100ft) ↔20m (70ft)

Juniperus communis '**Compressa**'

This slow-growing, spindle-shaped, dwarf form of the common juniper bears deep to blue-green, aromatic, evergreen scale-like leaves, borne in whorls of three along the stems. Small, oval or spherical fruits remain on the plant for three years, ripening from green through to cloudy-blue to black. 'Hibernica' is another recommended narrowly upright juniper, slightly faster-growing.

CULTIVATION: *Grow in any well-drained soil, preferably in full sun or light dappled shade. No pruning is needed.*

☼ ☀ ◊ ♀ ✿ ✿ ✿
‡to 80cm (32in) ↔ 45cm (18in)

Juniperus x *pfitzeriana* '**Wilhelm Pfitzer**'

This spreading, dense, evergreen shrub has ascending branches of grey-green foliage which droop at the tips; it eventually forms a flat-topped, tiered bush. The flattened, scale-like leaves are borne in whorls of three. Spherical fruits are at first dark purple, becoming paler as they age. Looks well as a specimen plant or in a large rock garden.

CULTIVATION: *Grow in any well-drained soil, preferably in full sun or light dappled shade. Keep pruning to a minimum, in late autumn if necessary.*

☼ ☀ ◊ ✿ ✿ ✿
‡1.2m (4ft) ↔ 3m (10ft)

Juniperus procumbens 'Nana'

A compact, mat-forming conifer that is excellent for ground cover in a wide range of situations. The needle-like, aromatic, yellow-green or light green leaves are carried in groups of three. Bears berry-like, brown to black, fleshy fruits which take two or three years to ripen.

CULTIVATION: *Grow in any well-drained soil, including sandy, dry, or chalky conditions. Site in full sun or dappled shade. No pruning is required.*

☼ ☀ ◊ ♚ ❀❀❀
‡15–20cm (6–8in) ↔75cm (30in)

Juniperus squamata 'Blue Star'

This conifer is a low-growing, dense, compact, rounded bush with rust-coloured, flaky bark. The silvery-blue leaves are sharply pointed, and grouped in whorls of three. The ripe fruits are oval and black. Useful as ground cover or in a rock garden.

CULTIVATION: *Grow in any well-drained soil, in full sun or dappled shade. Very little pruning is required.*

☼ ☀ ◊ ♚ ❀❀❀
‡to 40cm (16in) ↔to 1m (3ft)

Kalmia angustifolia

The sheep laurel is a tough, rabbit-proof shrub grown for its spectacular, rounded clusters of small flowers, usually pink to deep red, but occasionally white. They appear in early summer amid the dark green leaves. Useful for a shrub border or rockery; naturally mound-forming, it tolerates trimming to a neat shape if necessary.

CULTIVATION: *Choose a part-shaded site in moist, acid soil, rich in organic matter. Grow in full sun only where the soil remains reliably moist. Mulch in spring with leaf mould or pine needles. Trim or prune hard after flowering.*

☼ ◐ ♀ ❀❀❀ ↕60cm (24in) ↔1.5m (5ft)

Kalmia latifolia

The calico bush is a dense evergreen shrub producing large clusters of flowers from late spring to midsummer. These are cup-shaped, pink or occasionally white, and open from distinctively crimped buds. The oval leaves are glossy and dark green. An excellent specimen shrub for woodland gardens, but flowers best in full sun. 'Ostbo Red' is a recommended cultivar.

CULTIVATION: *Grow in moist, humus-rich, acid soil, in sun or partial shade. Mulch each spring with pine needles or leaf mould. Requires very little pruning, although deadheading is worthwhile.*

☼ ☼ ◐ ❀❀❀ ↕↔3m (10ft)

Kerria japonica 'Golden Guinea'

This vigorous, suckering, deciduous shrub forms clumps of arching, cane-like shoots which arise from ground level each year. Large, single yellow flowers are borne in mid- and late spring along the previous year's growth. ('Flore Pleno' has double flowers.) The bright green leaves are oval and sharply toothed.

CULTIVATION: *Grow in well-drained, fertile soil, in full sun or partial shade. Cut flowered canes back to different levels to obtain flowers at different heights. Chop out unwanted canes and suckers with a spade to restrict spread.*

☼ ☀ ◊ ♀ ✤ ✤ ✤ ↕2m (6ft) ↔2.5m (8ft)

Kirengeshoma palmata

A handsome, upright perennial with broad, lobed, pale green leaves. In late summer and early autumn, these are topped by loose clusters of nodding, pale yellow flowers, which are sometimes called "yellow wax bells". Brings a gentle elegance to a shady border, poolside, or woodland garden.

CULTIVATION: *Thrives in moist, lime-free soil, enriched with leaf mould, in partial shade sheltered from wind. If necessary, divide clumps in spring.*

☀ ◊ ✤ ✤ ✤
↕60–120cm (24–48in) ↔75cm (30in)

Kniphofia 'Bees' Sunset'

This red-hot poker is a deciduous perennial grown for its elegant spikes of soft yellowish-orange flowers. These appear through summer above the clumps of arching, grass-like leaves. Very attractive to bees.

CULTIVATION: *Grow in deep, fertile, moist but well-drained soil, ideally sandy but enriched with organic matter. Choose a site in full sun or partial shade. Mulch young plants for their first winter, and divide mature, crowded clumps in late spring.*

☼ ☀ ◊ ◗ ♀ ❄ ❄ ❄
‡90cm (36in) ↔ 60cm (24in)

Kniphofia caulescens

This stately evergreen perennial bears tall spikes of flowers from late summer to mid-autumn; coral-red, they fade upwards as they age to pale yellow. They are carried well above basal rosettes of arching, grass-like, blue-green leaves. Good for a herbaceous border; tolerant of coastal exposure.

CULTIVATION: *Grow in deep, fertile, moist but well-drained soil, preferably sandy and enriched with organic matter. Position in full sun or partial shade. Mulch young plants for their first winter. Divide large clumps in late spring.*

☼ ☀ ◊ ◗ ❄ ❄
‡to 1.2m (4ft) ↔ 60cm (24in)

Kniphofia 'Little Maid'

This clump-forming, deciduous perennial has tall heads of tubular flowers which appear from late summer to early autumn. They are pale green in bud, opening to pale buff-yellow, then fading to ivory. The leaves are narrow and grass-like. Good for late displays in a mixed or herbaceous border.

CULTIVATION: *Grow in well-drained, deep, fertile, humus-rich soil, in full sun. Keep moist when in growth. In their first winter and in cold areas, provide a mulch of straw or leaves.*

☼ ◊ ❀❀

‡60cm (24in) ↔ 45cm (18in)

Kniphofia 'Royal Standard'

A clump-forming, herbaceous perennial of classic red-hot poker appearance that bears tall, conical flowerheads from mid- to late summer. The bright yellow, tubular flowers open from red buds, starting at the base and moving upwards. The arching, grass-like leaves are deciduous, dying back in winter.

CULTIVATION: *Grow in deep, moist but well-drained, humus-rich soil, in sun. Water freely when in growth. Provide a mulch in cold areas, especially for young plants in their first winter.*

☼ ◊ ◊ ♀ ❀❀❀

‡1m (3ft) ↔ 60cm (24in)

Koelreuteria paniculata

The golden rain tree is a beautiful, exotic-looking deciduous tree, which bears large, open clusters of small yellow flowers in midsummer. These are followed by unusual bronze-red, inflated pods. The leaf colour is attractive in its own right, with reddish leaves emerging in spring, turning to green, with bright yellow tints in autumn. The leaves of 'Coral Sun' are more intensely coloured.

CULTIVATION: *Grow in well-drained soil in full sun.*

☼ ◊ ❄❄❄   ↕↔10m (30ft)

Kolkwitzia amabilis 'Pink Cloud'

The beauty bush is a fast-growing, suckering, deciduous shrub with an arching habit. Dense clusters of bell-shaped pink flowers with yellow-flushed throats appear in abundance from late spring to early summer. The leaves are dark green and broadly oval. Excellent for a shrub border or as a specimen plant.

CULTIVATION: *Grow in any well-drained, fertile soil, in full sun. Let the arching habit of young plants develop without pruning, then thin out the stems each year after flowering, to maintain vigour.*

☼ ◊ ♀ ❄❄❄   ↕3m (10ft) ↔4m (12ft)

Laburnum x *watereri* 'Vossii'

This spreading, deciduous tree bears long, hanging clusters of golden-yellow, pea-like flowers in late spring and early summer. The dark green leaves are made up of three oval leaflets. A fine specimen tree for small gardens; it can also be trained on an arch, pergola, or tunnel frame-work. All parts are toxic if eaten.

CULTIVATION: *Grow in well-drained, moderately fertile soil, in full sun. Cut back badly placed growth in winter or early spring. Remove any suckers or buds at the base of the trunk.*

☼ ◊ ♀ ❀ ❀ ❀   ↕↔8m (25ft)

Lagerstroemia indica 'Hopi'

Crepe myrtle is a large shrub or small deciduous tree that produces trusses of white or pink flowers in summer. In autumn, its dark green leaves turn orange to deep red. Ornamental seed heads and pale gray peeling bark add winter interest. Grow it as a shrub in cold regions. 'Seminole' is a compact pink-flowered cultivar reaching only 10m (30ft).

CULTIVATION: *Grow in fertile, well-drained soil in full sun. Remove spent flowers to encourage re-blooming. Will withstand hard pruning.*

☼ ◊ ◐ ❀   ↕↔to 8m (25ft)

Lamium maculatum 'White Nancy'

This colourful dead-nettle is a semi-evergreen perennial which spreads to form mats; this makes it effective as ground cover between shrubs. Spikes of pure white, two-lipped flowers are produced in summer above triangular to oval, silver leaves which are edged with green.

CULTIVATION: *Grow in moist but well-drained soil, in partial or deep shade. Can be invasive, so position away from other small plants, and dig up invasive roots or shoots to limit spread.*

☼ ☀ ◊ ◖ ❀ ❀ ❀
↕ to 15cm (6in) ↔ to 1m (3ft) or more

Lapageria rosea

The Chilean bell flower is a long-lived, twining, evergreen climber. From summer to late autumn, it produces large, narrowly bell-shaped, waxy red flowers which are borne either singly or in small clusters. The leaves are oval and dark green. In cold areas, it needs the protection of a warm, but partially shaded, wall. 'Nash Court' has soft pink flowers.

CULTIVATION: *Grow in well-drained, moderately fertile soil, preferably in partial shade. In frost-prone areas, shelter from wind and provide a winter mulch. Keep pruning to a minimum, removing damaged growth in spring.*

☼ ◊ ♀ ❀ ❀   ↕ 5m (15ft)

Larix decidua

The European larch is unusual among conifers in that it sheds its soft, pale green needles in autumn, once they have faded to a pretty straw yellow. The trees have a roughly conical shade and smooth, scaly grey bark. Small, rounded cones appear in early summer, and usually persist on the branches. This easily grown specimen tree tolerates a wide range of conditions.

CULTIVATION: *Grow in well-drained soil in sun.*

☼ ◊ ❀❀❀
‡30m (100ft) ↔ 4–6m (12–20ft)

Lathyrus latifolius

The everlasting or perennial pea is a tendril-climbing, herbaceous perennial with winged stems, ideal for growing through shrubs or over a bank. Clusters of pea-like, pink-purple flowers appear during summer and early autumn, amid the deciduous, blue-green leaves which are divided into two oblong leaflets. The seeds are not edible. For white flowers, choose 'Albus' or 'White Pearl'.

CULTIVATION: *Grow in well-drained, fertile, humus-rich soil, in sun or semi-shade. Cut back to ground level in spring and pinch out shoot tips to encourage bushiness. Resents disturbance.*

☼ ☀ ◊ ♀ ❀❀❀   ‡2m (6ft) or more

Sweet Peas (*Lathyrus odoratus*)

The many cultivars of *Lathyrus odoratus* are annual climbers cultivated for their long display of beautiful and fragrant flowers which cut well and are available in most colours except yellow. The flowers are arranged in clusters during summer to early autumn. The seeds are not edible. Most look very effective trained on a pyramid of canes or trellis, or scrambling amid shrubs and perennials. Compact cultivars such as 'Patio Mixed' suit containers; some of these are self-supporting. Grow sweet peas, also, in a vegetable garden, because they attract pollinating bees and other beneficial insects.

CULTIVATION: *Grow in well-drained, fertile soil; for the best flowers, add well-rotted manure the season before planting. Site in full sun or partial shade. Feed with a balanced fertilizer fortnightly when in growth. Deadhead or cut flowers for the house regularly. Support the climbing stems.*

☼ ☀ ◐ ◊ ◊ ❁ ❁ ❁

1 ↕2–2.5m (6–8ft) **2** ↕2–2.5m (6–8ft) **3** ↕2–2.5m (6–8ft)

4 ↕1m (3ft) **5** ↕2–2.5m (6–8ft) **6** ↕2–2.5m (6–8ft)

1 *Lathyrus odoratus* 'Aunt Jane' **2** *L. odoratus* 'Evening Glow' ♀
3 *L. odoratus* 'Noel Sutton' ♀ **4** *L. odoratus* 'Patio Mixed'
5 *L. odoratus* 'Teresa Maureen' ♀ **6** *L. odoratus* 'White Supreme' ♀

Laurus nobilis

Bay, or bay laurel, is a conical, evergreen tree grown for its oval, aromatic, leathery, dark green leaves which are used in cooking. Clusters of small, greenish-yellow flowers appear in spring, followed by black berries in autumn. Effective when trimmed into formal shapes.

CULTIVATION: *Grow in well-drained but moist, fertile soil, in sun or semi-shade with shelter from cold, drying winds. Grow male and female plants together for a reliable crop of berries. Prune young plants to shape in spring; trim established plants lightly in summer.*

☼ ☀ ◊ ♀ ❀❀❀
‡12m (40ft) ↔10m (30ft)

Laurus nobilis 'Aurea'

This golden-yellow-leaved bay is a conical tree bearing aromatic, oval, evergreen leaves, which can be used in cooking. In spring, clusters of small, greenish-yellow flowers appear, followed in autumn by black berries on female plants. Good for topiary and in containers, where it makes a much smaller plant.

CULTIVATION: *Grow in well-drained but moist, fertile soil. Position in full sun or partial shade with shelter from cold winds. Prune young plants to shape in spring; once established, trim lightly in summer to encourage a dense habit.*

☼ ☀ ◊ ♀ ❀❀❀
‡12m (40ft) ↔10m (30ft)

Lavandula angustifolia 'Hidcote'

This compact lavender with thin, silvery-grey leaves and dark purple flowers is an evergreen shrub useful for edging. Dense spikes of fragrant, tubular flowers, borne at the ends of long, unbranched stalks, appear during mid- to late summer. Like all lavenders, the flowers dry best if cut before they are fully open.

CULTIVATION: *Grow in well-drained, fertile soil, in sun. Cut back flower stems in autumn, and trim the foliage lightly with shears at the same time. In frost-prone areas, leave trimming until spring. Do not cut into old wood.*

☼ ◊ ♀ ❀❀❀ ‡60cm (24in) ↔75cm (30in)

Lavandula angustifolia 'Twickel Purple'

This evergreen shrub is a close relative of 'Hidcote' (above) with a more spreading habit, paler flowers and greener leaves. Dense spikes of fragrant purple flowers are borne in midsummer above narrowly oblong, grey-green leaves. Suited to a shrub border; like all lavenders, the flower-heads are very attractive to bees.

CULTIVATION: *Grow in well-drained, fairly fertile soil, in full sun. Trim in autumn, or delay until spring in frost-prone areas. Do not cut into old wood.*

☼ ◊ ❀❀❀ ‡60cm (24in) ↔1m (3ft)

Lavandula x *intermedia* **Dutch Group**

A tall, robust, bushy lavender, with broad, silvery, aromatic leaves and tall, slender spikes of scented lavender-blue flowers. Suited to a sunny shrub border, or large rock, or scree garden; makes good low hedging if regularly trimmed. 'Grappenhall' is very similar, but not as hardy.

CULTIVATION: *Grow in well-drained, fertile soil, in sun. Cut back flower stems in autumn, and trim the foliage lightly with shears at the same time. In frost-prone areas, leave trimming until spring. Do not cut into old wood.*

☼ ◊ ❀❀❀   ↕↔1.2m (4ft)

Lavandula pedunculata **subsp.** *pedunculata*

French lavender is a compact, evergreen shrub which blooms from late spring to summer. Dense spikes of tiny, fragrant, dark purple flowers, each spike topped by distinctive, rose-purple bracts, are carried on long stalks well above the narrow, woolly, silvery-grey leaves. Effective in a sheltered shrub border or rock garden. May suffer in cold areas.

CULTIVATION: *Grow in well-drained, fairly fertile soil, in sun. Trim back in spring or, in frost-free climates, after flowering. Avoid cutting into old wood.*

☼ ◊ ❀❀   ↕↔60cm (24in)

Shrubby Lavateras (*Lavatera x clementii*)

These upright and showy, flowering shrubs with stiff stems and sage-green leaves usually bloom from midsummer to autumn in pinks and purples. Although they are short-lived, they grow quickly on any well-drained soil, including thin, dry soil, which makes them a welcome addition to any garden where quick results are desired. They also perform well in coastal areas, being able to tolerate salt-laden winds, but the shrubs will need staking if grown in a site exposed to wind.

In cold regions prone to severe frost, plant them against a warm, sunny wall.

CULTIVATION: *Grow in any poor to moderately fertile, well-drained to dry soil in full sun. Shelter from cold, drying winds in frost-prone areas. After cold winters, frost-damaged plants are best pruned right down to the base in spring to encourage new growth and tall, vigorous stems. In milder areas, trim to shape after flowering.*

☼ ◊ ❀ ❀

1 ↕↔2m (6ft)

2 ↕↔2m (6ft)

3 ↕↔2m (6ft)

4 ↕↔2m (6ft)

5 ↕↔2m (6ft)

1 *Lavatera* x *clementii* 'Barnsley' **2** *L.* x *clementii* 'Bredon Springs' ♀ **3** *L.* x *clementii* 'Burgundy Wine' ♀ **4** *L.* x *clementii* 'Candy Floss' ♀ **5** *L.* x *clementii* 'Rosea' ♀

Annual Lavateras (*Lavatera trimestris*)

These sturdy, bushy annuals are an excellent choice for planting in groups in a herbaceous border or for summer bedding, flowering continuously from midsummer into autumn. They are softly hairy or downy plants with shallowly lobed, soft green leaves; the open funnel-shaped, pink, reddish-pink, or purple blooms are nicely complemented by the foliage. Although annual lavateras are only a temporary visitor to the garden, with their cottage-garden appearance they seem to look as if they have been around for years. Excellent for a dry, sunny site and as a source of cut flowers.

CULTIVATION: *Grow in light, moderately fertile, well-drained soil in full sun. Young plants need a regular supply of water until they become established; after this, the plants are fairly drought-tolerant. Watch out for aphids, which often attack young growth.*

☼ ◊ ❀ ❀

‡ to 60cm (24in) ↔ 45cm (18in)   ‡ to 60cm (24in) ↔ 45cm (18in)   ‡ to 60cm (24in) ↔ 45cm (18in)

‡ to 75cm (30in) ↔ 45cm (18in)   ‡ to 60cm (24in) ↔ 45cm (18in)   ‡ to 75cm (30in) ↔ 45cm (18in)

1 *Lavatera trimestris* 'Beauty Formula Mixture' **2** *L. trimestris* 'Pink Beauty' ♀ **3** *L. trimestris* 'Salmon Beauty' ♀ **4** *L. trimestris* 'Silver Cup' ♀ **5** *L. trimestris* 'White Beauty' ♀ **6** *L. trimestris* 'White Cherub'

Leiophyllum buxifolium

Sand myrtle is an upright to mat-forming, evergreen perennial, grown for its glossy, dark green foliage and for its abundance of star-shaped, pinkish-white flowers in late spring and early summer. The leaves tint bronze in winter. Good, free-flowering underplanting for a shrub border or woodland garden.

CULTIVATION: *Grow in moist but well-drained, acid soil rich in organic matter. Choose a site in partial or deep shade with protection from cold, drying winds. Trim after flowering; it may spread widely if left unattended.*

☀ ☀ ◊ ◐ ♀ ❄ ❄ ❄
‡30–60cm (12–24in) ↔ 60cm (24in) or more

Leptospermum rupestre

This low-growing, evergreen shrub with dense foliage bears star-shaped white flowers from late spring to summer. The small, aromatic leaves are glossy, elliptic, and dark green. Native to coastal areas of Tasmania, it is useful in seaside gardens provided that the climate is mild. May be sold as *L. humifusum*.

CULTIVATION: *Grow in well-drained, fertile soil, in full sun or partial shade. Trim young growth in spring to promote bushiness, but do not cut into old wood.*

☀ ☀ ◊ ❄ ❄
‡0.3–1.5m (1–5ft) ↔ 1–1.5m (3–5ft)

Leptospermum scoparium 'Kiwi'

A compact shrub with arching shoots bearing an abundance of small, flat, dark crimson flowers during late spring and early summer. The small, aromatic leaves are flushed with purple when young, maturing to mid- or dark green. Suitable for a large rock garden, and attractive in an alpine house display; may not survive cold winters. 'Nicholsii Nanum' is similarly compact.

CULTIVATION: *Grow in well-drained, moderately fertile soil, in full sun or part-shade. Trim new growth in spring for bushiness; do not cut into old wood.*

☼ ☀ ◊ ♀ ❋ ❋   ↕↔1m (3ft)

Leucanthemum x *superbum* 'Wirral Supreme'

A robust, clump-forming, daisy-flowered perennial with, from early summer to early autumn, dense, double white flowerheads. These are carried singly at the end of long stems, above lance-shaped, toothed, dark green leaves. Good for cut flowers. 'Aglaia' and 'T.E. Killin' are other recommended cultivars.

CULTIVATION: *Grow in moist but well-drained, moderately fertile soil, in full sun or partial shade. May need staking.*

☼ ☀ ◊ ◊ ♀ ❋ ❋ ❋
↕90cm (36in) ↔75cm (30in)

Leucojum aestivum
'Gravetye giant'

This robust cultivar of summer snowflake is a spring-flowering, bulbous perennial with upright, strap-shaped, dark green leaves, to 40cm (16in) tall. The faintly chocolate-scented, drooping, bell-shaped white flowers with green petal-tips are borne in clusters. Good planted near water, or for naturalizing in grass.

CULTIVATION: *Grow in reliably moist, humus-rich soil, preferably near water. Choose a position in partial shade.*

☼ ◑ ♀ ❁❁❁   ↕1m (3ft) ↔8cm (3in)

Leucojum autumnale

A slender, late-summer-flowering, bulbous perennial bearing stems of two to four drooping, bell-shaped white flowers, tinged red at the petal bases. Narrow, upright, grass-like leaves appear at the same time as or just after the flowers. Suitable for a rock garden.

CULTIVATION: *Grow in any moist but well-drained soil. Choose a position in full sun. Divide and replant bulbs once the leaves have died down.*

☼ ◑ ◑ ❁❁❁
↕10–15cm (4–6in) ↔5cm (2in)

Lewisia cotyledon

This evergreen perennial produces tight clusters of open funnel-shaped, usually pinkish-purple flowers; they may be white, cream, yellow, or apricot. These are borne on long stems from spring to summer. The dark green, fleshy, lance-shaped leaves are arranged in basal rosettes. Suitable for growing in wall crevices. The Sunset Group are recommended garden forms.

CULTIVATION: *Grow in sharply drained, fairly fertile, humus-rich, neutral to acid soil. Choose a site in light shade where plants will be sheltered from winter wet.*

☀ ◊ ♀ ❀ ❀ ❀
‡15–30 cm (6–12in) ↔20–40cm (8–16in)

Leycesteria formosa

Himalayan honeysuckle is an attractive flowering shrub with bamboo-like, dark green young stems. Chains of white flowers with dark red bracts dangle from the branches in summer and early autumn; these are followed by maroon to purple-black fruits, which are attractive to birds. Trim or cut back hard in spring.

CULTIVATION: *Grow in moist but well-drained soil in full sun or partial shade.*

☀ ☀ ◊ ◊ ❀ ❀ ❀   ‡↔to 6m (20ft)

Ligularia 'Gregynog Gold'

A large, robust, clump-forming perennial with pyramidal spikes of daisy-like, golden-orange, brown-centred flowers from late summer to early autumn. These are carried on upright stems above the large, rounded, mid-green leaves. Excellent by water; it naturalizes readily in moist soils. Dark-leaved 'Desdemona', only 1m (3ft) tall, and 'The Rocket' are other recommended ligularias.

CULTIVATION: *Grow in reliably moist, deep, moderately fertile soil. Position in full sun with some midday shade, and shelter from strong winds.*

☼ ☀ ◊ ♀ ❁ ❁ ❁
‡to 2m (6ft) ↔1m (3ft)

Ligustrum lucidum

The Chinese privet is a vigorous, conical, evergreen shrub with glossy dark green, oval leaves. Loose clusters of small white flowers are produced in late summer and early autumn, followed by oval, blue-black fruits. Good as hedging, but also makes a useful, well-shaped plant for a shrub border.

CULTIVATION: *Best in well-drained soil, in full sun or partial shade. Cut out any unwanted growth in winter.*

☼ ☀ ◊ ♀ ❁ ❁ ❁   ‡↔10m (30ft)

Ligustrum lucidum 'Excelsum Superbum'

This variegated Chinese privet, with yellow-margined, bright green leaves, is a fast-growing, conical, evergreen shrub. Loose clusters of small, creamy-white flowers appear in late summer and early autumn, followed by oval, blue-black fruits.

CULTIVATION: *Grow in any well-drained soil, in full sun for the best leaf colour. Remove unwanted growth in winter. Remove any shoots that have plain green leaves as soon as seen.*

☼ ◊ ♀ ✿✿✿　　　　↕↔10m (30ft)

Lilium candidum

The Madonna lily is an upright, bulbous perennial, bearing sprays of up to 20 highly fragrant, trumpet-shaped flowers on each stiff stem in midsummer. The flowers have pure white petals with tinted yellow bases, and yellow anthers. The lance-shaped, glossy bright green leaves that appear after the flowers usually last over winter.

CULTIVATION: *Grow in well-drained, neutral to alkaline soil that is rich in well-rotted organic matter. Tolerates drier soil than most lilies. Position in full sun with the base in shade.*

☼ ◊ ♀ ✿✿✿　　　　↕1–2m (3–6ft)

Lilium formosanum var. *pricei*

An elegant, clump-forming perennial
bearing very fragrant, slender,
trumpet-shaped flowers. These are
borne singly or in clusters of up to
three during summer. The flowers
have curved petal tips, white
insides, and strongly purple-flushed
outsides. Most of the oblong, dark
green leaves grow at the base of
the stem. Lovely for an unheated
greenhouse or conservatory.

CULTIVATION: *Grow in moist, neutral
to acid, humus-rich soil or compost,
in sun with the base in shade.
Protect from excessively hot sun.*

☼ ◑ ❀❀❀   ‡0.6–1.5m (2–5ft)

Lilium henryi

A fast-growing, clump-forming
perennial that bears a profusion
of slightly scented, "turkscap" flowers
(with reflexed, or backward-bending
petals) in late summer. These are
deep orange with brown spots and
red anthers, carried on purple-
marked green stems above lance-
shaped leaves. Excellent for a wild
garden or woodland planting.

CULTIVATION: *Grow in well-drained,
neutral to alkaline soil with added leaf
mould or well-rotted organic matter.
Choose a position in partial shade.*

☼ ◑ ♀ ❀❀❀   ‡1–3m (3–10ft)

Lilium longiflorum

The Easter lily is a fast-growing perennial carrying short clusters of one to six pure white, strongly fragrant, trumpet-shaped flowers with yellow anthers. They appear during midsummer above scattered, lance-shaped, deep green leaves. One of the smaller, less hardy lilies, it grows well in containers and under glass.

CULTIVATION: *Best in well-drained soil or compost with added organic matter, in partial shade. Tolerates lime.*

☼ ◊ ♀ ❀   ‡40–100cm (16–39in)

Lilium martagon var. *album*

A clump-forming, vigorous perennial bearing sprays of up to 50 small, nodding, glossy white, "turkscap" flowers with strongly curled petals. The leaves are elliptic to lance-shaped, mostly borne in dense whorls. Unlike most lilies, it has an unpleasant smell and is better sited in a border or wild garden, for which it is well-suited. Looks good planted with var. *cattaniae*, with its deep maroon flowers.

CULTIVATION: *Best in almost any well-drained soil, in full sun or partial shade. Water freely when in growth.*

☼ ☀ ◊ ♀ ❀❀❀   ‡1–2m (3–6ft)

Lilium monadelphum

This stout, clump-forming perennial, also known as *L. szovitsianum*, bears up to 30 large, fragrant, trumpet-shaped flowers on each stiff stem in early summer. The blooms are pale yellow, flushed brown-purple on the outsides and flecked purple-maroon on the insides. The scattered, bright green leaves are narrowly oval. Good for containers, as it tolerates drier conditions than most lilies.

CULTIVATION: *Grow in any well-drained soil, in full sun. Tolerates fairly heavy soils and lime.*

☼ ◊ ❀❀❀   ↕1–1.5m (3–5ft)

Lilium Pink Perfection Group

These stout-stemmed lilies bear clusters of large, scented, trumpet-shaped flowers with curled petals in midsummer. Flower colours range from deep purplish-red to purple-pink, all with bright orange anthers. The mid-green leaves are strap-like. Excellent for cutting.

CULTIVATION: *Grow in well-drained soil that is enriched with leaf mould or well-rotted organic matter. Choose a position in full sun with the base in shade.*

☼ ◊ ♀ ❀❀❀   ↕1.5–2m (5–6ft)

Lilium pyrenaicum

A relatively short, bulbous lily that produces up to 12 nodding, green-yellow or yellow, purple-flecked flowers per stem, in early to midsummer. Their petals are strongly curved back. The green stems are sometimes spotted with purple, and the lance-shaped, bright green leaves often have silver edges. Not a good lily for patio planting, since its scent is unpleasant.

CULTIVATION: *Grow in well-drained, neutral to alkaline soil with added leaf mould or well-rotted organic matter. Position in full sun or partial shade.*

☼ ☀ ◊ ❀ ❀ ❀   ‡30–100cm (12–39in)

Lilium regale

The regal lily is a robust, bulbous perennial with very fragrant, trumpet-shaped flowers opening during midsummer. They can be borne in sprays of up to 25, and are white, flushed with purple or purplish-brown on the outsides. The narrow leaves are numerous and glossy dark green. A bold statement in a mixed border. Suitable for growing in pots.

CULTIVATION: *Grow in well-drained soil enriched with organic matter. Dislikes very alkaline conditions. Position in full sun or partial shade.*

☼ ☀ ◊ ♀ ❀ ❀ ❀   ‡0.6m–2m (2–6ft)

Limnanthes douglasii

The poached egg plant is an upright
to spreading annual which produces
a profusion of white-edged, yellow
flowers from summer to autumn.
The deeply toothed, glossy, bright
yellow-green leaves are carried on
slender stems. Good for brightening
up a rock garden or path edging, and
attractive to hoverflies which help
control aphids.

CULTIVATION: *Grow in moist but well-*
drained, fertile soil, in full sun. Sow
seed outdoors during spring or autumn.
After flowering, it self-seeds freely.

☼ ◊ ♀ ✿✿✿ ↕↔to 15cm (6in) or more

Limonium sinuatum 'Forever Gold'

An upright perennial bearing tightly
packed clusters of bright yellow flowers
from summer to early autumn. The stiff
stems have narrow wings. Most of the
dark green leaves are arranged in
rosettes around the base of the plant.
Suitable for a sunny border or in a
gravel garden; the flowers dry well for
indoor arrangements. Like other statice,
this is usually grown as an annual since
it may not survive cold winters.

CULTIVATION: *Grow in well-drained,*
preferably sandy soil, in full sun.
Tolerates dry and stony conditions.

☼ ◊ ✿✿✿
↕to 60cm (24in) ↔30cm (12in)

Liquidambar styraciflua '**Worplesdon**'

Sweetgum trees are often mistaken for maples because of their similar foliage and wonderful autumn colours, which range from yellow to orange, purple, and crimson. 'Worplesdon' is one of the most reliable forms for its intense red, orange, and yellow autumn leaves. Liquidambars make narrowly conical, deciduous trees and also have attractive bark.

CULTIVATION: *Grow in well-drained soil in sun.*

☼ ◐ ◑ ♀ ❀❀❀
‡25m (80ft) ↔12m (40ft)

Liriodendron tulipifera

The stately tulip tree has a broadly columnar habit, spreading with age. The deciduous, squarish, lobed leaves are dark green, turning butter-yellow in autumn. Tulip-shaped, pale green flowers, tinged orange at the base, appear in summer and are followed by cone-like fruits in autumn. An excellent specimen tree for a large garden.

CULTIVATION: *Grow in moist but well-drained, moderately fertile, preferably slightly acid soil. Choose a site in full sun or partial shade. Keep pruning of established specimens to a minimum.*

☼ ☀ ◐ ♀ ❀❀❀
‡30m (100ft) ↔15m (50ft)

Liriope muscari

This stout, evergreen perennial forms dense clumps of dark green, strap-like leaves. Spikes of small, violet-purple flowers open in autumn amid the foliage, and may be followed by black berries. Good in a woodland border, or use as drought-tolerant ground cover for shady areas.

CULTIVATION: *Grow in light, moist but well-drained, moderately fertile soil. Prefers slightly acid conditions. Position in partial or full shade with shelter from cold, drying winds. Tolerates drought.*

☼ ☀ ◊ ◊ ♀ ❀ ❀ ❀
‡30cm (12in) ↔ 45cm (18in)

Lithodora diffusa 'Heavenly Blue'

A spreading, evergreen shrub, sometimes sold as *Lithospermum* 'Heavenly Blue', that grows flat along the ground. Deep azure-blue, funnel-shaped flowers are borne in profusion over long periods from late spring into summer. The leaves are elliptic, dark green, and hairy. Suits an open position in a rock garden or raised bed.

CULTIVATION: *Grow in well-drained, humus-rich, acid soil, in full sun. Trim lightly after flowering.*

☼ ◊ ♀ ❀ ❀ ❀
‡15cm (6in) ↔ 60cm (24in) or more

Lobelia cardinalis 'Queen Victoria'

This short-lived, clump-forming perennial bears almost luminous spikes of vivid red, two-lipped flowers from late summer to mid-autumn. Both the leaves and stems are deep purple-red. Effective in a mixed border or waterside planting. May not survive in cold areas. 'Bee's Flame' is a very similar plant, enjoying the same conditions.

CULTIVATION: *Grow in deep, reliably moist, fertile soil, in full sun. Short-lived, but can be easily propagated by division in spring.*

☀ ◐ ♡ ✲ ✲ ‡1m (3ft) ↔30cm (12in)

Lobelia 'Crystal Palace'

A compact, bushy perennial that is almost always grown as an annual, with vibrant clusters of two-lipped, dark blue flowers during summer to autumn. The tiny leaves are dark green and bronzed. Useful for edging and to spill over the edges of containers.

CULTIVATION: *Grow in deep, fertile soil or compost that is reliably moist, in full sun or partial shade. Plant out, after the risk of frost has passed, in spring.*

☀ ☀ ◐ ✲ ✲ ‡to 10cm (4in) ↔10–15cm (4–6in)

Lonicera x *italica*

This vigorous, deciduous, woody-stemmed honeysuckle is a free-flowering, twining, or scrambling climber. In summer and early autumn, it bears large whorls of tubular, very fragrant, soft flesh-pink flowers, flushed red-purple with yellow insides; red berries follow later in the season. The leaves are oval and dark green. Train onto a wall, or up into a small tree.

CULTIVATION: *Grow in moist but well-drained, fertile, humus-rich soil, in full sun or partial shade. Once established, cut back shoots by up to one-third after they have flowered.*

☼ ☀ ◊ ◑ ❀❀❀   ↕7m (22ft)

Lonicera nitida 'Baggesen's Gold'

A dense, evergreen shrub bearing tiny, oval, bright yellow leaves on arching stems. Inconspicuous yellow-green flowers are produced in spring, occasionally followed by small, blue-purple berries. Excellent for hedging or topiary in urban gardens, as it is pollution-tolerant.

CULTIVATION: *Grow in any well drained soil, in full sun or partial shade. Trim hedges at least 3 times a year, between spring and autumn. Plants that become bare at the base will put out renewed growth if cut back hard.*

☼ ☀ ◊ ♀ ❀❀❀   ↕↔1.5m (5ft)

Lonicera periclymenum 'Graham Thomas'

This long-flowering form of English honeysuckle is a woody, deciduous, twining climber bearing abundant, very fragrant, tubular white flowers. These mature to yellow over a long period in summer, without the red flecking seen on *L. periclymenum* 'Belgica'. The leaves are mid-green and oval in shape.

CULTIVATION: *Grow in moist but well-drained, fertile, humus-rich soil. Thrives in full sun, but prefers shade at the base. Once established, cut back shoots by up to one-third after flowering.*

☼ ☼ ◊ ◊ ♀ ✿ ✿ ✿   ↕7m (22ft)

Lonicera periclymenum 'Serotina'

The late Dutch honeysuckle is a fast-growing, deciduous, twining climber with very fragrant, rich red-purple flowers which appear in abundance during mid- and late summer. These may be followed by red berries. The leaves are oval and mid-green. If given plenty of space, it scrambles naturally with little pruning needed.

CULTIVATION: *Grow in well-drained but moist, humus-rich, fertile soil, in sun with shade at the base. To keep trained specimens within bounds, prune shoots back by one-third after flowering.*

☼ ☼ ◊ ◊ ♀ ✿ ✿ ✿   ↕7m (22ft)

Lonicera sempervirens

A semi-evergreen, woody-stemmed, twining honeysuckle with oval leaves and salmon-red to orange tubular flowers in summer and early autumn, followed by bright red berries in winter. Suitable for a large container with a trellis, or cascading over an arbour or fence. *L. sempervirens* f. *sulphurea* is a yellow-flowered variety.

CULTIVATION: *Grow in fertile, moist but well-drained soil in sun or partial shade. Prune back young growth soon after flowering.*

☼ ☀ ◊ ◐ ✿✿✿   ‡to 4m (12ft)

Lonicera x *tellmanniana*

A twining, deciduous, woody-stemmed climber that bears clusters of coppery-orange, tubular flowers which open from late spring to midsummer. The deep green, elliptic leaves have blue-white undersides. Train onto a wall or fence, or up into a large shrub. Summer-flowering *L. tragophylla*, one of its parents, is similar and is also recommended.

CULTIVATION: *Grow in moist but well-drained, fertile, humus-rich soil. Will tolerate full sun, but produces better flowers in a more shaded position. After flowering, trim shoots by one-third.*

☼ ☀ ◊ ◐ ♀ ✿✿✿   ‡5m (15ft)

Lotus hirsutus

The hairy canary clover is a silvery shrub with greyish green, very downy leaves. In summer and early autumn, clusters of pink-flushed, creamy white, pea-like flowers appear. These mature into reddish brown seed pods. Good for a sheltered rock garden or Mediterranean-style border, where the shrub can be trimmed lightly in spring if required.

CULTIVATION: *Needs well-drained soil in sun as it dislikes winter wet.*

☼ ◊ ♀ ❈❈
‡to 60cm (24in) ↔ to 1m (3ft)

Lupinus arboreus

The tree lupin is a fast-growing, sprawling, semi-evergreen shrub grown for its spikes of fragrant, clear yellow flowers which open through the summer. The divided leaves are bright and grey-green. Native to scrub of coastal California, it is tolerant of seaside conditions. May not survive in cold, frosty winters.

CULTIVATION: *Grow in light or sandy, well-drained, fairly fertile, slightly acid soil, in full sun. Cut off seedheads to prevent self-seeding and trim after flowering to keep compact.*

☼ ◊ ♀ ❈❈❈
‡↔2m (6ft)

Lychnis chalcedonica

The Jerusalem cross is a clump-forming perennial which produces slightly domed, brilliant red flowerheads in early to midsummer. These are borne on upright stems above the oval, mid-green, basal leaves. The flowers are small and cross-shaped. Good for a sunny border or wild garden, but needs some support. It self-seeds freely.

CULTIVATION: *Grow in moist but well-drained, fertile, humus-rich soil, in full sun or light dappled shade.*

☼ ☀ ◊ ◊ ♀ ❀❀❀
↕1–1.2m (3–4ft) ↔30cm (12in)

Lycoris radiata

A late-summer flowering, bulbous perennial, red spider lily sends up a leafless stalk topped with a head of bright, spidery flowers. In mild climates, strappy leaves follow and last until spring, when it becomes dormant. In areas with dry summers, grow in a sunny border or rock garden. In cold or wet regions, grow in a container and overwinter in a frost-free area. Cut flowers last for several days.

CULTIVATION: *Grow in fertile, well-drained soil that dries out in summer, in full sun. Dig and divide every three years.*

☼ ◊ ❀❀
↕30–40cm (12–16in) ↔to 15cm (6in)

Lysichiton americanus

Yellow skunk cabbage is a striking, colourful perennial, which flowers in early spring and is ideal for a waterside planting. Each dense spike of tiny, greenish-yellow flowers is hooded by a bright yellow spathe and has an unpleasant, slightly musky scent. The large, dark green leaves, 50–120cm (20–48in) long, emerge from the base of the plant.

CULTIVATION: *Grow in moist, fertile, humus-rich soil. Position in full sun or partial shade, allowing plenty of room for the large leaves to develop.*

☀ ◐ ◊ ♀ ❋❋❋ ↕1m (3ft) ↔1.2m (4ft)

Lysichiton camtschatcensis

White skunk cabbage is a bold waterside perennial with a slightly musky scent. In early spring, dense spikes of tiny green flowers emerge, cloaked by pointed white spathes. The large, dark green leaves, up to 1m (3ft) long, grow from the base. Ideal beside a stream or a pool.

CULTIVATION: *Grow in moist, waterside conditions, in fertile, humus-rich soil. Position in full sun or partial shade, allowing room for the leaves to develop.*

☀ ◐ ◊ ♀ ❋❋❋ ↕↔to 1m (3ft)

Lysimachia clethroides

This spreading, clump-forming, herbaceous perennial is grown for its tapering spikes of tiny, star-shaped white flowers. These droop when in bud, straightening up as the flowers open in mid- to late summer. The narrow leaves are yellow-green when young, maturing to mid-green with pale undersides. Suitable for naturalizing in a wild woodland or bog garden.

CULTIVATION: *Grow in reliably moist soil that is rich in humus, in full sun or partial shade. May need staking.*

☼ ☀ ◊ ♀ ✽✽✽
‡1m (3ft) ↔60cm (24in)

Lysimachia nummularia '**Aurea**'

Golden creeping Jenny is a rampant, sprawling, evergreen perennial, which makes an excellent ground-cover plant. Its golden-yellow leaves are broadly oval in shape, with heart-shaped bases. The bright yellow, cup-shaped, summer flowers further enhance the foliage colour.

CULTIVATION: *Grow in reliably moist soil that is enriched with well-rotted organic matter. Site in full sun or partial shade.*

☼ ☀ ◊ ♀ ✽✽✽
‡to 5cm (2in) ↔indefinite

Macleaya x *kewensis* 'Kelway's Coral Plume'

A clump-forming perennial grown for its foliage and large, graceful plumes of tiny, coral-pink to deep-buff flowers. These open from pink buds from early summer, appearing to float above large, olive-green leaves. Grow among shrubs, or group to form a hazy screen. *M. cordata*, with paler flowers, can be used similarly.

CULTIVATION: *Best in moist but well-drained, moderately fertile soil, in sun or light shade. Shelter from cold winds. May be invasive; chop away roots at the margins of the clump to confine.*

☼ ☀ ◊ ◗ ❀ ❀ ❀
‡2.2m (7ft) ↔1m (3ft) or more

Magnolia grandiflora 'Exmouth'

This hardy cultivar of the bull bay is a dense, evergreen tree bearing glossy dark green leaves with russet-haired undersides. *M. grandiflora* is quite distinct from other magnolias, since it blooms sporadically from late summer to autumn, producing flowers that are large, very fragrant, cup-shaped, and creamy-white.

CULTIVATION: *Best in well-drained but moist, humus-rich, acid soil, in full sun or light shade. Tolerates dry, alkaline conditions. Mulch with leaf mould in spring. Keep pruning to a minimum.*

☼ ☀ ◊ ◗ ❀ ❀
‡6–18m (20–60ft) ↔to 15m (50ft)

Magnolia grandiflora 'Goliath'

This cultivar of bull bay is slightly less hardy than 'Exmouth' (see facing page, below), but has noticeably larger flowers, to 30cm (12in) across. It is a dense, conical, evergreen tree with slightly twisted, dark green leaves. The cup-shaped, very fragrant, creamy-white flowers appear from late summer to autumn.

CULTIVATION: *Best in well-drained but moist, humus-rich, acid soil, in full sun or light shade. Tolerates dry, alkaline conditions. Mulch with leaf mould in spring. Keep pruning to a minimum.*

☼ ☼ ◊ ◊ ❀❀
↕6–18m (20–60ft) ↔to 15m (50ft)

Magnolia liliiflora 'Nigra'

A dense, summer-flowering shrub bearing goblet-shaped, deep purple-red flowers. The deciduous leaves are elliptic and dark green. Plant as a specimen or among other shrubs and trees. Unlike many magnolias, it begins to flower when quite young.

CULTIVATION: *Grow in moist but well-drained, rich, acid soil, in sun or semi-shade. Provide a mulch in early spring. Prune young shrubs in midsummer to encourage a good shape; once mature, very little other pruning is needed.*

☼ ☼ ◊ ◊ ♀ ❀❀❀
↕3m (10ft) ↔2.5m (8ft)

Spring-flowering Magnolias

Spring-flowering magnolias are handsome deciduous trees and shrubs valued for their elegant habit and beautiful, often fragrant blooms which emerge just before the leaves. Flowers range from the tough but delicate-looking, star-shaped blooms of *M. stellata*, to the exotic, goblet-like blooms of hybrids such as 'Ricki'. Attractive red fruits form in autumn. The architectural branch framework of the bare branches makes an interesting display for a winter garden, especially when grown as free-standing specimens. *M. x soulangeana* 'Rustica Rubra' can be trained against a wall.

CULTIVATION: *Best in deep, moist but well-drained, humus-rich, neutral to acid soil. M. wilsonii tolerates alkaline conditions. Choose a position in full sun or partial shade. After formative pruning when young, restrict pruning to the removal of dead or diseased branches after flowering.*

☼ ☀ ◊ ◊

‡↔10m (30ft)

‡10m (30ft) ↔8m (25ft) ‡15m (50ft) ↔10m (30ft) ‡10m (30ft) ↔6m (20ft)

1 *Magnolia x loebneri* 'Merrill' ♀ ❀❀❀ **2** *M. campbellii* 'Charles Raffill' ❀❀
3 *M. denudata* ♀ ❀❀❀ **4** *M.* 'Elizabeth' ♀ ❀❀❀

‡↔ 4m (12ft)

‡ 8m (25ft) ↔ 6m (20ft)

‡ 10m (30ft) ↔ 5m (15ft)

‡↔ 6m (20ft)

‡ 9m (28ft) ↔ 6m (20ft)

‡ 8m (25ft) ↔ 6m (20ft)

‡ 3m (10ft) ↔ 4m (12ft)

5 *M.* 'Ricki' ❀❀❀ **6** *M.* × *loebneri* 'Leonard Messel' ♀ ❀❀❀ **7** *M. salicifolia* ❀❀
8 *M.* × *soulangeana* 'Rustica Rubra' ❀❀ **9** *M. salicifolia* 'Wada's Memory' ♀ ❀❀❀
10 *M. wilsonii* ♀ ❀❀❀ **11** *M. stellata* ❀❀❀

Mahonia aquifolium 'Apollo'

Oregon grape is a low-growing, evergreen shrub with dark green leaves; these are divided into several spiny leaflets and turn brownish-purple in winter. Dense clusters of deep golden flowers open in spring, followed by small, blue-black fruits. Can be grown as groundcover.

CULTIVATION: *Grow in moist but well-drained, rich, fertile soil, in semi-shade; tolerates sun if the soil remains moist. Every 2 years after flowering, shear ground-cover plants close to the ground.*

☼ ☀ ◊ ◊ ♀ ❀❀❀
‡60cm (24in) ↔1.2m (4ft)

Mahonia japonica

A dense, upright, winter-flowering, evergreen shrub carrying large, glossy dark green leaves divided into many spiny leaflets. Long, slender spikes of fragrant, soft yellow flowers are borne from late autumn into spring, followed by purple-blue fruits. Suits a shady border or woodland garden.

CULTIVATION: *Grow in well-drained but moist, moderately fertile, humus-rich soil. Prefers shade, but will tolerate sun if soil remains moist. Limit pruning to removal of dead wood, after flowering.*

☼ ☀ ◊ ◊ ♀ ❀❀❀ ‡2m (6ft) ↔3m (10ft)

Mahonia x *media* 'Buckland'

A vigorous, upright, evergreen shrub bearing dense and sharply spiny, dark green foliage. Small, fragrant, bright yellow flowers are produced in arching spikes from late autumn to early spring. A good vandal-resistant shrub for a boundary or front garden.

CULTIVATION: *Best in moist but well-drained, fairly fertile, humus-rich soil. Thrives in semi-shade, but will become leggy in deep shade. Little pruning is needed, but over-long stems can be cut back to a low framework after flowering.*

☀ ◊ ◐ ♀ ❀❀❀ ‡5m (15ft) ↔4m (12ft)

Mahonia x *media* 'Charity'

A fast-growing, evergreen shrub, very similar to 'Buckland' (above), but has more upright, densely-packed flower spikes. The dark green leaves are spiny, making it useful for barrier or vandal-proof planting. Fragrant yellow flowers are borne from late autumn to spring.

CULTIVATION: *Grow in moist but well-drained, moderately fertile, rich soil. Prefers partial shade, and will become leggy in deep shade. After flowering, bare, leggy stems can be pruned hard to promote strong growth from lower down.*

☀ ◊ ◐ ❀❀❀ ‡5m (15ft) ↔4m (12ft)

Malus floribunda

The Japanese crab apple is a dense, deciduous tree with a long season of interest. Graceful, arching branches, bearing dark green foliage, flower during mid- to late spring to give a glorious display of pale pink blossom. The flowers are followed by small yellow crab apples; these often persist, providing a valuable source of winter food for garden wildlife.

CULTIVATION: *Grow in moist but well-drained, moderately fertile soil, in sun or light shade. Prune to shape in the winter months when young; older specimens require little pruning.*

☼ ☀ ◐ ◊ ♥ ❋ ❋ ❋   ↕↔10m (30ft)

Malus 'John Downie'

This vigorous, deciduous tree is upright when young, becoming conical with age. Large, cup-shaped white flowers, which open from pale pink buds in late spring, are followed by egg-shaped, orange and red crab apples. The oval leaves are bright green when young, maturing to dark green. An ideal small garden tree.

CULTIVATION: *Grow in well-drained but moist, fairly fertile soil. Flowers and fruits are best in full sun, but tolerates some shade. Remove damaged or crossing shoots when dormant, to form a well-spaced crown. Avoid hard pruning of established branches.*

☼ ☀ ◐ ◊ ❋ ❋ ❋
↕10m (30ft) ↔ 6m (20ft)

Malus tschonoskii

This upright, deciduous tree with upswept branches produces pink-flushed white blossom in late spring, followed in autumn by red-flushed yellow crab apples. The leaves turn from green to a vibrant gold, then red-purple in autumn. It is taller than many crab apples, but is still a beautiful specimen tree which can be accommodated in small gardens.

CULTIVATION: *Grow in well-drained, moderately fertile soil. Best in full sun, but tolerates some shade. Forms a good shape with little or no pruning; does not respond well to hard pruning.*

☼ ☀ ◊ ❀❀❀ ↕12m (40ft) ↔7m (22ft)

Malus x zumi 'Golden Hornet'

A broadly pyramidal, deciduous tree bearing a profusion of large, pink-flushed white flowers which open from deep pink buds in late spring. Small yellow crab apples follow, and persist well into winter. The display of golden fruit is further enhanced when the dark green foliage turns yellow in autumn.

CULTIVATION: *Grow in any but waterlogged soil, in full sun for best flowers and fruit. To produce a well-spaced crown, remove damaged or crossing shoots on young plants when dormant. Do not prune older specimens.*

☼ ☀ ◊ ◑ ❀❀❀
↕10m (30ft) ↔8m (25ft)

Malva moschata **f. *alba***

This white- to very light pink-flowered musk mallow is a bushy, upright perennial suitable for wild-flower gardens or borders. The very attractive and showy flowers are borne in clusters from early to late summer, amid the slightly musk-scented, mid-green foliage.

CULTIVATION: *Grow in moist but well-drained, moderately fertile soil, in full sun. Taller plants may need staking. Often short-lived, but self-seeds readily.*

☼ ◊ ♀ ❀ ❀ ❀ ↕1m (3ft) ↔60cm (24in)

Matteuccia struthiopteris

The shuttlecock fern forms clumps of upright or gently arching, pale green, deciduous fronds. In summer, smaller, dark brown fronds form at the centre of each clump, which persist until late winter. No flowers are produced. An excellent foliage perennial for a damp, shady border, and woodland or waterside plantings.

CULTIVATION: *Grow in moist but well-drained, humus-rich, neutral to acid soil. Chose a site in light dappled shade.*

☼ ◊ ◊ ♀ ❀ ❀ ❀
↕1–1.5m (3–5ft) ↔45–75cm (18–30in)

Meconopsis betonicifolia

The Tibetan blue poppy is a clump-forming perennial bearing upright stems of large, saucer-shaped flowers which are clear blue or often purple-blue or white. These appear in early summer, above the oval and bluish-green leaves. Naturalizes well in a woodland garden.

CULTIVATION: *Best in moist but well-drained, rich, acid soil. Site in partial shade with shelter from cold winds. May be short-lived, especially in hot or dry conditions. Divide clumps after flowering to maintain vigour.*

☀ ◊ ◊ ✿✿✿
‡1.2m (4ft) ↔45cm (18in)

Meconopsis grandis

The Himalayan blue poppy is an upright, clump-forming perennial, similar to *M. betonicifolia* (see above), but with larger, less clustered, rich blue to purplish-red flowers. These are carried above the mid- to dark green foliage in early summer. Appealing when grown in large groups in a woodland setting.

CULTIVATION: *Grow in moist, leafy, acid soil. Position in partial shade with shelter from cold, drying winds. Mulch generously and water in dry spells; may fail to flower if soil becomes too dry.*

☀ ◊ ✿✿✿
‡1–1.2m (3–4ft) ↔60cm (24in)

Melianthus major

The honey bush is an excellent foliage shrub of upright to spreading habit. From late spring to midsummer, spikes of blood-red flowers may appear above the grey-green to bright blue-green, divided leaves. In cold areas, treat as a herbaceous perennial, but in milder climates it is ideal for a coastal garden.

CULTIVATION: *Grow in moist but well-drained, fertile soil. Choose a sunny site protected from cold, drying winds and winter wet. In cold areas, cut back the last year's growth in spring; in warm areas, cut out flowered stems in autumn.*

☼ ◊ ♀ ❀❀
‡2–3m (6–10ft) ↔1–3m (3–10ft)

Metasequoia glyptostroboides

The dawn redwood is an elegant conifer with a neat, narrowly conical outline, and it looks effective planted as an architectural specimen in a waterside location. The soft, fern-like leaves, which are emerald green when young, are set off by the shaggy, cinnamon-brown bark. The leaves turn a warm russet gold before falling in autumn. 'Gold Rush' is a variety with golden leaves.

CULTIVATION: *Ideal in any moist soil, in full sun.*

☼ ☀ ◊ ♦ ♀ ❀❀❀
‡20–40m (70–130ft) ↔5m (15ft)

Mimulus cardinalis

The scarlet monkey flower is a creeping perennial with tubular, scarlet, sometimes yellow-marked flowers. They appear throughout summer amid the oval, light green leaves, on hairy stems. Good for adding colour to a warm border. May not survive in cold climates.

CULTIVATION: *Grow in any well-drained, fertile, humus-rich soil; tolerates quite dry conditions. Position in sun or light dappled shade. May be short-lived, but easily propagated by division in spring.*

☼ ☀ ◊ ♀ ❀❀ ‡1m (3ft) ↔60cm (24in)

Miscanthus sinensis 'Flamingo'

This variety is one of the best of the miscanthus grasses, notable for its soft pink flowerheads that droop over the gracefully arching foliage from late summer. Miscanthus make good garden plants for sunny positions, either in large pots or in a border, where they will give structural interest well into winter. Cut back plants in early spring.

CULTIVATION: *Best in moist but well-drained soil in full sun.*

☼ ☀ ◊ ◑ ♀ ❀❀❀
‡to 2m (6ft) ↔1.5m (5ft)

Molinia caerulea 'Variegata'

The purple moor grass is a tufted perennial forming clumps of cream-striped, narrow, dark green leaves. Dense, purple flowering spikes are produced over a long period from spring to autumn on tall, ochre-tinted stems. A good structural plant for a border or a woodland garden.

CULTIVATION: *Grow in any moist but well-drained, preferably acid to neutral soil, in full sun or partial shade.*

☼ ☀ ◊ ♀ ❀❀❀
‡to 60cm (24in) ↔40cm (16in)

Monarda 'Cambridge Scarlet'

This bergamot hybrid is a clump-forming perennial bearing shaggy heads of rich scarlet-red, tubular flowers from midsummer to early autumn. They appear in profusion above the aromatic leaves, and are very attractive to bees; a common name for Monarda is "bee-balm". A colourful addition to any mixed or herbaceous border.

CULTIVATION: *Prefers moist but well-drained, moderately fertile, humus-rich soil, in full sun or light dappled shade. Keep moist in summer, but protect from excessive wet in winter.*

☼ ☀ ◊ ◊ ❀❀❀
‡1m (3ft) ↔45cm (18in)

Monarda 'Croftway Pink'

A clump-forming, herbaceous perennial with shaggy heads of tubular pink flowers carried above the small, aromatic, light green leaves from midsummer to early autumn. Suits mixed or herbaceous borders; the flowers are attractive to bees. 'Beauty of Cobham' is very similar, but with mauve outer bracts surrounding the pink flowers.

CULTIVATION: *Grow in well-drained, fairly fertile, humus-rich soil that is reliably moist in summer. Site in full sun or light shade, with protection from excessive winter wet.*

☼ ☀ ◊ ❀❀❀   ‡1m (3ft) ↔45cm (18in)

Muscari armeniacum

A vigorous, bulbous perennial that bears dense spikes of tubular, rich blue flowers in early spring. The mid-green leaves are strap-like, and begin to appear in autumn. Plant massed together in borders or allow to spread and naturalize in grass, although it can be invasive.

CULTIVATION: *Grow in moist but well-drained, fairly fertile soil, in full sun. Divide clumps of bulbs in summer.*

☼ ◊ ♀ ❀❀❀   ‡20cm (8in) ↔5cm (2in)

Muscari aucheri

This bulbous perennial, less invasive than *M. armeniacum* (see previous page, bottom), bears dense spikes of small, bright blue spring flowers. The flower spikes, often topped with paler blue flowers, are carried above the basal clumps of strap-shaped, mid-green leaves. Suitable for a rock garden. Sometimes known as *M. tubergianum*.

CULTIVATION: *Grow in moist but well-drained, moderately fertile soil. Choose a position in full sun.*

☼ ◊ ♀ ❋ ❋ ❋
‡10–15cm (4–6in) ↔5cm (2in)

Myrtus communis

The common myrtle is a rounded shrub with dense, evergreen foliage. From midsummer to early autumn, amid the small, glossy dark green, aromatic leaves, it bears great numbers of fragrant white flowers with prominent tufts of stamens. Purple-black berries appear later in the season. Can be grown as an informal hedge or a specimen shrub. In areas with cold winters, grow against a warm, sunny wall.

CULTIVATION: *Best in well-drained but moist, fertile soil, in full sun. Protect from cold, drying winds. Trim in spring; tolerates close clipping.*

☼ ◊ ♀ ❋ ❋
‡↔3m (10ft)

Myrtus communis subsp. *tarentina*

This dense, evergreen shrub is more compact and rounded than the species (see facing page, below), with smaller leaves and pink-tinted cream flowers. These appear during mid-spring to early autumn and are followed by white berries. Grow in a border or as an informal hedge; may not survive in cold climates.

CULTIVATION: *Grow in moist but well-drained, moderately fertile soil. Choose a site in full sun with shelter from cold, drying winds. Trim back in spring; tolerates close clipping.*

☼ ◊ ♡ ❈❈ ↕↔1.5m (5ft)

Nandina domestica

Heavenly bamboo is an upright, evergreen or semi-evergreen shrub with fine spring and autumn colour. The divided leaves are red when young, maturing to green, then flushing red again in late autumn. Conical clusters of small white flowers with yellow centres appear in midsummer, followed, in warm climates, by long-lasting, bright red fruits. May not survive cold winters.

CULTIVATION: *Grow in moist but well-drained soil, in full sun. Cut back on planting, then prune in mid-spring to keep the plant tidy. Deadhead regularly.*

☼ ◊ ❈❈ ↕2m (6ft) ↔1.5m (5ft)

Small Daffodils (*Narcissus*)

Small and miniature daffodils make good spring-flowering, bulbous perennials for both indoor and outdoor displays. They are cultivated for their elegant, mostly yellow or white flowers, of which there is great variety in shape. The blooms are carried either singly or in clusters above basal clumps of long, strap-shaped leaves on upright, leafless stems. All small daffodils are suitable for a rock garden, and they can look effective when massed together to form drifts. Because of their manageable size, these daffodils are useful for indoor displays. Some, such as *N. bulbocodium*, will naturalize well in short, fine grass.

CULTIVATION: *Best in well-drained, fertile soil that is moist during growth, preferably in sun. Feed with a balanced fertilizer after flowering to encourage good flowers the following year. Deadhead as flowers fade, and allow the leaves to die down naturally; do not tie them into bunches.*

☼ ◊

1 ‡35cm (14in) ↔ 8cm (3in)

2 ‡10–15cm (4–6in) ↔ 5–8cm (2–3in)

3 ‡30cm (12in) ↔ 8cm (3in)

4 ‡15–20cm (6–8in) ↔ 5–8cm (2–3in)

5 ‡30cm (12in) ↔ 8cm (3in)

6 ‡30cm (12in) ↔ 8cm (3in)

1 *Narcissus* 'Avalanche' ♀ ❀❀ 2 *N. bulbocodium* ❀❀ 3 *N.* 'Charity May' ❀❀❀
4 *N. cyclamineus* ♀ ❀❀❀ 5 *N.* 'Dove Wings' ❀❀❀ 6 *N.* 'February Gold' ♀ ❀❀❀

7
‡17cm (7in) ↔ 5–8cm (2–3in)

8
‡20cm (8in) ↔ 8cm (3in)

9
‡20cm (8in) ↔ 8cm (3in)

10
‡17cm (7in) ↔ 5–8cm (2–3in)

11
‡10–15cm (4–6in) ↔ 5–8cm (2–3in)

12
‡15cm (6in) ↔ 5–8cm (2–3in)

13
‡10–25cm (4–10in) ↔ 5–8cm (2–3in)

7 *N.* 'Hawera' ♀ ✿✿✿ **8** *N.* 'Jack Snipe' ♀ ✿✿✿ **9** *N.* 'Jetfire' ♀ ✿✿✿ **10** *N.* 'Jumblie'
♀ ✿✿✿ **11** *N. minor* ✿✿✿ **12** *N.* 'Tête-à-tête' ♀ ✿✿✿ **13** *N. triandrus* ♀ ✿✿

Large Daffodils (*Narcissus*)

Large daffodils are tall, bulbous perennials, easily cultivated for their showy, mostly white or yellow flowers in spring. These are borne singly or in clusters on upright, leafless stems above long, strap-shaped, mid-green foliage arising from the bulb. A great diversity of elegant flower shapes is available, and all of the cultivars illustrated here are excellent for cutting; 'Sweetness' has blooms that last particularly well when cut. Most look very effective flowering in large groups between shrubs or in a border. Some naturalize easily in grass, or under deciduous trees and shrubs in a woodland garden.

CULTIVATION: *Best in well-drained, fertile soil, preferably in full sun. Keep soil reliably moist during the growing season, and feed with a balanced fertilizer after flowering to ensure good blooms the following year. Deadhead as flowers fade, and allow leaves to die down naturally.*

☼ ◊ ❀ ❀ ❀

1 ‡45cm (18in) ↔15cm (6in)

2 ‡40cm (16in) ↔12cm (5in)

3 ‡40cm (16in) ↔12cm (5in)

4 ‡40cm (16in) ↔15cm (6in)

1 *N.* 'Actaea' ♀ **2** *N.* 'Empress of Ireland' **3** *N.* 'Ceylon' **4** *N.* 'Cheerfulness' ♀

‡40cm (16in) ↔15cm (6in)

‡45cm (18in) ↔15cm (6in)

‡35cm (14in) ↔15cm (6in)

‡45cm (18in) ↔15cm (6in)

‡40cm (16in) ↔15cm (6in)

‡40cm (16in) ↔8cm (3in)

‡45cm (18in) ↔15cm (6in)

‡45cm (18in) ↔15cm (6in)

‡40cm (16in) ↔8cm (3in)

‡45cm (18in) ↔15cm (6in)

5 *N.* 'Ice Follies' ♀ **6** *N.* 'Kingscourt' ♀ **7** *N.* 'Merlin' ♀ **8** *N.* 'Mount Hood' ♀
9 *N.* 'Passionale' ♀ **10** *N.* 'Suzy' ♀ **11** *N.* 'Saint Keverne' ♀ **12** *N.* 'Tahiti' ♀
13 *N.* 'Sweetness' ♀ **14** *N.* 'Yellow Cheerfulness' ♀

Nepeta racemosa

This catmint is a fairly low, spreading perennial with strongly aromatic, greyish green foliage and spires of vivid blue flowers through summer. It is a good, drought-tolerant plant for the edge or middle of a sunny border, or spilling over a low wall. Cats and butterflies find the plant irresistible. Cut back in autumn when the foliage is looking tired. 'Walker's Low' is a very good variety.

CULTIVATION: *Grow in well-drained soil in sun or partial shade.*

☼ ☀ ◊ ♀ ❀❀❀
‡to 30cm (12in) ↔50cm (20in) or more

Nerine bowdenii

This robust perennial is one of the best late flowering bulbs. In autumn, it bears open sprays of five to ten trumpet-shaped, faintly scented, bright pink flowers with curled, wavy-edged petals. The strap-like, fresh green leaves appear after the flowers, at the base of the plant. Despite its exotic appearance, it will survive winter cold with protection at the base of a warm, sunny wall. The flowers are good for cutting.

CULTIVATION: *Grow in well-drained soil, in a sunny, sheltered position. Provide a deep, dry mulch in winter.*

☼ ◊ ♀ ❀❀
‡45cm (18in) ↔12–15cm (5–6in)

Nicotiana 'Lime Green'

This striking tobacco plant is an upright, free-flowering, bushy annual, ideal for a summer border. It bears loose clusters of night-scented, yellow-green flowers, with long throats and flattened faces, from mid- to late summer. The leaves are mid-green and oblong. Also good on patios, where the evening scent can be appreciated. For tobacco flowers in mixed colours that include lime green, the Domino Series is recommended.

CULTIVATION: *Grow in moist but well-drained, fertile soil. Choose a position in full sun or partial shade.*

☼ ☀ ◊ ❀❀
‡60cm (24in) ↔25cm (10in)

Nigella 'Miss Jekyll'

This tall, slender annual bears pretty, sky-blue flowers during summer, which are surrounded by a feathery "ruff" of bright green foliage. These are followed later in the season by attractive, inflated seed pods. The blooms last well when cut, and the seed pods can be dried for indoor flower arrangements. There is a white version, also recommended: 'Miss Jekyll Alba'.

CULTIVATION: *Grow in well-drained soil, in full sun. Like all love-in-a-mists, it self-seeds freely, although seedlings may differ from the parent.*

☼ ◊ ♀ ❀❀❀
‡to 45cm (18in) ↔to 23cm (9in)

Water Lilies (*Nymphaea*)

These aquatic perennials are cultivated for their showy, sometimes fragrant summer flowers and rounded, floating leaves. The shade cast by the leaves is useful in reducing growth of pond algae. Flowers are mostly white, yellow, pink, or red with yellow stamens in the centres. Spread varies greatly, so choose carefully to match the size of your water feature. Planting depths may be between 45cm and 1m (18–30in), though 'Pygmaea Helvola' can grow in water no deeper than 15cm (6in). Using aquatic plastic mesh planting baskets makes lifting and dividing much easier.

CULTIVATION: *Grow in still water in full sun. Plant in aquatic compost with rhizomes just below the surface, anchored with a layer of grit. Stand young plants on stacks of bricks so that shoot tips reach the water surface; lower plants as stems lengthen. Remove yellow leaves regularly. Divide in spring, giving plants an aquatic fertilizer.*

☼ ❀❀❀

1 ↔1.2–1.5m (4–5ft) 2 ↔1.5–2.5m (5–8ft) 3 ↔0.9–1.2m (3–4ft) 4 ↔0.9–1.2m (3–4ft) 5 ↔1.2–1.5m (4–5ft) 6 ↔25–40cm (10–16in)

1 *Nymphaea* 'Escarboucle' ♀ **2** *N.* 'Gladstoniana' ♀ **3** *N.* 'Gonnère' ♀
4 *N.* 'James Brydon' ♀ **5** *N.* 'Marliacea Chromatella' ♀ **6** *N.* 'Pygmaea Helvola' ♀

Nyssa sinensis

The Chinese tupelo is a deciduous tree, conical in form, grown for its lovely foliage. The elegant leaves are bronze when young, maturing to dark green, then becoming brilliant shades of orange, red, and yellow in autumn before they fall. The flowers are inconspicuous. Ideal as a specimen tree near water.

CULTIVATION: *Grow in fertile, reliably moist but well-drained, neutral to acid soil in sun or partial shade, with shelter from cold, drying winds. Thin out crowded branches in late winter.*

☼ ☀ ◊ ◑ ❀ ❀ ❀   ↕↔10m (30ft)

Nyssa sylvatica

A taller tree than *N. sinensis* (above), the tupelo is similar in form, with drooping lower branches. Its dark green leaves change to a glorious display of orange, yellow, or red in autumn. A brilliant tree for autumn colour, with brownish-grey bark that breaks up into large pieces on mature specimens. It tolerates acid soil.

CULTIVATION: *Grow in moist but well-drained, fertile, neutral to acid soil in sun or partial shade. Shelter from cold, drying winds, and prune in late winter, if necessary.*

☼ ☀ ◊ ◑ ❀ ❀ ❀
↕20m (70ft) ↔10m (30ft)

Oenothera fruticosa 'Fyrverkeri'

An upright, clump-forming perennial carrying clusters of short-lived, cup-shaped, bright yellow flowers which open in succession from late spring to late summer. The lance-shaped leaves are flushed red-purple when young, contrasting beautifully with the red stems. It associates well with bronze-or copper-leaved plants. *O. macrocarpa* subsp. *glauca* is quite similar, with slightly paler flowers.

CULTIVATION: *Grow in sandy, well-drained soil that is well-fertilized. Choose a site in full sun.*

☼ ◊ ❀❀❀
‡30–100cm (12–39in) ↔30cm (12in)

Oenothera macrocarpa

This vigorous perennial has similar flowers to *O. fruticosa* 'Fyrverkeri' (above), but its trailing habit makes this plant more suitable for border edging. The golden-yellow blooms appear from late spring to early autumn, amid the lance-shaped, mid-green leaves. Can also be used in a scree bed or rock garden. Also known as *O. missouriensis*.

CULTIVATION: *Grow in well-drained, poor to moderately fertile soil. Position in full sun, in a site that is not prone to excessive winter wet.*

☼ ◊ ♀ ❀❀❀
‡15cm (6in) ↔to 50cm (20in)

Olearia macrodonta

A summer-flowering, evergreen shrub or small tree that forms an upright, broadly columnar habit. Large clusters of fragrant, daisy-like white flowers with reddish-brown centres are borne amid the sharply toothed, glossy dark green leaves. A good hedging plant or windbreak for coastal areas with mild climates; in colder regions it is better grown in a warm, sheltered shrub border.

CULTIVATION: *Grow in well-drained, fertile soil, in full sun with shelter from cold, drying winds. Prune unwanted or frost-damaged growth in late spring.*

☼ ◊ ♀ ❀❀ ‡6m (20ft) ↔5m (15ft)

Omphalodes cappadocica

This clump-forming, shade-loving, evergreen perennial bears sprays of small, azure-blue, forget-me-not-like flowers with white centres. These appear in early spring above the pointed, mid-green leaves. Effective planted in groups through a woodland garden.

CULTIVATION: *Grow in moist, humus-rich, moderately fertile soil. Choose a site in partial shade.*

☀ ◊ ♀ ❀❀❀ ‡to 25cm (10in) ↔to 40cm (16in)

Omphalodes cappadocica '**Cherry Ingram**'

This clump-forming, evergreen perennial is very similar to the species (see previous page, bottom), but with larger, deep blue flowers which have white centres. These appear in early spring, above the pointed, finely hairy, mid-green leaves. Suits a woodland garden.

CULTIVATION: *Best in reliably moist, moderately fertile, humus-rich soil. Choose a site in partial shade.*

☀ ◊ ♀ ❀❀❀
‡to 25cm (10in) ↔ to 40cm (16in)

Onoclea sensibilis

The sensitive fern forms a beautifully textured mass of arching, finely divided, broadly lance-shaped, deciduous fronds. The foliage is pinkish-bronze in spring, maturing to pale green. There are no flowers. Thrives at the edge of water, or in a damp, shady border.

CULTIVATION: *Grow in moist, humus-rich, preferably acid soil. Site in partial shade, as fronds will scorch in sun.*

☀ ◊ ◊ ♀ ❀❀❀
‡60cm (24in) ↔ indefinite

Ophiopogon planiscapus 'Nigrescens'

This evergreen, spreading perennial forms clumps of grass-like, curving, almost black leaves. It looks very unusual and effective when planted in gravel-covered soil. Spikes of small, tubular, white to lilac flowers appear in summer, followed by round, blue-black fruits in autumn.

CULTIVATION: *Grow in moist but well-drained, fertile, slightly acid soil that is rich in humus, in full sun or partial shade. Top-dress with leaf mould in autumn, where practicable.*

☼ ☀ ◊ ◑ ♀ ❀ ❀ ❀
‡20cm (8in) ↔30cm (12in)

Origanum laevigatum

A woody-based, bushy perennial bearing open clusters of small, tubular, purplish-pink flowers from late spring to autumn. The oval, dark green leaves are aromatic, powerfully so when crushed. Suits a rock garden or scree bed; the flowers are attractive to bees. May not survive in cold climates.

CULTIVATION: *Grow in well-drained, poor to moderately fertile, preferably alkaline soil. Position in full sun. Trim back flowered stems in early spring.*

☼ ◊ ♀ ❀ ❀
‡to 60cm (24in) ↔45cm (18in)

Origanum laevigatum 'Herrenhausen'

A low-growing perennial that is more hardy than the species (see previous page, bottom), with purple-flushed young leaves and denser whorls of pink flowers in summer. The mature foliage is dark green and aromatic. Suits a Mediterranean-style planting or rock garden.

CULTIVATION: *Grow in very well-drained, poor to fairly fertile, preferably alkaline soil, in full sun. Trim back flowered stems in early spring.*

☼ ◊ ♀ ✳✳✳ ‡↔45cm (18in)

Origanum vulgare 'Aureum'

Golden marjoram is a colourful, bushy perennial with tiny, golden-yellow leaves that age to greenish-yellow. Short spikes of tiny, pretty pink flowers are occasionally produced in summer. The highly aromatic leaves can be used in cooking. Good for ground cover on a sunny bank, or in a herb garden, although it tends to spread.

CULTIVATION: *Grow in well-drained, poor to moderately fertile, alkaline soil. Position in full sun. Trim back after flowering to maintain a compact form.*

☼ ◊ ♀ ✳✳✳ ‡↔30cm (12in)

Osmanthus x *burkwoodii*

A dense and rounded, evergreen shrub, sometimes known as x *Osmarea burkwoodii*, carrying oval, slightly toothed, leathery, dark green leaves. Profuse clusters of small, very fragrant white flowers, with long throats and flat faces, are borne in spring. Ideal for a shrub border, or as a hedge.

CULTIVATION: *Grow in well-drained, fertile soil, in sun or partial shade with shelter from cold, drying winds. Prune to shape after flowering, giving hedges a trim in summer.*

☼ ☀ ◊ ♀ ❋ ❋ ❋　　　$\updownarrow\leftrightarrow$3m (10ft)

Osmanthus heterophyllus 'Goshiki'

This slow-growing, rounded shrub has much to offer. The multi-coloured, holly-like leaves are speckled in pink, coral, cream, and yellow. Inconspicuous, fragrant white flowers are produced in autumn followed by black berries. Grow as a low hedge, specimen, or container plant.

CULTIVATION: *Grow in fertile, well-drained soil in sun or partial shade, with shelter from cold, drying winds.*

☼ ☀ ◊ ♀ ❋ ❋　　　$\updownarrow\leftrightarrow$2m (8ft)

Osmunda regalis

The royal fern is a stately, clump-forming perennial with bright green, finely divided foliage. Distinctive, rust-coloured fronds are produced at the centre of each clump in summer. There are no flowers. Excellent in a damp border, or at the margins of a pond or stream. There is an attractive version of this fern with crested fronds, 'Cristata', growing slightly less tall, to 1.2m (4ft).

CULTIVATION: *Grow in very moist, fertile, humus-rich soil, in semi-shade. Tolerates full sun if conditions are reliably damp.*

☼ ◐ ◊ ◑ ♀ ✿✿✿ ‡2m (6ft) ↔4m (12ft)

Osteospermum jucundum

A neat, clump-forming, woody-based perennial bearing large, daisy-like, mauve-pink flowers which are flushed bronze-purple on the undersides. The blooms open in succession from late spring until autumn. Ideal for wall crevices or at the front of a border. Also known as *O. barberae*. 'Blackthorn Seedling', with dark purple flowers, is a striking cultivar.

CULTIVATION: *Best in light, well-drained, fairly fertile soil. Choose a site in full sun. Deadhead to prolong flowering.*

☼ ◊ ♀ ✿✿
‡10–50cm (4–20in) ↔50–100cm (20–39in)

Osteospermum Hybrids

These evergreen subshrubs are grown primarily for their daisy-like, bright and cheerful flowerheads, sometimes with pinched petals or centres in contrasting colours. They are borne singly or in open clusters over a long season, which begins in late spring and ends in autumn. Numerous cultivars have been named, varying from deep magenta through to white, pink, or yellow. Osteospermums are ideal for a sunny border – the flowers close in dull conditions. The half-hardy types are best grown as annuals in frost-prone areas, or in containers so that they can easily be moved into a greenhouse or conservatory during winter.

CULTIVATION: *Grow in light, moderately fertile, well-drained soil in a warm, sheltered site in full sun. In frost-prone areas, overwinter half-hardy types in frost-free conditions. Regular deadheading will encourage more flowers.*

☼ ◊

‡↔ 60cm (24in)

‡30cm (12in) ↔ 45cm (18in)

‡↔ 60cm (24in)

‡↔ 45cm (18in)

‡35cm (14in) ↔ 45cm (18in)

‡↔ 60cm (24in)

1 *Osteospermum* 'Buttermilk' ❀ **2** *O.* 'Hopleys' ♀ ❀❀ **3** *O.* 'Pink Whirls' ❀
4 *O.* 'Stardust' ❀❀ **5** *O.* 'Weetwood' ♀ ❀❀ **6** *O.* 'Whirligig' ❀

Oxalis adenophylla

A bulbous perennial forming clumps of pretty, grey-green leaves which are divided into many heart-shaped leaflets. In late spring, widely funnel-shaped, purple-pink flowers contrast beautifully with the foliage. Native to the Andes, it suits a well-drained rock garden, trough, or raised bed. Pink-flowered *O. enneaphylla* and the blue-flowered *O.* 'Ione Hecker' are very similar, though they grow from rhizomatous roots rather than bulbs.

CULTIVATION: *Grow in any moderately fertile soil with good drainage. Choose a position in full sun.*

☼ ◊ ♀ ❀❀❀
‡10cm (4in) ↔ to 15cm (6in)

Pachysandra terminalis

This freely spreading, bushy, evergreen foliage perennial makes a very useful ground cover plant for a shrub border or woodland garden. The oval, glossy dark green leaves are clustered at the tips of the stems. Spikes of small white flowers are produced in early summer. There is a less vigorous version with white-edged leaves, 'Variegata'.

CULTIVATION: *Grow in any but very dry soil that is rich in humus. Choose a position in partial or full shade.*

☼ ☼ ◊ ❀❀❀
‡20cm (8in) ↔ indefinite

Paeonia delavayi

An upright, sparsely branched, deciduous shrub bearing nodding, bowl-shaped, rich dark red flowers in early summer. The dark green leaves are deeply cut into pointed lobes and have blue-green undersides. A tall bush peony, good in a shrub border.

CULTIVATION: *Grow in deep, moist but well-drained, fertile soil that is rich in humus. Position in full sun or partial shade with shelter from cold, drying winds. Occasionally cut an old, leggy stem back to ground level in autumn, but avoid regular or hard pruning.*

☼ ☀ ◊ ◊ ❀ ❀ ❀   ↕2m (6ft) ↔1.2m (4ft)

Paeonia lactiflora '*Bowl of Beauty*'

A herbaceous, clump-forming perennial bearing very large, bowl-shaped flowers in early summer. These have carmine-pink, red-tinted petals arranged around a dense cluster of creamy-white stamens. The leaves are mid-green and divided into many leaflets. Ideal for a mixed or herbaceous border.

CULTIVATION: *Grow in deep, moist but well-drained, fertile, humus-rich soil, in full sun or partial shade. Provide support. Resents being disturbed.*

☼ ☀ ◊ ◊ ♀ ❀ ❀ ❀
↕↔80–100cm (32–39in)

Paeonia lactiflora 'Sarah Bernhardt'

With its large, fragrant, fully double flowers, this peony is indispensable in a herbaceous border. The flowers are light pink with ruffled, silver-margined inner petals and appear in early summer above the clumps of mid-green, deeply divided leaves.

CULTIVATION: *Grow in moist but well-drained, deep, humus-rich, fertile soil. Position in full sun or partial shade and provide support for the flowering stems. Does not respond well to root disturbance.*

☼ ☀ ◊ ◊ ♀ ❀ ❀ ❀　　　↕↔to 1m (3ft)

Paeonia ludlowii

This vigorous, deciduous shrub has an upright and open form. In late spring, large, bright yellow, nodding flowers open amid the bright green foliage. The leaves are deeply divided into several pointed leaflets. Good in a shrub border or planted on its own.

CULTIVATION: *Best in deep, well-drained but moist, fertile soil that is rich in humus. Site in sun or semi-shade with shelter from cold, drying winds. Avoid hard pruning, but occasionally cut old, leggy stems to ground level in autumn.*

☼ ☀ ◊ ◊ ❀ ❀ ❀　　　↕↔1.5m (5ft)

Paeonia officinalis 'Rubra Plena'

This long-lived, clump-forming, herbaceous peony makes a fine early-summer-flowering addition to any border display. The large, fully double, vivid crimson flowers, with ruffled, satiny petals, contrast well with the deep green, divided leaves. 'Rosea Plena', very similar, is also recommended.

CULTIVATION: *Grow in well-drained but moist, deep, fertile, humus-rich soil. Choose a site in full sun or partial shade. Support the flowering stems.*

☼ ☀ ◊ ◑ ♀ ✿ ✿ ✿
‡↔70–75cm (28–30in)

Panicum virgatum 'Shenandoah'

This perennial switch grass forms vase-shaped clumps of upright, strap-like, green leaves, which turn red at the tips in late summer when it also produces delicate, airy pink flower panicles. In autumn, the foliage turns a deep burgundy. Plant in small groups or as a specimen in a mixed border.

CULTIVATION: *Grow in moderately fertile, well-drained soil in full sun. Cut back in late winter and divide as needed. May self-seed.*

☼ ◊ ◑ ✿ ✿ ✿
‡1m (3ft) ↔75cm (30in)

Papaver orientale 'Beauty of Livermere'

This tall oriental poppy with crimson-scarlet flowers is an upright, clump-forming perennial. The flowers open during late spring to midsummer, and develop into large seed pods. Each petal has a bold, black mark at the base. The mid-green, divided leaves are borne on upright, bristly stems. Looks spectacular in a border.

CULTIVATION: *Grow in well-drained, poor to moderately fertile soil. Choose a position in full sun.*

☼ ◊ ❀❀❀
‡1–1.2m (3–4ft) ↔90cm (36in)

Papaver orientale 'Black and White'

This oriental poppy, white-flowered with crimson-black markings at the petal bases, is a clump-forming perennial. The flowers are borne above the mid-green foliage at the tips of white-bristly, upright stems during early summer; they are followed by distinctive seed pods. Makes a good border perennial.

CULTIVATION: *Best in deep, moderately fertile soil with good drainage. Choose a position in full sun.*

☼ ◊ ♀ ❀❀❀
‡45–90cm (18–36in) ↔60–90cm (24–36in)

Papaver orientale 'Cedric Morris'

This oriental poppy has very large, soft pink flowers with black-marked bases that are set off well against the grey-hairy foliage. It forms upright clumps which are well-suited to a herbaceous or mixed border. Distinctive seed pods develop after the flowers have faded.

CULTIVATION: *Grow in deep, moderately fertile soil with good drainage. Choose a position in full sun.*

☼ ◊ ♀ ✿✿✿
‡45–90cm (18–36in) ↔60–90cm (24–36in)

Papaver rhoeas Shirley Mixed

These field poppies are summer-flowering annuals with single, semi-double or double, bowl-shaped flowers in shades of yellow, orange, pink, and red. These appear on upright stems above the finely divided, bright green leaves. Can be naturalized in a wildflower meadow.

CULTIVATION: *Best in well-drained, poor to moderately fertile soil, in full sun. Divide and replant clumps in spring.*

☼ ◊ ✿✿✿
‡to 1m (3ft) ↔to 30cm (12in)

Parahebe perfoliata

Digger's speedwell is a spreading, evergreen perennial bearing short spikes of blue, saucer-shaped flowers in late summer. The blue- or grey-green, overlapping leaves are oval and slightly leathery. Suitable for gaps in old walls or a rock garden.

CULTIVATION: *Grow in poor to fairly fertile soil with good drainage, in full sun. In frost-prone climates, provide shelter from cold, drying winds.*

☼ ◊ ❀❀
‡60–75cm (24–30in) ↔45cm (18in)

Parrotia persica

Persian ironwood is a slow-growing, deciduous, short-trunked tree with flaking grey and fawn bark. Rich green leaves turn spectacular shades of yellow, orange, and red-purple in autumn. Small red flowers are borne on bare wood in early spring. An excellent specimen tree. The cultivar 'Vanessa' has a more upright habit.

CULTIVATION: *Grow in deep, fertile, moist but well-drained soil in full sun or partial shade. Is lime-tolerant, but colours best in acid soil. Drought tolerant once established.*

☼ ☀ ◊ ❀❀❀
‡8m (25ft) ↔10m (30ft)

Parthenocissus henryana

Chinese Virginia creeper is a woody, twining, deciduous climber with colourful foliage in autumn. The insignificant summer flowers are usually followed by blue-black berries. The conspicuously white-veined leaves, made up of three to five leaflets, turn bright red late in the season. Train over a wall or a strongly built fence.

CULTIVATION: *Grow in well-drained but moist, fertile soil. Tolerates sun, but leaf colour is best in deep or partial shade. Young plants may need some support. Prune back unwanted growth in autumn.*

☼ ☀ ◊ ◑ ♀ ❋ ❋ ❋   ‡10m (30ft)

Parthenocissus tricuspidata

Even more vigorous than Virginia creeper (*P. quinquefolia*), Boston ivy is a woody, deciduous climber with foliage that turns a beautiful colour in autumn. The variably lobed, bright green leaves flush a brilliant red, fading to purple before they fall. Creates a strong textural effect on cold, featureless walls.

CULTIVATION: *Best in moist but well-drained, fertile soil that is rich in humus. Position in partial or full shade. Young plants may need some support before they are established. Remove any unwanted growth in autumn.*

☼ ☀ ◊ ◑ ❋ ❋ ❋   ‡20m (70ft)

Passiflora caerulea

The blue passion flower is a fast-growing, evergreen climber valued for its large, exotic flowers, crowned with prominent blue- and purple-banded filaments. These are borne from summer to autumn amid the dark green, divided leaves. Of borderline hardiness; in cold climates grow in the protection of a warm wall.

CULTIVATION: *Best in moist but well-drained, moderately fertile soil, in a sunny, sheltered site. Remove crowded growth in spring, cutting back flowered shoots at the end of the season.*

☼ ◊ ◑ ♀ ❀❀ ‡10m (30ft) or more

Passiflora caerulea 'Constance Elliott'

A fast-growing, evergreen climber, resembling the species (above), but with white flowers, borne from summer to autumn. The leaves are dark green and deeply divided into three to nine lobes. Good for a pergola in warm climates, but needs the shelter of a warm wall in areas with cold winters.

CULTIVATION: *Grow in moist but well-drained, moderately fertile soil. Choose a sheltered site, in full sun. Remove weak growth in spring, cutting back flowered shoots at the end of the season.*

☼ ◊ ◑ ♀ ❀❀ ‡10m (30ft) or more

Passiflora racemosa

The red passion flower is a vigorous climber noted for its large, bright red flowers borne in hanging clusters in summer and autumn. They are followed by deep green fruits. The leathery leaves are glossy and mid-green. It cannot be grown outdoors in cool climates, but suits a green-house that is heated over winter, or makes a striking conservatory plant.

CULTIVATION: *Grow in a greenhouse border or large tub, in loam-based potting compost. Provide full light, with shade from hot sun. Water sparingly in winter. Prune in early spring. Minimum temperature 16˚C (61˚F)*

☼ ☀ ◊ ◊ ♀ ❀　　　　　　‡5m (15ft)

Patrinia scabiosifolia

An upright, herbaceous perennial, golden lace forms tidy mounds of attractive foliage. Its airy clusters of tiny yellow flowers bloom on tall stems above the leaves from late summer into autumn. Grow in a mixed border where it combines well with other late-season perennials.

CULTIVATION: *Although it prefers well-drained, moist, humus-rich soil and full sun, it tolerates dry soil, as well as heat and humidity. Flower stems may need some support.*

☼ ◊ ❀❀❀❀
‡to 1m (3ft) ↔ to 60cm (24in)

Scented-Leaved Pelargoniums

Pelargoniums, often incorrectly called geraniums (see pages 226–229), are tender, evergreen perennials. Many are grown specifically for their pretty, scented foliage. Their flowers, generally in pinks and mauves, are less showy and more delicate in form than those of the zonal and regal pelargoniums, bred more specifically for floral display. Leaf fragrance varies from sweet or spicy through to the citrus of *P. crispum* 'Variegatum' or the peppermint-like *P. tomentosum*.

They will perfume a greenhouse or conservatory, or use as summer edging along a path or in pots on a patio, where they will be brushed against and release their scent.

CULTIVATION: *Grow in well-drained, fertile, neutral to alkaline soil or compost, in sun. Deadhead regularly. In cold areas, over-winter in frost-free conditions, cutting back top growth by one-third. Repot as growth resumes. Minimum temperature 2˚C (36˚F).*

☼ ◊ ⌂

1 ‡50cm (20in) ↔25-30cm (10-12in)

2 ‡ ↔60cm (24in)

3 ‡60-90cm (24-36in) ↔60cm (24in)

4 ‡to 1.2m (4ft) ↔30cm (12in)

5 ‡45cm (18in) ↔20-25cm (8-10in)

6 ‡35-45cm (30-36in) ↔15m (6in)

1 *Pelargonium* 'Attar of Roses' ♀ **2** *P.* 'Bolero' ♀ **3** *P.* 'Charity' ♀ **4** *P.* 'Citriodorum' ♀
5 *P.* 'Copthorne' ♀ **6** *P. crispum* 'Variegatum' ♀

‡↔ 60cm (24in)

‡↔ 60cm (24in)

‡30–40cm (12–16in) ↔ 20cm (8in)

‡↔ 60cm (24in)

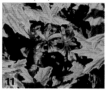

‡30–35cm (12–14in) ↔ 15cm (6in)

‡90cm (36in) ↔ 30cm (12in)

‡30–35cm (12–14in) ↔ 25cm (10in)

‡60cm (24in) ↔ 30cm (12in)

‡30–40cm (12–16in) ↔ 30cm (12in)

‡45–50cm (18–20in) ↔ 25cm (10in)

‡75–90cm (30–36in) ↔ to 75cm (30in)

7 *P.* 'Gemstone' ♀ **8** *P.* 'Grace Thomas' ♀ **9** *P.* 'Lady Plymouth' ♀
10 *P.* 'Lara Starshine' ♀ **11** *P.* 'Mabel Grey' ♀ **12** *P.* 'Nervous Mabel' ♀ **13** *P.* 'Orsett' ♀
14 *P.* 'Peter's Luck' ♀ **15** *P.* 'Royal Oak' ♀ **16** *P.* 'Sweet Mimosa' ♀ **17** *P. tomentosum* ♀

Flowering Pelargoniums

Pelargoniums grown for their bold flowers are tender, evergreen perennials: most cultivars are bushy, but there are also trailing types for window boxes and hanging baskets. Flower forms vary from the tightly frilled 'Apple Blossom Rosebud' to the delicate, narrow-petalled 'Bird Dancer'; colours range from shades of orange through pink and red to rich purple, and some have coloured foliage.

A popular choice as bedding plants, flowering from spring into summer, although many will flower throughout the year if kept above 7°C (45°F), making them excellent for a conservatory and as houseplants.

CULTIVATION: *Grow in well-drained, fertile, neutral to alkaline soil or compost, in sun or partial shade. Deadhead regularly to prolong flowering. In cold areas, overwinter in frost-free conditions, cutting back by one-third. Repot in late winter as new growth resumes. Minimum temperature 2°C (36°F).*

☼ ☀ ◊ ❀

1
‡25–30cm (10–12in) ↔ to 25cm (10in)

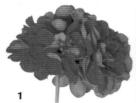

2
‡to 30cm (12in) ↔ 25cm (10in)

3
‡30–40cm (12–16in) ↔ 20–25cm (8–10in)

---

1 *Pelargonium* 'Alice Crousse' ♥ **2** *P.* AMETHYST 'Fisdel' ♥ **3** *P.* 'Apple Blossom Rosebud' ♥

‡15–20cm (6–8in) ↔15cm (6in)

‡25–30cm (10–12in) ↔15cm (6in)

‡45–60cm (18–24in) ↔to 25cm (10in)

‡10–12cm (4–5in) ↔7–10cm (3–4in)

‡40–45cm (16–18in) ↔25cm (10in)

4 *P.* 'Bird Dancer' ♀ **5** *P.* 'Dolly Varden' ♀ **6** *P.* 'Flower of Spring'
7 *P.* 'Francis Parrett' ♀ **8** *P.* 'Happy Thought' ♀

‡40–45cm 16–18in) ↔ to 30cm (12in)

‡25–30cm (10–12in) ↔12cm (5in)

‡to 30cm (12in) ↔30cm (12in)

‡40–45cm (16–18in) ↔20cm (8in)

‡to 60cm (24in) ↔ to 25cm (10in)

‡to 60cm (24in) ↔25cm (10in)

‡20cm (8in) ↔18cm (7in)

9 _P._ 'Irene' **10** _P._ 'Mr Henry Cox' ♀ **11** _P._ Multibloom Series **12** _P._ 'Paton's Unique' ♀
13 _P._ 'The Boar' ♀ **14** _P._ 'Voodoo' ♀ **15** _P._ Video Series

Regal Pelargoniums

These bushy, half-hardy perennials and shrubs have gloriously showy, blowsy flowers, earning them the sobriquet "the queen of pelargoniums". Their main flowering period is from spring to early summer, with blooms carried in clusters in reds, pinks, and purples, orange, white, and reddish-black; colours are combined in some cultivars. Often grown as container plants in frost-prone areas both for indoor and outdoor display, although they may also be used in summer bedding schemes.

CULTIVATION: *Best in good-quality, moist but well-drained potting compost in full light with shade from strong sun. Water moderately during growth, feeding every two weeks with a liquid fertilizer. Water sparingly if overwintering in a frost-free greenhouse. Cut back by one-third and repot in late winter. Outdoors, grow in fertile, neutral to alkaline, well-drained soil in full sun. Deadhead regularly. Minimum temperature 7°C (45°F).*

☼ ◊ ◐ ◊ ❀

‡45cm (18in) ↔ to 25cm (10in)

‡30–40cm (12–16in) ↔ to 20cm (8in)

‡45cm (18in) ↔ to 25cm (10in)

‡45cm (18in) ↔ to 25cm (10in)

‡30–40cm (12–16in) ↔ to 20cm (8in)

MORE CHOICES

'Chew Magna' Pink petals with a red blaze.
'Leslie Judd' Salmon-pink and wine-red.
'Lord Bute' Dark reddish-black flowers.
'Spellbound' Pink with wine-red petal markings.

1 *P.* 'Ann Hoystead' ♥ **2** *P.* 'Bredon' ♥ **3** *P.* 'Carisbrooke' ♥
4 *P.* 'Lavender Grand Slam' ♥ **5** *P.* 'Sefton' ♥

Penstemons

Penstemons are elegant, semi-evergreen perennials valued for their spires of tubular, foxglove-like flowers, in white and shades of pink, red, and purple, held above lance-shaped leaves. Smaller penstemons, such as *P. newberryi*, are at home in a rock garden or as edging plants, while larger cultivars make reliable border perennials which flower throughout summer and into autumn. 'Andenken an Friedrich Hahn' and 'Schoenholzeri' are among the hardiest cultivars, although most types benefit from a dry winter mulch in areas that are prone to heavy frosts. Penstemons tend to be short-lived, and are best replaced after a few seasons.

CULTIVATION: *Grow border plants in well-drained, humus-rich soil, and dwarf cultivars in sharply drained, gritty, poor to moderately fertile soil. Choose a site in full sun or partial shade. Deadhead regularly to prolong the flowering season.*

☼ ☀ ◊

‡1.2m (4ft) ↔ 45cm (18in)

‡↔ 45–60cm (18–24in)

‡75cm (30in) ↔ 60cm (24in)

1 *Penstemon* 'Alice Hindley' ♀ ❀❀ **2** *P.* 'Apple Blossom' ♀ ❀❀
3 *P.* 'Andenken an Friedrich Hahn' ♀ ❀❀❀ (syn. *P.* 'Garnet')

‡90cm (36in) ↔ 75cm (30in)

‡45–60cm (18–24in) ↔ 30cm (12in)

‡25cm (10in) ↔ 30cm (12in)

‡90cm (36in) ↔ 60cm (24in)

‡60cm (24in) ↔ 45cm (18in)

4 *P.* 'Chester Scarlet' ♀ ❀❀❀ **5** *P.* 'Evelyn' ♀ ❀❀ **6** *P. newberryi* ♀ ❀❀❀
7 *P.* 'Schoenholzeri' ♀ ❀❀❀ (syn. *P.* 'Firebird', *P.* 'Ruby') **8** *P.* 'White Bedder' ♀ ❀❀

Perilla frutescens var. *crispa*

An upright, bushy annual grown for its frilly, pointed, mid-green leaves which are flecked with deep purple. Spikes of tiny white flowers appear in summer. The dark foliage makes a good contrasting background to the bright flowers of most summer bedding plants.

CULTIVATION: *Grow in moist but well-drained, fertile soil. Position in sun or partial shade. Plant in spring after the danger of frost has passed.*

☼ ◐ ◇ ◆ ❄❄
↕to 1m (3ft) ↔to 30cm (12in)

Perovskia 'Blue Spire'

This upright, deciduous subshrub, grown for its foliage and flowers, suits a mixed or herbaceous border. Branching, airy spikes of tubular, violet-blue flowers are borne in profusion during late summer and early autumn, above the silvery-grey, divided leaves. Tolerates coastal conditions.

CULTIVATION: *Best in poor to moderately fertile soil that is well-drained. Tolerates chalky soil. For vigorous, bushy growth, prune back hard each spring to a low framework. Position in full sun.*

☼ ◇ 🝖 ❄❄❄ ↕1.2m (4ft) ↔1m (3ft)

Persicaria affinis '**Superba**'

A vigorous, evergreen perennial, formerly in the genus *Polygonum*, that forms mats of lance-shaped, deep green leaves which turn rich brown in autumn. Dense spikes of long-lasting, pale pink flowers, ageing to dark pink, are borne from midsummer to mid-autumn. Plant in groups at the front of a border or use as groundcover. 'Darjeeling Red' is similar and equally good.

CULTIVATION: *Grow in any moist soil, in full sun or partial shade. Dig out invasive roots in spring or autumn.*

☼ ☀ ◊ ♀ ❀❀
↕to 25cm (10in) ↔60cm (24in)

Persicaria amplexicaulis '**Firetail**'

Red bistort is a valuable border perennial, producing late summer flowers that persist well. These begin to appear from midsummer and last until the first frosts in autumn. Individually, the bright red flowers are quite small, but they are carried in numbers on long spikes on top of tall stems above the pointed foliage. 'Alba' has white flowers.

CULTIVATION: *Best in moist soil in full sun or partial shade.*

☼ ☀ ◊ ❀❀❀
↕↔1.2m (4ft)

Persicaria bistorta 'Superba'

A fast-growing, semi-evergreen perennial, formerly in the genus *Polygonum*, that makes good groundcover. Dense, cylindrical, soft pink flowerheads are produced over long periods from early summer to mid-autumn, above the clumps of mid-green foliage.

CULTIVATION: *Best in any well-drained, reliably moist soil, in full sun or partial shade. Tolerates dry soil.*

☼ ☀ ◊ ◊ ♀ ❀❀❀
‡75cm (30in) ↔ 90cm (36in)

Persicaria vacciniifolia

This creeping, evergreen perennial, formerly in the genus *Polygonum*, bears glossy mid-green leaves which flush red in autumn. In late summer and autumn, spikes of deep pink flowers appear on branching, red-tinted stems. Suits a rock garden by water, or at the front of a border. Good for ground cover.

CULTIVATION: *Grow in any moist soil, in full sun or semi-shade. Control spread by digging up invasive roots in spring or autumn.*

☼ ☀ ◊ ♀ ❀❀❀
‡20cm (8in) ↔ 50cm (20in) or more

Petrorhagia saxifraga

The creeping, wiry stems of the tunic flower root to form mats of grass-like, rich green leaves that are studded throughout summer with many tiny, white or pink flowers. The plant thrives in poor soils and makes good ground cover or neat edging to sunny borders. It also suits rock gardens.

CULTIVATION: *Grow in any poor to moderately fertile, well-drained soil in sun.*

☼ ◊ ♀ ❁ ❁ ❁
‡10cm (4in) ↔ to 20cm (8in)

Phalaris arundinacea 'Picta'

Gardeners' garters is an evergreen, clump-forming perennial grass with narrow, white-striped leaves. Tall plumes of pale green flowers, fading to buff as they mature, are borne on upright stems during early to midsummer. Good for ground cover, but can be invasive.

CULTIVATION: *Grow in any soil, in full sun or partial shade. Cut down all but the new young shoots in early summer to encourage fresh growth. To control spread, lift and divide regularly.*

☼ ☼ ◊ ❁ ❁ ❁
‡to 1m (3ft) ↔ indefinite

Phlomis fruticosa

Jerusalem sage is a mound-forming, spreading, evergreen shrub, carrying sage-like and aromatic, grey-green leaves with woolly undersides. Short spikes of hooded, dark golden-yellow flowers appear from early to midsummer. Effective when massed in a border.

CULTIVATION: *Best in light, well-drained, poor to fairly fertile soil, in sun. Prune out any weak or leggy stems in spring.*

☼ ◊ ♡ ❁❁❁　　　‡1m (3ft) ↔1.5m (5ft)

Phlomis russeliana

An upright, evergreen border perennial, sometimes known as either *P. samia* or *P. viscosa*, bearing pointed, hairy, mid-green leaves. Spherical clusters of hooded, pale yellow flowers appear along the stems from late spring to autumn. The flowers last well into winter.

CULTIVATION: *Grow in any well-drained, moderately fertile soil, in full sun or light shade. May self-seed.*

☼ ☼ ◊ ♡ ❁❁❁
‡to 1m (3ft) ↔75cm (30in)

Phlox divaricata 'Chattahoochee'

A short-lived, semi-evergreen border perennial bearing many flat-faced, long-throated, lavender-blue flowers with red eyes. These are produced over a long period from summer to early autumn, amid the lance-shaped leaves which are carried on purple-tinted stems.

CULTIVATION: *Grow in moist but well-drained, humus-rich, fertile soil. Chose a site in partial shade.*

☀ ◊ ◊ ♀ ❀ ❀ ❀
‡15cm (6in) ↔ 30cm (12in)

Phlox douglasii 'Boothman's Variety'

A low and creeping, evergreen perennial that forms mounds of narrow, dark green leaves. Dark-eyed, violet-pink flowers with long throats and flat faces appear in late spring or early summer. Good in a rock garden or wall, or as edging in raised beds.

CULTIVATION: *Grow in well-drained, fertile soil, in full sun. In areas with low rainfall, position in dappled shade.*

☀ ☀ ◊ ♀ ❀ ❀ ❀
‡20cm (8in) ↔ 30cm (12in)

Phlox drummondii cultivars

The Greek name *Phlox* means "a flame", referring to the very bright colours of the blooms. *P. drummondii*, the annual phlox, is the parent of many named cultivars used primarily for bold summer bedding. They are upright to spreading, bushy annuals, which bear bunched clusters of hairy purple, pink, red, lavender-blue, or white flowers in late spring. The flowers are often paler at the centres with contrasting marks at the bases of the petal lobes. The stem-clasping leaves are mid-green. They are useful in rock gardens, herbaceous borders, flower beds, or in containers; if raised in a greenhouse, they may come into bloom earlier.

CULTIVATION: *Grow in reliably moist but well-drained, sandy soil in full sun. Enrich the soil with plenty of organic matter before planting and feed once a week with a diluted liquid fertilizer. Slugs and snails are attracted by young plants.*

☼ ◐ ◊ ❀ ❀ ❀

1 ‡10–45cm (4–18in) ↔25cm (10in) 2 ‡10–45cm (4–18in) ↔25cm (10in) 3 ‡10–45cm (4–18in) ↔25cm (10in)

4 ‡10–45cm (4–18in) ↔25cm (10in) 5 ‡10–45cm (4–18in) ↔25cm (10in) 6 ‡10–45cm (4–18in) ↔25cm (10in)

1 *Phlox drummondii* 'Beauty Mauve' **2** 'Beauty Pink' **3** 'Brilliancy Mixed' **4** 'Brilliant'
5 'Buttons Salmon with Eye' **6** 'Phlox of Sheep'

Phlox 'Kelly's Eye'

A vigorous, evergreen, mound-forming perennial producing a colourful display of long-throated, flat-faced, pale pink flowers with red-purple centres in late spring and early summer. The leaves are dark green and narrow. Suitable for a rock garden or wall crevices.

CULTIVATION: *Grow in fertile soil that has good drainage, in full sun. In low-rainfall areas, site in partial shade.*

☼ ☀ ◊ ♀ ✿ ✿ ✿
‡15cm (6in) ↔20cm (8in)

Phlox maculata 'Alpha'

A cultivar of meadow phlox with tall, fat clusters of lilac-pink flowers, which appear above the foliage in the first half of summer. It is an upright herbaceous perennial with wiry stems. Ideal in a moist border, and a good source of fragrant flowers for cutting.

CULTIVATION: *Grow in fertile, moist soil in full sun or partial shade. Remove spent flowers to prolong flowering, and cut back to the ground in autumn. May need staking.*

☼ ☀ ◊ ♀ ✿ ✿ ✿
‡to 90cm (36in) ↔45cm (18in)

Phlox paniculata cultivars

Cultivars of the perennial phlox, *P. paniculata*, are herbaceous plants bearing dome-shaped or conical clusters of flowers above lance-shaped, toothed, mid-green leaves. Appearing throughout summer and into autumn, the flat-faced, long-throated, delicately fragranced flowers are white, pink, red, purple, or blue, often with contrasting centres; they are long-lasting when cut. Larger flowers can be encouraged by removing the weakest shoots in spring when the plant is still quite young.

All *Phlox paniculata* cultivars are well-suited to a herbaceous border; most will need staking, but others, such as 'Fujiyama', have particularly sturdy stems.

CULTIVATION: *Grow in any reliably moist, fertile soil. Choose a position in full sun or partial shade. Feed with a balanced liquid fertilizer in spring and deadhead regularly to prolong flowering. After flowering, cut back all foliage to ground level.*

☼ ☀ ◊ ❁ ❁ ❁

‡1.2m (4ft) ↔ 60cm (24in)

‡90cm (36in) ↔ 45cm (18in)

‡1.2m (4ft) ↔ 60cm (24in)

‡1.1m (3½ft) ↔ to 1m (3ft)

‡1m (3ft) ↔ 45cm (18in)

‡1.2m (4ft) ↔ 60-100cm (24-39in)

1 *Phlox paniculata* 'Brigadier' ♀ **2** 'Eventide' ♀ **3** 'Mount Fuji' ♀ (syn *P.* 'Fujiyama')
4 'Le Mahdi' ♀ **5** 'Mother of Pearl' ♀ **6** 'Windsor' ♀

Phlox stolonifera 'Blue Ridge'

Evergreen and prostrate, creeping phlox creates an arresting swath of saucer-shaped, lavender-blue flowers in early summer. It spreads quickly without becoming invasive. Plant at the edge of a border, in a rockery, or grow in a woodland garden.

CULTIVATION: *Grow in humus-rich, fertile, moist but well-drained soil in partial shade.*

☀ ◊ ♀ ✿ ✿ ✿
‡10–15cm (4–6in) ↔30cm (12in)

Phlox subulata 'Marjorie'

An evergreen, mat-forming perennial with unusual hairy, needle-like leaves and a profusion of vivid pink, early-summer flowers with darker pink lines radiating out from the yellow eyes. Good for a sunny rock garden, or planted along the edge of a wall.

CULTIVATION: *Grow in well-drained, fertile soil in full sun. Trim after flowering.*

☀ ◊ ✿ ✿ ✿
‡10cm (4in) ↔20cm (8in)

Phormium cookianum subsp. *hookeri* 'Cream Delight'

This mountain flax forms a clump of broad, arching leaves, to 1.5m (5ft) long, each with broad vertical bands of creamy yellow. Tall, upright clusters of tubular, yellow-green flowers appear in summer.
An unusual plant, often used as a focal point. Good in a large container, and tolerant of coastal exposure.

CULTIVATION: *Grow in fertile, moist but well-drained soil in full sun. In frost-prone areas, provide a deep, dry winter mulch. Divide crowded clumps in spring.*

☼ ◊ ◑ ♡ ✿✿ ↕to 2m (6ft) ↔3m (10ft)

Phormium cookianum subsp. *hookeri* 'Tricolor'

This striking perennial, very useful as a focal point in a border, forms arching clumps of broad, light green leaves, to 1.5m (5ft) long; they are boldly margined with creamy-yellow and red stripes. Tall spikes of tubular, yellow-green flowers are borne in summer. In cold climates, grow in a container during the summer, and overwinter under glass.

CULTIVATION: *Grow in moist but well-drained soil, in sun. Provide a deep, dry winter mulch in frost-prone areas.*

☼ ◊ ◑ ♡ ✿✿ ↕0.6–2m (2–6ft) ↔0.3–3m (1–10ft)

Phormium **'Sundowner'**

A clump-forming, evergreen perennial valued for its form and brilliant colouring. It has broad, bronze-green leaves, to 1.5m (5ft) long, with creamy rose-pink margins. In summer, tall, upright clusters of tubular, yellow-green flowers are borne, to be followed by decorative seedheads that persist through winter. Suits coastal gardens.

CULTIVATION: *Grow in any deep soil that does not dry out, but best in a warm, wet, sheltered site in full sun. Provide a deep mulch in winter, and divide overcrowded clumps in spring.*

☼ ◊ ◐ ♈ ❀ ❀   ↕↔to 2m (6ft)

Phormium tenax

The New Zealand flax is an evergreen perennial that forms clumps of very long, tough leaves, to 3m (10ft) in length in ideal conditions. They are dark green above, and blue-green beneath. Dusky red flowers appear in summer on very tall, stout and upright spikes. One of the largest phormiums – a striking plant with decorative seedheads in winter.

CULTIVATION: *Best in deep, reliably moist but well-drained soil in a warm, sheltered site in full sun. Mulch deeply for winter and divide overcrowded clumps in spring.*

☼ ◊ ◐ ❀ ❀ ❀   ↕4m (12ft) ↔2m (6ft)

Phormium tenax Purpureum Group

An evergreen perennial that forms clumps of long, stiff, sword-shaped, deep copper to purple-red leaves. Large spikes of dark red, tubular flowers on blue-purple stems appear in summer. Ideal for coastal gardens, but will suffer in areas with cold winters. Can be container-grown, or choose the similar but much smaller 'Nanum Purpureum'.

CULTIVATION: *Grow in deep, fertile, humus-rich soil that is reliably moist. Position in full sun with shelter from cold winds. In frost-prone areas, provide a deep, dry mulch in winter.*

☼ ◊ ♀ ✿✿ ↕2–2.5m (6–8ft) ↔1m (3ft)

Phormium 'Yellow Wave'

An evergreen perennial, forming clumps of broad, arching, yellow-green leaves with mid-green vertical stripes; they take on chartreuse tones in autumn. Spikes of tubular red flowers emerge from the centre of each leaf clump in summer. Especially good for seaside gardens, it will add a point of interest to any border, especially in winter when it remains bold and attractive.

CULTIVATION: *Best in fertile, moist but well-drained soil in full sun. Provide a deep, dry mulch for winter in frost-prone areas. Divide crowded clumps in spring.*

☼ ◊ ◊ ♀ ✿✿ ↕3m (10ft) ↔2m (6ft)

Photinia x fraseri 'Red Robin'

An upright, compact, evergreen shrub often grown as a formal or semi-formal hedge for its bright red young foliage, the effect of which is prolonged by clipping. The mature leaves are leathery, lance-shaped, and dark green. Clusters of small white flowers appear in mid-spring. 'Robusta' is another recommended cultivar.

CULTIVATION: *Grow in moist but well-drained, fertile soil, in full sun or semi-shade. Clip hedges 2 or 3 times a year to perpetuate the colourful foliage.*

☼ ☀ ◊ ◐ ♚ ❉ ❉ ❉   ↕↔5m (15ft)

Photinia villosa

A spreading, shrubby tree grown for its foliage, flowers, and fruits. Flat clusters of small white flowers appear in late spring, and develop into attractive red fruits. The dark green leaves are bronze when young, turning orange and red in autumn before they fall. Attractive all year round, and tolerant of permanently damp soil.

CULTIVATION: *Grow in fertile, moist but well-drained, neutral to acid soil in full sun or partial shade. Remove any congested or diseased growth in late winter.*

☼ ☀ ◊ ◐ ❉ ❉ ❉   ↕↔5m (15ft)

Phygelius aequalis 'Yellow Trumpet'

An upright, evergreen shrub forming loose spikes of hanging, tubular, pale cream-yellow flowers during summer. The leaves are oval and pale green. Suits a herbaceous or mixed border; in cold areas, grow against a warm, sunny wall.

CULTIVATION: *Best in moist but well-drained soil, in sun. In cold climates, shelter from wind and cut frost-damaged stems back to the base in spring. Dead-head to prolong flowering. Dig up unwanted shoots to contain spread.*

☼ ◊ ◊ ♀ ❄❄   ↕↔1m (3ft)

Phygelius capensis

The Cape figwort is an evergreen shrub valued for its summer display of upright spikes of orange flowers. The foliage is dark green, and the plant is often mistaken for a novel-coloured fuchsia. Grow near the back of a herbaceous border, against a warm, sunny wall in cold climates. Birds may find the flowers attractive.

CULTIVATION: *Grow in fertile, moist but well-drained soil in full sun with shelter from cold, drying winds. Remove spent flower clusters to encourage more blooms, and provide a dry winter mulch in frost-prone areas. Cut back to the ground in spring.*

☼ ◊ ◊ ❄❄   ↕1.2m (4ft) ↔1.5m (5ft)

Phygelius x *rectus* 'African Queen'

An upright, evergreen border shrub that bears long spikes of hanging, tubular, pale red flowers with orange to yellow mouths. These appear in summer above the oval, dark green leaves. Best against a warm wall in cold areas.

CULTIVATION: *Grow in moist but well-drained, fertile soil, in sun with shelter from cold, drying winds. Cut frost-damaged stems back to the base in spring. Deadhead to prolong flowering.*

☼ ◊ ◑ ♀ ❀ ❀   ‡1m (3ft) ↔1.2m (4ft)

Phygelius x *rectus* 'Devil's Tears'

An upright, evergreen shrub with dark green foliage. In summer, it carries spikes of hanging, red-pink flowers with yellow throats. A reasonably compact shrub with abundant flowers, ideal for a herbaceous or mixed border.

CULTIVATION: *Best in moist but well-drained, reasonably fertile soil in full sun. Remove spent flower clusters to encourage further blooming, and cut the plant back to ground level in spring, if damaged over winter; otherwise trim to shape.*

☼ ◊ ◑ ♀ ❀ ❀ ❀   ‡↔1.5m (5ft)

Phygelius x *rectus* 'Salmon Leap'

This upright shrub is very similar to 'Devil's Tears' (see previous page), but with orange flowers, which turn slightly back toward the stems. They appear in large sprays above the dark green foliage. Good in a mixed or herbaceous border.

CULTIVATION: *Grow in moist but well-drained, fertile soil in full sun. Remove spent flower clusters to encourage more blooms. Cut the plant back to ground level in spring if damaged by winter weather, otherwise trim to shape.*

☼ ◊ ◊ ♈ ✿✿✿
‡1.2m (4ft) ↔1.5m (5ft)

Phyllostachys nigra

Black bamboo is an arching, clump-forming, evergreen shrub. The gentle, lance-shaped, dark green leaves are produced on slender green canes which turn black in their second or third year. Use as a screen or as a large feature plant.

CULTIVATION: *Grow in moist but well-drained soil, in sun or partial shade with shelter from cold winds. Mulch over winter. Cut out damaged and overcrowded canes in spring or early summer. Confine spread by burying a barrier around the roots.*

☼ ☀ ◊ ◊ ♈ ✿✿✿
‡3–5m (10–15ft) ↔2–3m (6–10ft)

Phyllostachys nigra **f. henonis**

This clump-forming, evergreen bamboo is similar to the black bamboo in habit (see facing page, below), but has bright green canes that mature to yellow-green in the second or third year. The lance-shaped, dark green leaves are downy and rough when young.

CULTIVATION: *Grow in well-drained but moist soil, in sun or semi-shade. Shelter from cold winds and mulch over winter. Thin crowded clumps in late spring. Bury a barrier around the roots to confine spread.*

☼ ☀ ◊ ◊ ♀ ❀❀❀
‡3–5m (10–15ft) ↔2–3m (6–10ft)

Physocarpus opulifolius **'Diabolo'**

Ninebark is named for its peeling, multi-hued bark, but the real show-stealer is its deep purple, maple-like foliage, which turns bronze in autumn. Clusters of small white or pink flowers bloom in summer. This deciduous shrub looks best as a specimen or at the back of the border, creating a dark background for pale-flowered and silver-leaved plants.

CULTIVATION: *Prefers moist, well-drained, acid soil in full sun; it does not grow well on shallow, chalky soil. Remove some of the oldest stems each year to promote vigorous new growth.*

☼ ◊ ◊ ♀ ❀❀❀   ‡↔2.5m (8ft)

Physostegia virginiana 'Vivid'

An upright, densely clump-forming border perennial bearing spikes of bright purple-pink, hooded flowers from midsummer to early autumn above the narrow, mid-green leaves. The flowers are good for cutting and will remain in a new position if they are moved on the stalks; because of this, it is known as the obedient plant. Looks good mixed with the white cultivar, 'Summer Snow'.

CULTIVATION: *Grow in reliably moist, fertile soil that is rich in humus. Position in full sun or partial shade.*

☼ ☀ ◑ ◊ ♀ ❀❀❀
‡30–60cm (12–24in) ↔30cm (12in)

Picea glauca var. *albertiana* 'Conica'

A conical, slow-growing, evergreen conifer carrying dense, blue-green foliage. The short, slender needles are borne on buff-white to ash-grey stems. Oval cones appear during summer, green at first, then maturing to brown. Makes an excellent neat specimen tree for a small garden.

CULTIVATION: *Grow in deep, moist but well-drained, preferably neutral to acid soil, in full sun. Prune in winter if necessary, but keep to a minimum.*

☼ ◊ ◑ ♀ ❀❀❀
‡2–6m (6–20ft) ↔1–2.5m (3–8ft)

Picea mariana 'Nana'

This dwarf, low-growing form of black spruce is a mound-forming conifer with scaly, grey-brown bark. The evergreen, bluish-grey needles are short, soft, and slender. Useful in a rock garden or conifer bed, or as an edging plant.

CULTIVATION: *Best in deep, moist but well-drained, fertile, humus-rich soil, in partial shade. Completely remove any vigorous upright growth as soon as they are seen.*

☼ ◑ ♀ ✻✻✻   ↕↔50cm (20in)

Picea pungens 'Koster'

This conical evergreen conifer, with attractive horizontal branches, becomes more columnar with age, and the young growth is clothed in silvery-blue foliage. The long, sharp-pointed needles turn greener as they mature. Cylindrical green cones are borne during the summer, aging to pale brown. Good in large gardens as a prominent specimen tree. 'Hoopsii' is very similar indeed, with blue-white foliage.

CULTIVATION: *Grow in well-drained, fertile, neutral to acid soil, in full sun. Prune in late autumn or winter if necessary, but keep to a minimum.*

☼ ◊ ✻✻✻   ↕15m (50ft) ↔5m (15ft)

Pieris 'Forest Flame'

An upright, evergreen shrub valued for its slender, glossy, lance-shaped leaves that are bright red when young; they mature through pink and creamy-white to dark green. Upright clusters of white flowers enhance the effect in early to mid-spring. Ideal for a shrub border or a peaty, lime-free soil; it will not thrive in alkaline soil.

CULTIVATION: *Grow in moist but well-drained, fertile, humus-rich, acid soil, in full sun or partial shade. In frost-prone areas, shelter from cold, drying winds. Trim lightly after flowering.*

☼ ☀ ◊ ◑ ♀ ❀❀❀
‡4m (12ft) ↔2m (6ft)

Pieris formosa var. *forrestii* 'Wakehurst'

This upright, evergreen, acid-soil-loving shrub has brilliant red young foliage that matures to dark green. The large, slightly drooping clusters of small, fragrant white flowers from mid-spring are also attractive. Grow in a woodland garden or shrub border; may not survive in areas with cold winters. 'Jermyns' is very similar, with darker red young leaves.

CULTIVATION: *Best in well-drained but moist, fertile, humus-rich, acid soil. Site in full sun or partial shade with shelter from cold winds in frost-prone climates. Trim lightly after flowering.*

☼ ☀ ◊ ◑ ♀ ❀❀
‡5m (15ft) ↔4m (12ft)

Pieris japonica 'Blush'

A rounded, evergreen shrub that bears small, pink-flushed white flowers in late winter and early spring. These are carried in long, drooping clusters amid the glossy dark green foliage. A good early-flowering border shrub, although it will not thrive in alkaline soil.

CULTIVATION: *Grow in well-drained but moist, fertile, humus-rich, acid soil. Site in full sun or partial shade. Trim lightly after flowering, removing any dead, damaged or diseased shoots.*

☼ ◑ ◊ ◔ ♀ ❀ ❀ ❀
‡4m (12ft) ↔ 3m (10ft)

Pileostegia viburnoides

A slow-growing, woody, evergreen climber that is dusted with feathery clusters of tiny, creamy-white flowers in late summer and autumn. The glossy dark green, leathery leaves look very attractive against a large tree trunk or shady wall.

CULTIVATION: *Grow in well-drained, fertile soil, in full sun or shade. Shorten stems after flowering as the plant begins to outgrow the allotted space.*

☼ ◑ ◊ ❀ ❀ ❀   ‡6m (20ft)

Pinus mugo 'Mops'

This dwarf pine is an almost spherical conifer with scaly, grey bark and thick, upright branches. The shoots are covered with long, well-spaced, dark to bright green needles. The dark brown, oval cones take a few years to ripen. Effective in a large rock garden or, where space allows, planted in groups.

CULTIVATION: *Grow in any well-drained soil, in full sun. Very little pruning is required since growth is slow.*

☼ ◊ ♀ ❀❀❀
‡to 1m (3ft) ↔ to 2m (6ft)

Pinus sylvestris Aurea Group

Unlike most Scots pine trees, pines in this group grow slowly, making them suited to smaller gardens, particularly when winter turns the blue-green, evergreen pine needles golden yellow in response to the cooler weather. Pine trees tolerate a range of conditions, including coastal sites and drought. Allow the tree to develop its shape naturally without pruning.

CULTIVATION: *Grow in well-drained soil in sun.*

☼ ◊ ❀❀❀
‡to 10m (30ft) ↔ 5m (15ft)

Pittosporum tenuifolium

A columnar, evergreen shrub, much valued for its glossy green leaves with wavy edges. Fast growing at first, it then broadens out into a tree. Tiny, honey-scented, purple-black, bell-shaped flowers open from late spring. Makes a good hedge, but may not survive cold winters. There is a gold-leaved version, 'Warnham Gold'.

CULTIVATION: *Grow in well-drained but moist, fertile soil, in full sun or partial shade. In frost-prone areas, shelter from cold winds. Trim to shape in spring; avoid pruning after midsummer.*

☼ ☀ ◊ ◑ ❀❀
‡4–10m (12–30ft) ↔2–5m (6–15ft)

Pittosporum tenuifolium 'Tom Thumb'

A compact, rounded, evergreen foliage shrub that would suit a mixed border designed for year-round interest. The glossy bronze-purple leaves are elliptic and wavy-edged. Tiny, honey-scented, purple flowers are borne in late spring and early summer. Shelter against a warm wall in cold areas.

CULTIVATION: *Grow in well-drained but moist, fertile soil, in full sun for best colour. Shelter from cold winds in frost-prone areas. Trim to shape in spring; established plants need little pruning.*

☼ ☀ ◊ ◑ ♡ ❀❀
‡to 1m (3ft) ↔60cm (24in)

Platycodon grandiflorus

The balloon flower is a clump-forming perennial producing clusters of large purple to violet-blue flowers that open from balloon-shaped buds in late summer, above bluish-green, oval leaves. For a rock garden or herbaceous border, 'Apoyama', with deep-coloured flowers, and 'Mariesii' are recommended cultivars.

CULTIVATION: *Best in deep, well-drained, loamy, fertile soil that does not dry out, in full sun or partial shade. Flower stems may need staking. Established plants dislike root disturbance.*

☼ ☀ ◊ ♀ ❀ ❀ ❀
‡to 60cm (24in) ↔30cm (12in)

Pleioblastus variegatus

This upright, evergreen bamboo is much shorter than *P. viridistriatus* (see facing page, above), with cream-and green-striped foliage. The lance-shaped leaves, borne on hollow, pale green canes, are covered in fine white hairs. Suits a sunny border backed by shrubs, and where it has space to spread; it will swamp less vigorous neighbours unless confined.

CULTIVATION: *Grow in moist but well-drained, fertile, humus-rich soil, in sun with shelter from cold winds. Thin out clumps in late spring. Bury a barrier around the roots to confine spread.*

☼ ◊ ◊ ♀ ❀ ❀ ❀
‡75cm (30in) ↔1.2m (4ft)

Pleioblastus viridistriatus

This upright bamboo is an evergreen shrub grown for its brilliant green, yellow-striped foliage. The bristly-edged, lance-shaped leaves are carried on purple-green canes. Effective in an open glade in a woodland garden. Good in sun, backed by trees or tall shrubs.

CULTIVATION: *Grow in moist but well-drained, fertile, humus-rich soil, in full sun for best leaf colour. Provide shelter from cold, drying winds. Thin over-crowded clumps in late spring or early summer. Confine spread by burying a barrier around the roots.*

☼ ◐ ◊ ◊ ♀ ❀ ❀ ❀　　　　↔to 1.5m (5ft)

Plumbago auriculata

Cape leadwort is a scrambling, semi-evergreen, frost-tender shrub often trained as a climber. It bears dense trusses of long-throated, sky-blue flowers from summer to late autumn, amid the oval leaves. In cold climates, plants overwintered under glass can be moved outside in summer. Often sold as *P. capensis.*

CULTIVATION: *Grow in well-drained, fertile soil or compost, in full sun or light shade. Pinch out the tips of young plants to promote bushiness, and tie climbing stems to a support. Cut back to a permanent framework in early spring.*

☼ ☀ ◊ ♀ ☙
↕3–6m (10–20ft) ↔1–3m (3–10ft)

Polemonium 'Lambrook Mauve'

This clump-forming perennial, a garden variety of Jacob's ladder, forms rounded mounds of neat, divided, mid-green leaves. An abundance of funnel-shaped, sky-blue flowers cover the foliage from late spring to early summer. Good in any border, or in a wild garden.

CULTIVATION: *Grow in well-drained but moist, moderately fertile soil, in full sun or partial shade. Deadhead regularly.*

☼ ◑ ◇ ◆ ❄❄❄ ↕↔to 45cm (18in)

Polygonatum x *hybridum*

Solomon's seal is a perennial for a shady border, bearing hanging clusters of tubular, small white flowers with green mouths, along slightly arching stems. These appear in late spring among elliptic, bright green leaves. Round blue-black fruits develop after the flowers. For a woodland garden. *P.* x *odoratum* 'Flore Pleno' is a very similar but smaller plant.

CULTIVATION: *Grow in moist but well-drained, fertile soil that is rich in humus. Position in partial or full shade.*

◑ ● ◇ ◆ ♀ ❄❄❄

↕to 1.5m (5ft) ↔30cm (12in)

Polystichum acrostichoides

This attractive, glossy fern earned the common name Christmas fern because its fronds stay green through the winter holidays. In spring, silvery fiddleheads (croziers) emerge. Use this clump-former as ground cover or edging, or naturalize in a woodland setting. It is excellent for cutting.

CULTIVATION: *Plant in moist, humus-rich, well-drained soil in deep or partial shade. Remove dead fronds before new ones unfurl. Protect the crowns from winter wet.*

☀ ☀ ◊ ❋❋❋
‡60cm (24in) ↔ 45cm (18in)

Polystichum aculeatum

The prickly shield fern is an elegant, evergreen perennial producing a shuttlecock of finely divided, dark green fronds. There are no flowers. An excellent foliage plant for shady areas in a rock garden or well-drained border.

CULTIVATION: *Grow in fertile soil that has good drainage, in partial or deep shade. Choose a site sheltered from excessive winter wet. Remove the previous year's dead fronds before the new growth unfurls in spring.*

☀ ☀ ◊ ♀ ❋❋❋
‡60cm (24in) ↔ 1m (3ft)

Potentilla fruticosa

Cultivars of *P. fruticosa* are compact and rounded, deciduous shrubs that produce an abundance of flowers over a long period from late spring to mid-autumn. The leaves are dark green and composed of several oblong leaflets. Flowers are saucer-shaped and wild rose-like, sometimes borne singly, but often in clusters of three. Most cultivars are yellow-flowered, but blooms may also be white, as with 'Abbotswood', or flushed with pink, as in 'Daydawn'. These are undemanding shrubs that make invaluable additions to mixed or shrub borders; they can also be grown as attractive low hedges.

CULTIVATION: *Grow in well-drained, poor to moderately fertile soil. Best in full sun, but many tolerate partial shade. Trim lightly after flowering, cutting older wood to the base and removing weak, twiggy growth. Old shrubs sometimes respond well to renovation, but may be better replaced.*

☼ ◊ ❋❋❋

‡75cm (30in) ↔ 1.2m (4ft)

‡1m (3ft) ↔ 1.5m (5ft)

‡1m (3ft) ↔ 1.2m (4ft)

‡1m (3ft) ↔ 1.5m (5ft)

1 *Potentilla fruticosa* 'Abbotswood' ♥ **2** *P. fruticosa* 'Elizabeth' **3** *P. fruticosa* 'Daydawn'
4 *P. fruticosa* 'Primrose Beauty' ♥

Potentilla 'Gibson's Scarlet'

A dense, clump-forming herbaceous perennial grown for its very bright scarlet flowers, borne in succession throughout summer. The soft green leaves are divided into five leaflets. Good for a rock garden or for a bold summer colour in a mixed or herbaceous border. 'William Rollisson' is a similar plant, with more orange-red, semi-double flowers.

CULTIVATION: *Grow in well-drained, poor to moderately fertile soil. Choose a position in full sun.*

☼ ◊ ♀ ❀ ❀ ❀
‡to 45cm (18in) ↔ 60cm (24in)

Potentilla megalantha

A compact, clump-forming perennial bearing a profusion of upright, cup-shaped, rich yellow flowers during mid- to late summer. The slightly hairy, mid-green leaves are divided into three coarsely scalloped leaflets. Suits the front of a border.

CULTIVATION: *Grow in poor to fairly fertile soil that has good drainage. Choose a position in full sun.*

☼ ◊ ❀ ❀ ❀
‡15–30cm (6–12in) ↔ 15cm (6in)

Potentilla nepalensis 'Miss Wilmott'

A summer-flowering perennial that forms clumps of mid-green, divided leaves on wiry, red-tinged stems. The small pink flowers, with cherry-red centres, are borne in loose clusters. Good for the front of a border or in cottage-style plantings.

CULTIVATION: *Grow in any well-drained, poor to moderately fertile soil. Position in full sun or light dappled shade.*

☼ ☀ ◊ ✿✿✿
‡30–45cm (12–18in) ↔60cm (24in)

Primula denticulata

The drumstick primula is a robust, clump-forming perennial bearing spherical clusters of small purple flowers with yellow centres. They are carried on stout, upright stalks from spring to summer, above the basal rosettes of oblong to spoon-shaped, mid-green leaves. Thrives in damp, but not waterlogged, soil; ideal for a waterside planting.

CULTIVATION: *Best in moist, humus-rich, neutral to acid or peaty soil. Choose a site in partial shade, but tolerates full sun where soil is reliably damp.*

☼ ☀ ◊ ♀ ✿✿✿   ‡↔45cm (18in)

Primula elatior

The oxlip is a semi-evergreen perennial wildflower that can vary in appearance. Clusters of tubular yellow flowers emerge from the basal rosettes of scalloped, mid-green leaves on stiff, upright stems in spring and summer. Plant in groups to naturalize in a moist meadow.

CULTIVATION: *Grow in moderately fertile, deep, moist but well-drained soil. Site in partial shade, but full sun is tolerated as long as the soil remains moist at all times.*

☼ ☀ ◊ ◑ ♥ ❀❀❀
‡30cm (12in) ↔25cm (10in)

Primula flaccida

A rosette-forming, deciduous perennial for an open woodland garden or alpine house. In summer, tall flowering stems bear conical clusters of funnel-shaped and downward-pointing, floury, lavender-blue flowers, carried above the pale to mid-green leaves.

CULTIVATION: *Grow in deep or partial shade in peaty, gritty, moist but sharply drained, acid soil. Protect from excessive winter wet.*

☀ ☀ ◊ ◑ ❀❀❀
‡50cm (20in) ↔30cm (12in)

Primula florindae

The giant cowslip is a deciduous, summer-flowering perennial wildflower that grows naturally by pools and streams. It forms clumps of oval, toothed, mid-green leaves which are arranged in rosettes at the base of the plant. Drooping clusters of up to 40 funnel-shaped, sweetly-scented yellow flowers are borne well above the foliage on upright stems. Good in a bog garden or waterside setting.

CULTIVATION: *Grow in deep, reliably moist, humus-rich soil, in partial shade. Tolerates full sun if soil remains moist.*

☼ ☀ ◊ ♈ ✿✿✿
‡to 1.2m (4ft) ↔1m (3ft)

Primula frondosa

This deciduous border perennial, with its rosettes of spoon-shaped, mid-green leaves, carries yellow-eyed, pink to purple flowers in late spring or early summer. They are carried in loose clusters of up to 30 at the top of upright stems, and each has a pale yellow eye at its centre.

CULTIVATION: *Best in deep, moist, neutral to acid loam or peaty soil enriched with organic matter. Site ideally in partial shade, but it tolerates full sun if the soil is reliably moist.*

☼ ☀ ◊ ♈ ✿✿✿
‡15cm (6in) ↔25cm (10in)

Primula Gold-Laced Group

A group of semi-evergreen primroses grown for their showy spring flowers in the border or in containers. Clusters of golden-eyed, very dark mahogany-red or black flowers, with a thin gold margin around each petal, are carried atop upright flowering stems above the sometimes reddish foliage.

CULTIVATION: *Grow in moist, deep, neutral to acid loam or peaty soil that is rich in organic matter. Choose a site in partial shade, or in full sun if the soil is reliably moist.*

☼ ☀ ◊ ❁ ❁ ❁
↕25cm (10in) ↔30cm (12in)

Primula 'Guinevere'

A fast-growing, evergreen, clump-forming perennial, sometimes called 'Garryarde Guinevere', that bears clusters of pale purplish-pink flowers. They have yellow centres, flat faces, and long throats, and are carried above the deep bronze, oval leaves in spring. Suits damp, shady places.

CULTIVATION: *Grow in moist, neutral to acid soil that is well-drained, in partial shade. Tolerates full sun, but only if the soil remains damp at all times.*

☼ ☀ ◊ ♀ ❁ ❁ ❁
↕12cm (5in) ↔25cm (10in)

Candelabra Primroses (*Primula*)

Candelabra primroses are robust, herbaceous perennials, so-called because their flowers are borne in tiered clusters, which rise above the basal rosettes of broadly oval leaves in late spring or summer. Depending on the species, the foliage may be semi-evergreen, evergreen, or deciduous. The flowers have flat faces and long throats; as with most cultivated primroses, there is a wide choice of colours, from the brilliant red 'Inverewe' to the golden-yellow

P. prolifera. Some flowers change colour as they mature; those of *P. bulleyana* fade from crimson to orange. Candelabra primroses look most effective when grouped together in a bog garden or waterside setting.

CULTIVATION: *Grow in deep, moist, neutral to acid soil that is rich in humus. Site in partial shade, although full sun is tolerated if the soil remains moist at all times. Divide and replant clumps in early spring.*

☼ ◑ ◊ ❀ ❀ ❀

‡↔ 60cm (24in)

‡↔ 60cm (24in)

‡75cm (30in) ↔ 60cm (24in)

‡to 1m (3ft) ↔ 60cm (24in)

‡to 1m (3ft) ↔ 60cm (24in)

1 *P. bulleyana* ♀ **2** *P. prolifera* ♀ **3** *P.* 'Inverewe' ♀ **4** *P. pulverulenta* ♀
5 *P. pulverulenta* Bartley Hybrids ♀

Primula rosea

A deciduous perennial that bears rounded clusters of glowing pink, long-throated flowers on upright stalks in spring. Clumps of oval, toothed, mid-green leaves emerge after the flowers; these are tinted red-bronze when young. Good for a bog garden or waterside planting.

CULTIVATION: *Grow in deep, reliably moist, neutral to acid or peaty soil that is rich in humus. Prefers partial shade, but tolerates full sun if the soil is moist at all times.*

☀ ◑ ♀ ❄❄❄   ‡↔20cm (8in)

Primula veris

The cowslip is a semi-evergreen, spring-flowering perennial wildflower with a variable appearance. Stout flower stems carry dense clusters of small, funnel-shaped, sweetly scented, nodding yellow flowers, above the clumps of lance-shaped, crinkled leaves. Lovely naturalized in damp grassed areas.

CULTIVATION: *Best in deep, moist but well-drained, fertile, peaty soil that is rich in humus, in semi-shade or full sun, if soil remains reliably damp.*

☀ ☀ ◑ ♀ ❄❄❄   ‡↔to 25cm (10in)

Primula 'Wanda'

A very vigorous, semi-evergreen perennial that bears clusters of flat-faced, claret-red flowers with yellow centres over a long period in spring. The oval, toothed, purplish-green leaves are arranged in clumps at the base of the plant. Good in a waterside setting.

CULTIVATION: *Best in deep, moist but well-drained, fertile soil that is enriched with humus. Prefers partial shade, but tolerates full sun if soil remains damp.*

☼ ◐ ◊ ◑ ♀ ✽✽✽
‡10–15cm (4–6in) ↔30–40cm (12–16in)

Prunella grandiflora 'Loveliness'

This vigorous, spreading perennial bears dense, upright spikes of light purple, tubular flowers in summer. The lance-shaped, deep-green leaves are arranged in clumps at ground level. Versatile ground cover when planted in groups; the flowers are attractive to beneficial insects.

CULTIVATION: *Grow in any soil, in sun or partial shade. May swamp smaller plants, so allow room to expand. Divide clumps in spring or autumn to maintain vigour. Deadhead to prevent self-seeding.*

☼ ◐ ◊ ◑ ✽✽✽
‡15cm (6in) ↔to 1m (3ft) or more

Prunus x *cistena*

An upright, slow-growing, deciduous
shrub valued in particular for its
foliage, which is red when young,
maturing to red-purple. Bowl-shaped,
pinkish-white flowers open from
mid- to late spring, sometimes
followed by small, cherry-like,
purple-black fruits. Good as a
windbreak hedge.

CULTIVATION: *Grow in any but water-
logged soil, in full sun. Prune back
overcrowded shoots after flowering.
To grow as a hedge, prune the shoot
tips of young plants then trim in
midsummer to encourage branching.*

☼ ◊ ◖ ♀ ✿ ✿ ✿   ↕↔1.5m (5ft)

Prunus glandulosa 'Alba Plena'

This small cherry is a neat, rounded,
deciduous shrub producing dense
clusters of pure white, bowl-shaped,
double flowers during late spring.
The narrowly oval leaves are pale
to mid-green. Brings beautiful spring
blossom to a mixed or shrub border.
The cultivar 'Sinensis' has double
pink flowers.

CULTIVATION: *Grow in any moist but
well-drained, moderately fertile soil, in
sun. Can be pruned to a low framework
each year after flowering to enhance the
flowering performance.*

☼ ◊ ◖ ✿ ✿ ✿   ↕↔1.5m (5ft)

Prunus 'Kiku-Shidare-Zakura'

Also known as 'Cheal's Weeping', this small deciduous cherry tree is grown for its weeping branches and clear pink blossom. Dense clusters of large, double flowers are borne in mid- to late spring, with or before the lance-shaped, mid-green leaves, which are flushed bronze when young. Excellent in a small garden.

CULTIVATION: *Best in any moist but well-drained, moderately fertile soil, in full sun. Tolerates chalk. After flowering, prune out only dead, diseased or damaged wood; remove any shoots growing from the trunk as they appear.*

☼ ◊ ◊ ❀❀❀ ↕↔3m (10ft)

Prunus laurocerasus 'Otto Luyken'

This compact cherry laurel is an evergreen shrub with dense, glossy dark green foliage. Abundant spikes of white flowers are borne in mid- to late spring and often again in autumn, followed by conical red fruits that ripen to black. Plant in groups as a low hedge, or to cover bare ground.

CULTIVATION: *Grow in any moist but well-drained, moderately fertile soil, in full sun. Prune in late spring or early summer to restrict size.*

☼ ◊ ◊ ♀ ❀❀❀ ↕1m (3ft) ↔1.5m (5ft)

Prunus lusitanica subsp. *azorica*

This Portugal laurel is a slow-growing, evergreen shrub bearing slender spikes of small, fragrant white flowers in early summer. The oval, glossy dark green leaves have red stalks. Purple berries appear later in the season. Attractive year-round as a dense screen or hedge, except where winds are very cold.

CULTIVATION: *Grow in any moist but well-drained, fairly fertile soil, in sun with shelter from cold, drying winds. In late spring, prune to restrict size, or to remove old or overcrowded shoots.*

☼ ◊ ◖ ❀❀ ‡↔to 20m (70ft)

Prunus serrula

A rounded, deciduous tree that is valued for its striking, glossy, copper-brown to mahogany-red bark, which peels with age. Small, bowl-shaped white flowers in spring are followed by cherry-like fruits in autumn. The leaves are lance-shaped and dark green, turning yellow in autumn. Best seen as a specimen tree.

CULTIVATION: *Best in moist but well-drained, moderately fertile soil, in full sun. Remove dead or damaged wood after flowering, and remove any shoots growing from the trunk as they appear.*

☼ ◊ ◖ ❀❀❀ ‡↔10m (30ft)

Flowering Cherry Trees (*Prunus*)

Ornamental cherries are cultivated primarily for their white, pink, or red flowers which create a mass of blossom, usually on bare branches, from late winter to late spring; cultivars of the rosebud cherry, *P. x subhirtella*, flower from late autumn. Most popular cultivars not only bear dense clusters of showy, double flowers but have other ornamental characteristics to extend their interest beyond the flowering season: *P. sargentii*, for example, has brilliant autumn foliage colour, and some have shiny, coloured bark. All of these attractive features make flowering cherries superb specimen trees for small gardens.

CULTIVATION: *Grow in any moist but well-drained, fairly fertile soil, in sun. Keep all pruning to an absolute minimum; restrict formative pruning to shape to young plants only. Remove any damaged or diseased growth in midsummer, and keep trunks clear of sprouting shoots.*

☼ ◐ ◊ ♦ ❀ ❀ ❀ ❀

‡20m (70ft) ↔ 10m (30ft)

‡↔ 12m (40ft)

‡10m (30ft) ↔ 8m (25ft)

‡10m (30ft) ↔ 8m (25ft)

‡15m (50ft) ↔ 10m (30ft)

‡10m (30ft) ↔ 8m (25ft)

1 *Prunus avium* **2** *P. avium* 'Plena' ♀ **3** *P.* 'Kanzan' ♀ **4** *P. x incam* 'Okamé' ♀
5 *P. padus* 'Colorata' ♀ **6** *P.* 'Pandora' ♀

‡↔8m (25ft)

‡↔8m (25ft)

9

‡to 20m (70ft) ↔15m (50ft)

‡8m (25ft) ↔10m (30ft)

‡5m (15ft) ↔8m (25ft)

7

‡15m (50ft) ↔10m (30ft)

12

‡↔8m (25ft)

‡10m (30ft) ↔6m (20ft)

14

‡8m (25ft) ↔10m (30ft)

15

‡8m (25ft) ↔10m (30ft)

‡to 15m (50ft) ↔10m (30ft)

7 *P. padus* 'Watereri' ♀ **8** *P.* 'Pink Perfection' ♀ **9** *P. sargentii* **10** *P.* 'Shirofugen' ♀
11 *P.* 'Shogetsu' ♀ **12** *P.* x *subhirtella* 'Autumnalis Rosea' **13** *P.* 'Spire' ♀
14 *P.* 'Tai-haku' ♀ **15** *P.* 'Ukon' ♀ **16** *P.* x *yedoensis*

Pseudopanax lessonii '**Gold Splash**'

This evergreen, upright to spreading shrub or tree bears yellow-splashed, deep green foliage. In summer, less conspicuous clusters of yellow-green flowers are carried amid toothed leaves which are divided into teardrop-shaped leaflets. Purple-black fruits appear later in the season. In frost-prone areas, grow in a container as a foliage plant for a conservatory.

CULTIVATION: *Best in well-drained, fertile soil or compost, in sun or partial shade. Prune to restrict spread in early spring. Minimum temperature 2°C (36°F).*

☼ ☀ ◊ ♀ ☻
‡3–6m (10–20ft) ↔2–4m (6–12ft)

Pterostyrax hispida

The fragrant epaulette tree is a small, deciduous tree or shrub valued for its very pretty, hanging clusters of perfumed white flowers that appear in the early part of summer amid the broad, bright green leaves. The peeling grey bark is aromatic. Remove wayward or crossing shoots in winter.

CULTIVATION: *Grow in deep, fertile, well-drained, neutral to acid soil in sun or partial shade.*

☼ ☀ ◊ ♀ ❀❀❀
‡15m (50ft) ↔12m (40ft)

Pulmonaria 'Lewis Palmer'

This lungwort, sometimes called *P.* 'Highdown', is a deciduous perennial that forms clumps of upright, flowering stems. These are topped by open clusters of pink then blue, funnel-shaped flowers in early spring. The coarse, softly hairy leaves, dark green with white spots, are arranged along the stems. Grow in a wild or woodland garden.

CULTIVATION: *Best in moist but not waterlogged, fertile, humus-rich soil, in deep or light shade. Divide and replant clumps, after flowering, every few years.*

☼ ☀ ◑ ♀ ❀❀❀
‡35cm (14in) ↔45cm (18in)

Pulmonaria rubra

This lovely clump-forming, evergreen perennial is a good ground cover plant for a shady position. It has attractive, bright green foliage, and brings early colour with its funnel-shaped, bright brick- to salmon-coloured flowers from late winter to mid-spring. They are attractive to bees and other beneficial insects.

CULTIVATION: *Grow in humus-rich, fertile, moist but not waterlogged soil, in full or partial shade. Remove old leaves after flowering. Divide large or crowded clumps after flowering or in autumn.*

☼ ☀ ◑ ❀❀❀
‡to 40cm (16in) ↔90cm (36in)

Pulmonaria saccharata
Argentea Group

This group of evergreen perennials
has almost completely silver leaves.
A striking colour contrast is seen
from late winter to early spring,
when funnel-shaped red flowers
appear; these age to a dark violet.
They make good clumps at the front
of a shady mixed border.

CULTIVATION: *Best in fertile, moist but
not waterlogged soil that is rich in humus,
sited in full or partial shade. After
flowering, remove old leaves and divide
congested clumps.*

☼ ☀ ◐ ♀ ❀❀❀
‡30cm (12in) ↔ 60cm (24in)

Pulmonaria
'Sissinghurst White'

A neat, clump-forming, evergreen
perennial valued for its pure white
spring flowers and white-spotted
foliage. The elliptic, hairy, mid- to
dark green leaves are carried on
upright stems, below funnel-shaped
flowers which open from pale pink
buds in early spring. Plant in groups
as ground cover in a shady position.

CULTIVATION: *Grow in moist but not
waterlogged, humus-rich soil. Best in
deep or light shade, but tolerates full sun.
Divide and replant clumps every 2 or
3 years, after flowering.*

☼ ☀ ◐ ♀ ❀❀❀
‡to 30cm (12in) ↔ 45cm (18in)

Pulsatilla halleri

This silky-textured herbaceous perennial, ideal for a rock garden, is densely covered in long silver hairs. It bears upright, bell-shaped, pale violet-purple flowers in late spring above finely divided, light green leaves. The first flowers often open before the new spring leaves have fully unfurled.

CULTIVATION: *Grow in fertile, very well-drained gritty soil in full sun. May resent disturbance, so leave established plants undisturbed.*

☼ ◊ ♡ ❀ ❀ ❀ ‡20cm (8in) ↔15cm (6in)

Pulsatilla vulgaris

The pasque flower is a compact perennial forming tufts of finely divided, light green foliage. Its bell-shaped, nodding, silky-hairy flowers are carried above the leaves in spring; they are deep to pale purple or occasionally white, with golden centres. Good in a rock garden, scree bed or trough, or between paving.

CULTIVATION: *Best in fertile soil with very good drainage. Site in full sun, where it will not be prone to excessive winter wet. Do not disturb once planted.*

☼ ◊ ♡ ❀ ❀ ❀
‡10–20cm (4–8in) ↔20cm (8in)

Pulsatilla vulgaris '**Alba**'

This clump-forming perennial, a white form of the pasque flower, bears nodding, bell-shaped, silky-hairy white flowers with bold yellow centres in spring. These are carried above the finely divided, light green foliage which is hairy when young. Very pretty in a rock garden or scree bed.

CULTIVATION: *Best in fertile soil with very good drainage. Position in full sun with protection from excessive winter wet. Resents disturbance once planted.*

☼ ◊ ❀❀❀
‡10–20cm (4–8in) ↔20cm (8in)

Pyracantha '**Orange Glow**'

An upright to spreading, spiny, evergreen shrub bearing profuse clusters of tiny white flowers in late spring. Orange-red to dark orange berries follow in autumn, and persist well into winter. The leaves are oval and glossy dark green. Excellent as a vandal-resistant barrier hedge, which may also attract nesting birds.

CULTIVATION: *Grow in well-drained, fertile soil, in full sun to deep shade. Shelter from cold, drying winds. Prune in mid-spring, and trim new leafy growth again in summer to expose the berries.*

☼ ☀ ◊ ♀ ❀❀❀
‡↔3m (10ft)

Pyracantha 'Watereri'

A vigorous, upright, spiny shrub that forms a dense screen of evergreen foliage, ornamented by its abundance of white spring flowers and bright red berries in autumn. The leaves are elliptic and dark green. Good as a barrier hedge or in a shrub border; can also be trained against a shady wall. Attractive to nesting birds. For yellow berries, look for 'Soleil d'Or' or *P. rogersiana* 'Flava'.

CULTIVATION: *Grow in well-drained, fertile soil, in sun or shade with shelter from cold winds. Cut back unwanted growth in mid-spring, and trim leafy growth in summer to expose the berries.*

☼ ☀ ◊ ❀ ❀ ❀   ↕↔2.5m (8ft)

Pyrus calleryana 'Chanticleer'

This very thorny ornamental pear tree has a narrowly conical shape and makes a good specimen tree for a small garden. Attractive sprays of small white flowers in mid-spring are followed by spherical brown fruits in autumn. The oval, finely scalloped leaves are glossy dark green and deciduous; they turn red before they fall. Tolerates urban pollution.

CULTIVATION: *Grow in any well-drained, fertile soil, in full sun. Prune in winter to maintain a well-spaced crown.*

☼ ◊ ❀ ❀ ❀   ↕15m (50ft) ↔6m (20ft)

Pyrus salicifolia '**Pendula**'

This weeping pear is a deciduous tree with silvery-grey, willow-like leaves that are downy when young. Dense clusters of small, creamy-white flowers appear during spring, followed by pear-shaped green fruits in autumn. A fine, pollution-tolerant tree for an urban garden.

CULTIVATION: *Grow in fertile soil with good drainage, in sun. Prune young trees in winter to create a well-spaced, balanced framework of branches.*

☼ ◊ ♀ ❀❀❀ ‡8m (25ft) ↔6m (20ft)

Quercus palustris

The pin oak is fast-growing, deciduous tree with an attractive pyramidal shape and smooth grey bark. In autumn, its leaves are ablaze in scarlet and bronze, with some lingering into winter. Makes a striking specimen tree; particularly fine in parkland or lakeside settings.

CULTIVATION: *Thrives in moist, well-drained, neutral to slightly acid soil in full sun. It tolerates wet soil, pollution, and even drought, but develops chlorosis in alkaline soils.*

☼ ◊ ◊ ♀ ❀❀❀
‡20m (70ft) ↔12m (40ft)

Ramonda myconi

A tiny, neat, evergreen perennial with basal rosettes of dark green, slightly crinkled, broadly oval leaves. In late spring and early summer, deep violet-blue flowers are borne above the foliage on short stems. Pink-and white-flowered variants also occur. Grow in a rock garden or on a dry wall.

CULTIVATION: *Plant in moist but well-drained, moderately fertile, humus-rich soil, in partial shade. Set plants at an angle to avoid water pooling in the rosettes and causing rot. Leaves wither if too dry, but recover with watering.*

☼ ◊ ◊ ♀ ❀❀❀
‡10cm (4in) ↔ to 20cm (8in)

Ranunculus aconitifolius 'Flore Pleno'

White bachelor's buttons is a clump-forming, herbaceous perennial bearing small, almost spherical, fully double white flowers which last for a long time during late spring and early summer. The toothed leaves are deeply lobed and glossy dark green. Good for a woodland garden.

CULTIVATION: *Grow in moist but well-drained soil that is rich in humus. Site in deep or partial shade.*

☼ ☀ ◊ ◊ ♀ ❀❀❀
‡60cm (24in) ↔ 45cm (18in)

Ranunculus calandrinioides

This clump-forming perennial produces clusters of up to three cup-shaped, white or pink-flushed flowers from late winter to early spring. The lance-shaped, blue-green leaves emerge from the base in spring and die down in summer. Grow in a rock garden, scree bed, or alpine house. Will not survive outside in frosty, wet winters.

CULTIVATION: *Best in gritty, sharply drained, humus-rich soil, in sun. Water sparingly when dormant in summer.*

☼ ◊ ♀ ❀ ‡20cm (8in) ↔15cm (6in)

Rhamnus alaternus 'Argenteovariegata'

This Italian buckthorn is a fast-growing, upright to spreading, evergreen shrub bearing oval, leathery, grey-green leaves with creamy-white margins. Clusters of tiny, yellow-green flowers are borne in spring, followed by spherical red fruits which ripen to black.

CULTIVATION: *Grow in any well-drained soil, in full sun. Prune out unwanted growth in early spring; remove any shoots with all-green leaves as seen.*

☼ ◊ ♀ ❀❀❀ ‡5m (15ft) ↔4m (12ft)

Rheum palmatum 'Atrosanguineum'

Ornamental rhubarb is a big perennial with large, rounded, toothed, dark green leaves to 90cm (36in) across, of architectural value in the garden. The leaves of 'Atrosanguineum' have a purplish tint when young. Tall, feathery, cherry pink flowerheads rise above the foliage in early summer. Ideal for a moist border, particularly beside water.

CULTIVATION: *Best in deep, moist soil, enriched with plenty of organic matter, in full sun or partial shade.*

☼ ☀ ◊ ◑ ❀❀❀
‡2.5m (8ft) ↔2m (6ft)

Rhodanthemum hosmariense

A spreading subshrub valued for its profusion of daisy-like flowerheads with white petals and yellow eyes. These are borne from early spring to autumn, covering the silver, softly hairy, finely divided leaves. Grow at the base of a warm wall or in a rock garden. In an alpine house, it will flower year-round if deadheaded.

CULTIVATION: *Grow in very well-drained soil, in a sunny position. Deadhead regularly to prolong flowering.*

☼ ◊ ♀ ❀❀❀
‡10–30cm (4–12in) ↔30cm (12in)

Evergreen Azaleas (*Rhododendron*)

Azaleas can be distinguished from true rhododendrons by their smaller dark green leaves and more tubular flowers. They also tend to make smaller, more spreading, twiggy shrubs. Botanically, rhododendrons have at least ten stamens per flower, and azaleas just five. Evergreen azaleas make beautiful, spring-flowering shrubs, blooming in almost every colour. They suit a variety of uses: dwarf or compact types are excellent in containers or tubs on shaded patios, and larger varieties will brighten up areas in permanent light shade. They do well in sun provided that the soil is not allowed to dry out.

CULTIVATION: *Ideal in moist but well-drained, humus-rich, acid soil in part-day shade. Shallow planting is essential. Little formative pruning is necessary. If older plants become congested, thin in early summer. Maintain a mulch of leaf mould, but do not cultivate around the root area.*

☼ ☀ ◊ ❀ ❀ ❀

↕↔ 1.2m (4ft)

↕↔ 1.3m (3½ft)

↕↔ 60cm (24in)

↕↔ 60cm (24in)

↕↔ 60cm (24in)

↕↔ 60cm (24in)

1 *Rhododendron* 'Azuma-kagami' **2** *R.* 'Beethoven' **3** *R.* 'Hatsugiri' **4** *R.* 'Hinodegiri'
5 *R.* (Obtusum Group) 'Hinomayo' ♥ **6** *R.* 'Irohayama' ♥

‡↔1.5m (5ft)

‡↔1.5m (5ft)

‡↔1.2m (4ft)

MORE CHOICES

'Addy Wery'
Vermilion-red flowers.

'Elsie Lee' Light
reddish-mauve flowers.

'Greeting' Red-orange.

'Gumpo' Pinky-white.

'Hexe' Crimson.

'Hino-crimson'
Brilliant red flowers.

'Kure-no-yuki' White,
also called 'Snowflake'.

'Louise Dowdle' Vivid
red-purple flowers.

'Vida Brown' Rose-red.

'Wombat' Pink flowers.

‡↔60–90cm (24–36in)

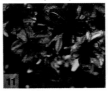

‡↔1.2m (4ft)

‡↔1.2m (4ft)

7 *R.* 'John Cairns' **8** *R.* 'Kirin' **9** *R.* 'Palestrina' ♀ **10** *R.* 'Rosebud'
11 *R.* 'Vuyk's Scarlet' ♀ **12** *R.* 'Vuyk's Rosyred' ♀

Deciduous Azaleas (*Rhododendron*)

A group of very hardy flowering shrubs, the only rhododendrons whose dark green leaves are deciduous, often colouring brilliantly before they fall. Deciduous azaleas are perhaps the most beautiful types of rhododendron, with large clusters of sometimes fragrant, white to yellow, orange, pink, or red flowers in spring and early summer. There is quite a variety in size, shape, and growth habit, but they suit most garden uses well, especially in light shade. *R. luteum* thrives in sun where the soil is reliably moist. Grow them in tubs if you do not have suitable soil.

CULTIVATION: *Grow in moist but well-drained, acid soil enriched with plenty of organic matter, ideally in partial shade. Shallow planting is essential; maintain a thick mulch of leaf mould, which will nourish the plant. Little or no pruning is necessary. Do not cultivate around the base of the plant as this will damage the roots.*

☼ ☀ ◊ ◊ ✤ ✤ ✤

1 ↕↔2.5m (8ft)

2 ↕↔3m (10ft)

3 ↕↔2.2m (7ft)

4 ↕↔1.5–2.5m (5–8ft)

5 ↕↔1.5m (5ft)

1 *R. albrechtii* **2** *R. austrinum* **3** *R.* 'Cecile' ♀ **4** *R.* 'Corneille' **5** *R.* 'Homebush' ♀

6

‡↔2m (6ft)

‡↔4m (12ft)

7

‡↔1.5–2.5m (5–8ft)

8

‡↔2m (6ft)

9

‡↔2.5m (8ft)

10

‡↔2m (6ft)

11

6 *R.* 'Irene Koster' ♀ **7** *R. luteum* ♀ **8** *R.* 'Narcissiflorum' ♀ **9** *R.* 'Persil' ♀
10 *R.* 'Spek's Orange' **11** *R.* 'Strawberry Ice' ♀

Large Rhododendrons

Large, woodland-type rhododendrons, which can reach tree-like proportions, are grown primarily for their bright, sometimes fragrant, mostly spring flowers which are available in a wide spectrum of shapes and colours. They are ideal for adding colour to shaded areas or woodland gardens. Most leaves are oval and dark green, although the attractive young foliage of *R. bureaui* is light brown. Some cultivars, like 'Cynthia' or 'Purple Splendour', are tolerant of direct sun (in reliably moist soil), making them more versatile than others; they make glorious, spring-flowering screens or hedges for a large garden.

CULTIVATION: *Grow in moist but well-drained, humus-rich, acid soil. Most prefer dappled shade in sheltered woodland. Shallow planting is essential. Little formative pruning is necessary, although most can be renovated after flowering to leave a balanced framework of old wood.*

☼ ☀ ◊ ◊ ❀ ❀ ❀

↕↔3m (10ft)

↕↔3m (10ft)

↕↔3.5m (11ft)

1 *R.* 'Blue Peter' ♀ **2** *R. bureaui* **3** *R.* 'Crest' ♀

$\updownarrow\leftrightarrow$ 6m (20ft)

$\updownarrow$ to 12m (40ft) $\leftrightarrow$ 5m (15ft)

$\updownarrow\leftrightarrow$ 4m (12ft)

$\updownarrow\leftrightarrow$ 3m (10ft)

$\updownarrow\leftrightarrow$ 4m (12ft)

$\updownarrow\leftrightarrow$ 3m (10ft)

$\updownarrow\leftrightarrow$ 3m (10ft)

$\updownarrow\leftrightarrow$ 3m (10ft)

4 _R._ 'Cynthia' ♀ **5** _R. falconeri_ ♀ **6** _R._ 'Fastuosum Flore Pleno' ♀
7 _R._ 'Furnivall's Daughter' ♀ **8** _R._ 'Loderi King George' ♀ **9** _R._ 'Purple Splendour'
10 _R._ 'Sappho' **11** _R._ 'Susan' J.C. Williams

Medium-sized Rhododendrons

These evergreen rhododendrons, between 1.5 and 3m (5 and 10ft) tall, are much-valued for their attractive, often scented blooms; the flowers are carried amid dark green foliage throughout spring. 'Yellow Hammer' will often produce an early show of flowers in autumn, and the foliage of 'Winsome' is unusual for its bronze tints when young. A vast number of different medium-sized rhododendrons are available, all suitable for shrub borders or grouped together in mass plantings. Some sun-tolerant varieties, especially low-growing forms such as 'May Day', are suitable for informal hedging.

CULTIVATION: *Grow in moist but well-drained, humus-rich, acid soil. Most prefer light dappled shade. Shallow planting is essential. 'Fragrantissimum' requires extra care in cold areas; provide a thick winter mulch and avoid siting in a frost-pocket. Trim after flowering, if necessary.*

☼ ☀ ◊ ◖

↕↔2m (6ft)

↕↔1.5m (5ft)

1 *R.* 'Fabia' ♀ ❀❀❀ **2** *R.* 'Golden Torch' ♀ ❀❀❀

‡↔2m (6ft)

‡↔1.5m (5ft)

‡↔1.5m (5ft)

‡↔2m (6ft)

‡↔2m (6ft)

‡↔1.5m (5ft)

3 *R.* 'Fragrantissimum' ♀ ✿ **4** *R.* 'Hydon Dawn' ♀ ✿✿✿ **5** *R.* 'May Day' ♀ ✿✿✿
6 *R.* 'Titian Beauty' ✿✿✿ **7** *R.* 'Yellow Hammer' ♀ ✿✿✿ **8** *R.* 'Winsome' ♀ ✿✿✿

Dwarf Rhododendrons

Dwarf rhododendrons are low-lying, evergreen shrubs with mid- to dark green, lance-shaped leaves. They flower throughout spring, in a wide variety of showy colours and flower forms. If soil conditions are too alkaline for growing rhododendrons in the open garden, these compact shrubs are ideal in containers or tubs on shaded patios; 'Ptarmigan' is particularly suited to this kind of planting since it is able to tolerate periods without water. Dwarf rhododendrons are also effective in rock gardens. In areas with cold winters, the earliest spring flowers may be vulnerable to frost.

CULTIVATION: *Grow in moist but well-drained, leafy, acid soil that is enriched with well-rotted organic matter. Site in sun or partial shade, but avoid the deep shade directly beneath a tree canopy. Best planted in spring or autumn; shallow planting is essential. No pruning is necessary.*

☼ ☀ ◊ ◐

1 ↕↔ 1.1m (3½ft)

2 ↕↔ 1.2m (4ft)

3 ↕↔ 60cm (24in)

4 ↕↔ 45–90cm (18–36in)

1 *R.* 'Cilpinense' ♀ ✿✿✿ **2** *R.* 'Doc' ✿✿✿ **3** *R.* 'Dora Amateis' ♀ ✿✿✿
4 *R.* 'Ptarmigan' ♀ ✿✿

Rhus typhina 'Dissecta'

Rhus typhina, stag's horn sumach (also seen as *R. hirta*), is an upright, deciduous shrub with velvety-red shoots that resemble antlers. This form has long leaves, divided into many finely cut leaflets, which turn a brilliant orange-red in autumn. Upright clusters of less significant, yellow-green flowers are produced in summer, followed by velvety clusters of deep crimson-red fruits.

CULTIVATION: *Grow in moist but well-drained, fairly fertile soil, in full sun to obtain best autumn colour. Remove any suckering shoots arising from the ground around the base of the plant.*

☼ ◊ ◑ ♀ ❀❀❀ ‡2m (6ft) ↔3m (10ft)

Ribes sanguineum 'Brocklebankii'

This slow-growing flowering currant is an upright, deciduous shrub with rounded, aromatic, yellow leaves, bright when young and fading in summer. The tubular, pale pink flowers, borne in hanging clusters in spring, are followed by small, blue-black fruits. 'Tydeman's White' is a very similar shrub, sometimes a little taller, with pure white flowers.

CULTIVATION: *Grow in well-drained, fairly fertile soil. Site in sun, with shade during the hottest part of the day. Prune some older stems after flowering. Cut back overgrown specimens in winter.*

☼ ☀ ◊ ❀❀❀ ‡↔1.2m (4ft)

Ribes sanguineum 'Pulborough Scarlet'

This vigorous flowering currant, larger than 'Brocklebankii', (see previous page, bottom) is an upright, deciduous shrub, bearing hanging clusters of tubular, dark red flowers with white centres in spring. The aromatic, dark green leaves are rounded with toothed lobes. Small, berry-like, blue-black fruits develop during the summer.

CULTIVATION: *Grow in well-drained, moderately fertile soil, in full sun. Cut out some older stems after flowering; overgrown specimens can be pruned hard in winter or early spring.*

☼ ◊ ◊ ♀ ❁❁❁   ↕2m (6ft) ↔2.5m (8ft)

Robinia hispida

The bristly locust is an upright and arching, deciduous shrub with spiny shoots, useful for shrub borders on poor, dry soils. Deep rose-pink, pea-like flowers appear in hanging spikes during late spring and early summer; these are followed by brown seed pods. The large, dark green leaves are divided into many oval leaflets.

CULTIVATION: *Grow in any but waterlogged soil, in sun. Provide shelter from wind to avoid damage to the brittle branches. No pruning is necessary.*

☼ ◊ ❁❁❁   ↕2.5m (8ft) ↔3m (10ft)

Robinia pseudoacacia 'Frisia'

The black locust, *R. pseudoacacia*, is a fast-growing, broadly columnar, deciduous tree, in this cultivar bearing gentle, yellow-green foliage which is golden-yellow when young, turning orange-yellow in autumn. Usually sparse, hanging clusters of fragrant, pea-like, small white flowers appear in midsummer. The stems are normally spiny.

CULTIVATION: *Grow in moist but well-drained, fertile soil, in full sun. When young, maintain a single trunk by removing competing stems as soon as possible. Do not prune once established.*

☼ ◊ ◑ ❀❀❀   15m (50ft) ↔8m (25ft)

Rodgersia pinnata 'Superba'

A clump-forming perennial that bears upright clusters of star-shaped, bright pink flowers. These are borne in mid- to late summer above bold, heavily veined, dark green foliage. The divided leaves, up to 90cm (36in) long, are purplish-bronze when young. Good near water, in a bog garden, or for naturalizing at a woodland margin. For creamy-white flowers, look for *R. podophylla*.

CULTIVATION: *Best in moist, humus-rich soil, in full sun or semi-shade. Provide shelter from cold, drying winds. Will not tolerate drought.*

☼ ☼ ◑ ♀ ❀❀❀
↕to 1.2m (4ft) ↔75cm (30in)

Large-flowered Bush Roses (*Rosa*)

Also known as hybrid tea roses, these deciduous shrubs are commonly grown in formal bedding displays, laid out with neat paths and edging. They are distinguished from other roses in that they carry their large flowers either singly or in clusters of two or three. The first blooms appear in early summer and repeat flushes continue into early autumn. In a formal bed, group five or six of the same cultivar together, and interplant with some standard roses to add some variation in height. These bush roses also combine well with herbaceous perennials and other shrubs in mixed borders.

CULTIVATION: *Grow in moist but well-drained, fertile soil in full sun. Cut spent flower stems back to the first leaf for repeat blooms. Prune main stems to about 25cm (10in) above ground level in early spring, and remove any dead or diseased wood as necessary at the base.*

☼ ◊ ◑ ❀❀❀

‡↔ 60cm (24in)

‡to 2m (6ft) ↔ 80cm (32in)

‡1.1m (3½ft) ↔ 75cm (30in)

‡1.1m (3½ft) ↔ 75cm (30in)

‡75cm (30in) ↔ 60cm (24in)

1 *Rosa* ABBEYFIELD ROSE 'Cocbrose' **2** *R.* ALEXANDER 'Harlex' ♥ **3** *R.* 'Blessings'
4 *R.* ELINA 'Dicjana' ♥ **5** *R.* FREEDOM 'Dicjem' ♥ **6** *R.* INDIAN SUMMER 'Peaperfume' ♥
7 *R.* INGRID BERGMAN 'Poulman' ♥ **8** *R.* 'Just Joey' ♥ **9** *R.* LOVELY LADY 'Dicjubell' ♥

‡55cm (22in) ↔ 60cm (24in)

‡80cm (32in) ↔ 65cm (26in)

‡75cm (30in) ↔ 70cm (28in)

‡75cm (30in) ↔ 60cm (24in)

‡1m (3ft) ↔ 75cm (30in)

‡1.2m (4ft) ↔ 1m (3ft)

‡1m (3ft) ↔ 60cm (24in)

‡1m (3ft) ↔ 75cm (30in)

‡80cm (32in) ↔ 60cm (24in)

‡1.1m (3½ft) ↔ 60cm (24in)

‡1m (3ft) ↔ 75cm (30in)

10 *R.* PAUL SHIRVILLE 'Harqueterwife' ♀ **11** *R.* PEACE 'Madame A. Meilland' ♀
12 *R.* REMEMBER ME 'Cocdestin' ♀ **13** *R.* ROYAL WILLIAM 'Korzaun' ♀
14 *R.* SAVOY HOTEL 'Harvintage' **15** *R.* 'Silver Jubilee' **16** *R.* TROIKA 'Poumidor'

Cluster-flowered Bush Roses (*Rosa*)

These very free-flowering bush roses are also known as floribunda roses. Like other roses, they come in a huge range of flower colours but are set apart from the large-flowered bush roses by their large, many-flowered trusses of relatively small blooms. Nearly all are fragrant, some more so than others. They lend themselves well to informal or cottage garden displays, mixing well with herbaceous perennials and other shrubs; remember to consider the colour and intensity of the flowers when choosing all neighbouring plants, since blooms will continue to appear from early summer through to early autumn.

CULTIVATION: *Best in moist but well-drained, fairly fertile soil in full sun. Deadhead for repeat blooms, unless the hips are wanted. Prune main stems to about 30cm (12in) above ground level in early spring, and remove any dead or diseased wood as necessary.*

☼ ◊ ◊ ❋ ❋ ❋

‡50cm (20in) ↔ 60cm (24in)

‡1m (3ft) ↔ 75cm (30in)

‡75cm (30in) ↔ 60cm (24in)

‡1m (3ft) ↔ 60cm (24in)

‡1.2m (4ft) ↔ 1m (3ft)

‡80cm (32in) ↔ 75cm (30in)

1 *R.* AMBER QUEEN 'Harroony' ♀ **2** *R.* ANISLEY DICKSON 'Dickimono' ♀ **3** *R.* ANNA LIVIA 'Kormetter' ♀
4 *R.* 'Arthur Bell' ♀ **5** *R.* 'Chinatown' ♀ **6** *R.* CITY OF LONDON 'Harukfore' ♀
7 *R.* ESCAPADE 'Harpade' ♀ **8** *R.* 'Fragrant Delight' ♀ **9** *R.* ICEBERG 'Korbin' ♀

‡75cm (30in) ↔ 60cm (24in)

‡1m (3ft) ↔ 75cm (30in)

‡80cm (32in) ↔ 65cm (26in)

10

‡↔75cm (30in)

‡80cm (32in) ↔ 60cm (24in)

‡1.2m (4ft) ↔ 1m (3ft)

‡70cm (28in) ↔ 60cm (24in)

‡75cm (30in) ↔ 60cm (24in)

‡to 2.2m (8ft) ↔ 1m (3ft)

10 *R.* MANY HAPPY RETURNS 'Harwanted' ♀ **11** *R.* MARGARET MERRIL 'Harkuly'
12 *R.* MOUNTBATTEN 'Harmantelle' ♀ **13** *R.* SEXY REXY 'Macrexy'
14 *R.* TANGO 'Macfirwal' **15** *R.* 'The Queen Elizabeth'

Climbing Roses (*Rosa*)

Climbing roses are often vigorous plants that will reach varying heights depending on the cultivar. All types have stiff, arching stems, usually with dense, glossy leaves divided into small leaflets. The frequently scented flowers are borne in summer, some in one exuberant flush, others having a lesser repeat flowering. They can be trained against walls or fences as decorative features in their own right, planted as a complement to other climbers, such as clematis, or allowed to scramble up into other wall-trained shrubs or even old trees. They are invaluable for disguising unsightly garden buildings, or as a backdrop to a summer border.

CULTIVATION: *Best in moist but well-drained, fairly fertile soil, in sun. Deadhead unless hips are wanted. As plants mature, prune back to within the allowed area, after flowering. Occasionally cut an old main stem back to the base to renew growth. Never prune in the first two years.*

☼ ◊ ◐ ❀❀❀

‡↔ to 6m (20ft)

‡3m (10ft) ↔2.5m (8ft)

‡↔2.2m (7ft)

‡ to 10m (30ft) ↔6m (20ft)

‡ to 5m (15ft) ↔4m (12ft)

1 *R. banksiae* 'Lutea' ♀ **2** *R.* 'Compassion' ♀ **3** *R.* DUBLIN BAY 'Macdub' ♀
4 *R. filipes* 'Kiftsgate' ♀ **5** *R.* 'Gloire de Dijon'

‡to 3m (10ft) ↔2m (6ft)

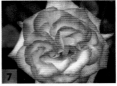

‡3m (10ft) ↔2.2m (7ft)

‡↔2.5m (8ft)

‡5m (15ft) ↔3m (10ft)

‡3m (10ft) ↔2.5m (8ft)

‡to 6m (20ft) ↔4m (12ft)

‡to 3m (10ft) ↔2m (6ft)

6 *R.* 'Golden Showers' **7** *R.* HANDEL 'Macha' **8** *R.* 'Maigold' ♀
9 *R.* 'Madame Alfred Carrière' ♀ **10** *R.* 'New Dawn' ♀
11 *R.* 'Madame Grégoire Staechelin' ♀ **12** *R.* 'Zéphirine Drouhin'

Rambling Roses (*Rosa*)

Rambling roses are very similar to climbers (see p.452), but with more lax, flexible stems. These are easier to train on to complex structures such as arches, tunnels and pergolas, or ropes and chains suspended between rigid uprights, provided they are solidly built; most ramblers are vigorous. Unlike climbers, they can succumb to mildew if trained flat against walls. All available cultivars have divided, glossy green leaves, borne on thorny or prickly stems.

Flowers are often scented, arranged singly or in clusters, and are borne during summer. Some bloom only once, others having a lesser repeat flowering later on.

CULTIVATION: *Best in moist but well-drained, fertile soil, in full sun. Train stems of young plants on to a support, to establish a permanent framework; prune back to this each year after flowering has finished, and remove any damaged wood as necessary.*

☼ ◊ ◊ ❀ ❀ ❀

‡to 5m (15ft) ↔ 3m (10ft)

‡to 5m (15ft) ↔ 4m (12ft)

1 *R.* 'Albéric Barbier' ♀ **2** *R.* 'Albertine' ♀

‡to 10m (30ft) ↔ 6m (20ft)

‡to 5m (15ft) ↔ to 4m (12ft)

‡↔ 6m (20ft)

‡to 6m (20ft) ↔ 4m (12ft)

‡↔ to 4m (12ft)

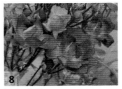

‡↔ 4m (12ft)

3 *R.* 'Bobbie James' ♥ **4** *R.* 'Félicité Perpétue' ♥ **5** *R.* 'Rambling Rector' ♥
6 *R.* 'Seagull' ♥ **7** *R.* 'Sander's White Rambler' ♥ **8** *R.* 'Veilchenblau' ♥

Patio and Miniature Roses (*Rosa*)

These small or miniature shrub roses, bred especially for their compact habit, greatly extend the range of garden situations in which roses can be grown. All have very attractive, scented flowers in a wide range of colours, blooming over long periods from summer to autumn amid deciduous, glossy green leaves. With the exception of 'Ballerina', which can grow to a height of about 1.5m (5ft), most are under 1m (3ft) tall, making them invaluable for confined, sunny spaces; planted in large tubs, containers or raised beds, they are also excellent for decorating patios and other paved areas.

CULTIVATION: *Grow in well-drained but moist, moderately fertile soil that is rich in well-rotted organic matter. Choose an open, sunny site. Remove all but the strongest shoots in late winter, then reduce these by about one-third of their height. Cut out any dead or damaged wood as necessary.*

☼ ◊ ◐ ❀ ❀ ❀

‡45cm (18in) ↔ 40cm (16in)

‡ to 1.5m (5ft) ↔ 1.2m (4ft)

‡75cm (30in) ↔ 60cm (24in)

‡45cm (18in) ↔ 30cm (12in)

‡50cm (20in) ↔ 40cm (16in)

‡75cm (30in) ↔ 60cm (30in)

1 *R.* ANNA FORD 'Harpiccolo' ♀ **2** *R.* 'Ballerina' ♀ **3** *R.* 'Cecile Brunner' ♀
4 *R.* CIDER CUP 'Dicladida' **5** *R.* GENTLE TOUCH 'Diclulu' **6** *R.* 'Mevrouw Nathalie Nypels'

‡1.2m (4ft) ↔ 1m (3ft)

‡40cm (16in) ↔ 60cm (24in)

‡40cm (16in) ↔ 35cm (14in)

‡25cm (10in) ↔ 30cm (12in)

‡↔35cm (14in)

‡↔60–90cm (24–30in)

‡↔1–1.5m (3–5ft)

7 *R.* 'Perle d'Or' ♀ **8** *R.* QUEEN MOTHER 'Korquemu' ♀ **9** *R.* 'Stacey Sue'
10 *R.* SWEET DREAM 'Fryminicot' ♀ **11** *R.* SWEET MAGIC 'Dicmagic' ♀ **12** *R.* 'The Fairy' ♀
13 *R.* 'Yesterday' ♀

Roses for Ground Cover (*Rosa*)

Ground cover roses are low-growing, spreading, deciduous shrubs, ideal for the front of a border, in both formal and informal situations. They produce beautiful, fragrant flowers over long periods from summer into autumn, amid divided, glossy green leaves, on thorny or prickly, sometimes trailing stems. Only those of really dense habit, like SWANY 'Meiburenac', will provide weed-smothering cover, and even these are only effective if the ground is weed-free to begin with.

Most give their best cascading over a low wall, or when used to clothe a steep bank which is otherwise difficult to plant.

CULTIVATION: *Best in moist but well-drained, reasonably fertile soil that is rich in humus, in full sun. Prune shoots back after flowering each year to well within the intended area of spread, removing any dead or damaged wood. Annual pruning will enhance flowering performance.*

☼ ◊ ◊ ❀❀❀

‡85cm (34in) ↔ 1.1m (3½ft)

‡45cm (18in) ↔ 1.2m (4ft)

‡75cm (30in) ↔ 1.2m (4ft)

‡1m (3ft) ↔ 1.2m (4ft)

1 *R.* BONICA 'Meidomonac' ♀ **2** *R.* 'Nozomi'
3 *R.* RED BLANKET 'Intercell' **4** *R.* ROSY CUSHION 'Interall'

5 ‡15cm (6in) ↔ 45cm (18in)

6 ‡50cm (20in) ↔ 1.5m (5ft)

7 ‡80cm (32in) ↔ 1.2m (4ft)

8 ‡to 75cm (30in) ↔ 1.7m (5½ft)

9 ‡75cm (30in) ↔ 1.2m (4ft)

5 *R.* SNOW CARPET 'Maccarpe' 6 *R.* SUMA 'Harsuma' 7 *R.* SURREY 'Korlanum' ♀
8 *R.* SWANY 'Meiburenac' 9 *R.* TALL STORY 'Dickooky' ♀

Old Garden Roses (*Rosa*)

The history of old garden roses extends back to Roman times, demonstrating their lasting appeal in garden design. They are deciduous shrubs, comprised of a very large number of cultivars categorized into many groups, such as the Gallica, Damask, and Moss roses. As almost all old garden roses flower in a single flush in early summer, they should be mixed with other flowering plants to maintain a lasting display. Try underplanting non-climbing types with spring bulbs,

and climbing cultivars can be interwoven with a late-flowering clematis, for example.

CULTIVATION: *Best in moist but well-drained, fertile soil in full sun. Little pruning is necessary; occasionally remove an old stem at the base to alleviate congested growth and to promote new shoots. Trim to shape in spring as necessary. Unless hips are wanted, remove spent flowers as they fade.*

☼ ◊ ◊ ❀❀❀

‡2.2m (7ft) ↔1.5m (5ft)

‡1.2m (4ft) ↔1m (3ft)

‡1m (3ft) ↔1.2m (4ft)

‡1.5m (5ft) ↔1.2m (4ft)

‡1.5m (5ft) ↔1.2m (4ft)

‡1.5m (5ft) ↔1.2m (4ft)

1 *R.* x *alba* 'Alba Maxima' **2** *R.* 'Belle de Crécy' **3** *R.* 'Cardinal de Richelieu'
4 *R.* 'Céleste' ♀ **5** *R.* x *centifolia* 'Cristata' ♀ **6** *R.* x *centifolia* 'Muscosa'
7 *R.* 'Charles de Mills' ♀ **8** *R.* 'De Rescht' **9** *R.* 'Duc de Guiche' ♀

$\updownarrow\leftrightarrow$1.2m (4ft) or more

$\updownarrow\leftrightarrow$ to 2m (6ft)

$\updownarrow\leftrightarrow$1.2m (4ft)

$\updownarrow$1.5m (5ft) $\leftrightarrow$1.2m (4ft)

$\updownarrow$1.3m (4½ft) $\leftrightarrow$1.2m (4ft)

$\updownarrow$1.5m (5ft) $\leftrightarrow$1.2m (4ft)

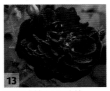

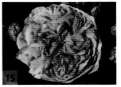

$\updownarrow$ to 2m (6ft) $\leftrightarrow$1.2m (4ft)

$\updownarrow$1.5m (5ft) $\leftrightarrow$1.2m (4ft)

$\updownarrow$1.5m (5ft) $\leftrightarrow$1.2m (4ft)

$\updownarrow$1.5m (5ft) $\leftrightarrow$1.2m (4ft)

$\updownarrow\leftrightarrow$1.2m (4ft)

$\updownarrow\leftrightarrow$1m (3ft)

10 *R.* 'Fantin-Latour' ♀ **11** *R.* 'Félicité Parmentier' ♀ **12** *R.* 'Ferdinand Pichard' ♀
13 *R.* 'Henri Martin' ♀ **14** *R.* 'Ispahan' ♀ **15** *R.* 'Königin von Dänemark' ♀
16 *R.* 'Madame Hardy' ♀ **17** *R.* 'Président de Sèze' ♀ **18** *R.* 'Tuscany Superb' ♀

Modern Shrub Roses (*Rosa*)

These roses are slightly larger and more spreading than most others, combining the stature of old garden roses with some of the benefits of modern types. Their general good health and vigour make them easy to grow and they repeat-flower over a long period, making ideal summer-flowering, deciduous shrubs for the back of a low-maintenance shrub border in a large garden. Like other roses, there is a wide choice in flower shape and colour, and several have a superb fragrance. Flowering begins in early summer, with repeat flushes up until early autumn.

CULTIVATION: *Grow in well-drained but moist, moderately fertile soil that is rich in organic matter. Choose an open, sunny site. To keep them at a manageable size, prune every year in early spring. It is often better to let them grow naturally, however, as their form is easily spoiled by severe or careless pruning.*

☼ ◊ ◗ ❈❈❈

‡1.5m (5ft) ↔1.1m (3½ft) ‡↔1.2m (4ft) ‡↔ to 3.5m (11ft)

‡2m (6ft) ↔1.5m (5ft) ‡↔1.5m (5ft) ‡1.5m (5ft) ↔1.2m (4ft)

1 *R.* 'Blanche Double de Coubert' ♀ **2** *R.* 'Buff Beauty' ♀ **3** *R.* 'Cerise Bouquet' ♀
4 *R.* 'Constance Spry' ♀ **5** *R.* 'Cornelia' ♀ **6** *R.* 'Felicia' ♀ **7** *R.* 'Fred Loads' ♀
8 *R.* GERTRUDE JEKYLL 'Ausbord' ♀

‡2m (6ft) ↔ 1m (3ft)

‡1.5m (5ft) ↔ 1m (3ft)

‡1.2m (4ft) ↔ 1.5m (5ft)

‡1.5m (5ft) ↔ 1.2m (4ft)

‡↔ 2.2m (7ft)

‡↔ 2.2m (7ft)

‡↔ 1.1m (3½ft)

‡ 2.2m (7ft) ↔ 2m (6ft)

‡2m (6ft) ↔ 1m (3ft)

‡↔ 1.1m (3½ ft)

‡2m (6ft) ↔ 1.2m (4ft)

9 R. GRAHAM THOMAS 'Ausmas' ♀ **10** R. JACQUELINE DU PRÉ 'Harwanna' ♀ **11** R. 'Marguerite Hilling' **12** R. 'Nevada' **13** R. 'Penelope' ♀ **14** R. 'Roseraie de l'Hay' ♀ **15** R. 'Sally Holmes' ♀ **16** R. 'The Lady' **17** R. WESTERLAND 'Korwest' ♀

Roses for Hedgerows and Wild Areas (*Rosa*)

The best types of rose for hedgerows or wild gardens are the species or wild roses, as many naturalize easily. Either shrubs or climbers, most have a natural-looking, scrambling or arching growth habit with single, five-petalled, often fragrant flowers, which appear in early summer on the previous year's growth. Although the main flowering season is fleeting, in many the flowers develop into beautiful rosehips, just as attractive as the flowers. Hips vary in colour from orange to red or black, and often persist into winter, providing valuable food for hungry wildlife.

CULTIVATION: *Best in moist but well-drained, reasonably fertile soil, rich in organic matter. In a wild garden, little pruning is needed; to control hedges, trim after flowering each year, removing any dead or damaged wood. The flowers are borne on the previous summer's stems, so do not remove too many older branches.*

☼ ◊ ◊ ❀ ❀ ❀

‡2.2m (7ft) ↔2.5m (8ft)

‡80cm (32in) ↔1m (3ft)

‡1m (3ft) ↔1.2m (4ft)

‡80cm (32in) ↔1m (3ft)

‡2m (6ft) ↔1.5m (5ft)

1 *R.* 'Complicata' **2** *R.* 'Fru Dagmar Hastrup' ♀ **3** *R. gallica* var. *officinalis* ♀
4 *R. gallica* 'Versicolor' ♀ **5** *R. glauca*

‡ to 3m (10ft) ↔ 2m (6ft)

‡1.1m (3½ft) ↔ 1.3m (4½ft)

‡1.5–2.5m (5–8ft) ↔ 1.2–2m (4–6ft)

‡1.2m (4ft) ↔ 1m (3ft)

‡2m (6ft) ↔ 1m (3ft)

‡↔1–2.5m (3–8ft)

‡2m (6ft) ↔ 1m (3ft)

13

‡2.5m (8ft) ↔ 1m (3ft)

6 *R.* 'Golden Wings' **7** *R. mulliganii* **8** *R. nutkana* 'Plena' ♀
9 *R.* x *odorata* 'Mutabilis' ♀ **10** *R. primula* **11** *R. rugosa* 'Rubra'
12 *R. xanthina* 'Canary Bird' ♀ **13** *R. xanthina* var. *hugonis*

Rosmarinus officinalis 'Miss Jessopp's Upright'

This vigorous, upright rosemary is an evergreen shrub with aromatic foliage which can be used in cooking. From mid-spring to early summer, whorls of small, purple-blue to white flowers are produced amid the narrow, dark green, white-felted leaves, often with a repeat show in autumn. A good hedging plant for a kitchen garden. For white flowers, try 'Sissinghurst White'.

CULTIVATION: *Grow in well-drained, poor to moderately fertile soil. Choose a sunny, sheltered site. After flowering, trim any shoots that spoil the symmetry.*

☼ ◊ ♀ ❀❀❀   ↕↔2m (6ft)

Rosmarinus officinalis Prostratus Group

These low-growing types of rosemary are aromatic, evergreen shrubs ideal for a rock garden or the top of a dry wall. Whorls of small, two-lipped, purple-blue to white flowers are produced in late spring, and often again in autumn. The dark green leaves have white-felted under-sides, and can be cut in sprigs for culinary use. Plant in a sheltered position and protect in cold winters.

CULTIVATION: *Grow in well-drained, poor to moderately fertile soil, in full sun. Trim or lightly cut back shoots that spoil the symmetry, after flowering.*

☼ ◊ ❀❀   ↕15cm (6in) ↔1.5m (5ft)

Rubus 'Benenden'

This flowering raspberry is an ornamental, deciduous shrub, which has spreading, arching, thornless branches and peeling bark. It is valued for its abundance of large, saucer-shaped, rose-like flowers with glistening, pure white petals in late spring and early summer. The lobed leaves are dark green. Suitable for a shrub border.

CULTIVATION: *Grow in any rich, fertile soil, in full sun or partial shade. After flowering, occasionally remove some old stems to the base to relieve overcrowding and to promote new growth.*

☼ ☀ ◊ ♀ ❀❀❀   ↕↔3m (10ft)

Rubus thibetanus

The ghost bramble is an upright, summer-flowering, deciduous shrub so named for its conspicuously white-bloomed, prickly stems in winter. The small, saucer-shaped, red-purple flowers are carried amid fern-like, white-hairy, dark green leaves, followed by spherical black fruits, also with a whitish bloom. *R. cockburnianus* is another bramble valued for its white winter stems.

CULTIVATION: *Grow in any fertile soil, in sun or partial shade. Each spring, cut all flowered stems back to the ground, leaving the previous season's new, unflowered shoots unpruned.*

☼ ☀ ◊ ♀ ❀❀❀   ↕↔2.5m (8ft)

Rudbeckia fulgida var. *sullivantii* 'Goldsturm'

This black-eyed Susan is a clump-forming perennial valued for its strongly upright form and large, daisy-like, golden-yellow flowerheads with cone-shaped, blackish-brown centres. These appear above the substantial clumps of lance-shaped, mid-green leaves during late summer and autumn. It is a bold addition to a late summer border, and the cut flowers last reasonably well in water.

CULTIVATION: *Grow in any moist but well-drained soil that does not dry out, in full sun or light shade.*

☼ ☀ ◊ ◊ ♀ ✿ ✿ ✿
‡to 60cm (24in) ↔ 45cm (18in)

Rudbeckia laciniata 'Goldquelle'

A tall but compact perennial that bears large, fully double, bright lemon-yellow flowers from midsummer to mid-autumn. These are carried above loose clumps of deeply-divided, mid-green leaves. The flowers are good for cutting.

CULTIVATION: *Grow in any moist but well-drained soil, in full sun or light dappled shade.*

☼ ☀ ◊ ◊ ✿ ✿ ✿
‡to 90cm (36in) ↔ 45cm (18in)

Salix babylonica var. *pekinensis* 'Tortuosa'

The dragon's claw willow is a fast-growing, upright, deciduous tree with curiously twisted shoots that are striking in winter. In spring, yellow-green catkins appear with the contorted, bright green leaves with grey-green undersides. Plant away from drains, as roots are invasive and water-seeking. Also sold as *S. matsudana* 'Tortuosa'.

CULTIVATION: *Grow in any but very dry or shallow, chalky soil. Choose a sunny site. Thin occasionally in late winter to stimulate new growth, which most strongly exhibits the fascinating growth pattern.*

☼ ◊ ✿✿✿   ↕15m (50ft) ↔8m (25ft)

Salix 'Boydii'

This tiny, very slow-growing, upright, deciduous shrub, with gnarled branches, is suitable for planting in a rock garden or trough. The small, almost rounded leaves are rough-textured, prominently veined and greyish-green. Catkins are only produced occasionally, in early spring.

CULTIVATION: *Grow in any deep, moist but well-drained soil, in full sun; willows dislike shallow, chalky soil. When necessary, prune in late winter to maintain a healthy framework.*

☼ ◊ ◊ ♀ ✿✿✿
↕30cm (12in) ↔20cm (8in)

Salix caprea 'Kilmarnock'

The Kilmarnock willow is a small, weeping, deciduous tree ideal for a small garden. It forms a dense, umbrella-like crown of yellow-brown shoots studded with silvery catkins in mid- and late spring, before the foliage appears. The broad, toothed leaves are dark green on top, and grey-green beneath.

CULTIVATION: *Grow in any deep, moist but well-drained soil, in full sun. Prune annually in late winter to prevent the crown becoming congested. Remove shoots that arise on the clear trunk.*

☼ ◊ ◊ ❀ ❀ ❀
‡1.5–2m (5–6ft) ↔2m (6ft)

Salix hastata 'Wehrhahnii'

This small, slow-growing, upright, deciduous shrub, with dark purple-brown stems and contrasting silvery-grey, early spring catkins, makes a beautiful specimen for winter colour displays. The leaves are oval and bright green.

CULTIVATION: *Grow in any moist soil, in sun; does not tolerate shallow, chalky soils. Prune in spring to maintain a balance between young stems, which usually have the best winter colour, and older wood with catkins.*

☼ ◊ ♀ ❀ ❀ ❀
‡↔1m (3ft)

Salix lanata

The woolly willow is a rounded,
slow-growing, deciduous shrub with
stout shoots which have an attractive
white-woolly texture when young.
Large, upright, golden to grey-yellow
catkins emerge on older wood in late
spring among the dark green, broadly
oval, silvery-grey woolly leaves.

CULTIVATION: *Grow in moist but well-
drained soil, in sun. Tolerates semi-shade,
but dislikes shallow, chalky soil. Prune
occasionally in late winter or early spring
to maintain a balance between old and
young stems.*

☼ ☀ ◊ ◑ ♀ ❉ ❉ ❉ ‡1m (3ft) ↔1.5m (5ft)

Salix reticulata

This little willow shrub has trailing
branches that bear rounded, deeply
veined, dark green leaves, which
are white-hairy beneath. In spring,
it produces upright, pink-tipped
catkins, and these are a useful source
of pollen for emerging bumblebees.
The plant makes good ground cover
for the front of a border.

CULTIVATION: *Grow in any poor to fertile,
well-drained soil in sun.*

☼ ◊ ♀ ❉ ❉ ❉
‡8–10cm (3–4in) ↔30cm (12in)

Salvia argentea

This short-lived perennial forms large clumps of soft, felty grey leaves around the base of the plant. Spikes of hooded, two-lipped, white or pinkish-white flowers are borne in mid- and late summer, on strong, upright stems. Suits a Mediterranean-style border, but will need shelter in areas with cold winters.

CULTIVATION: *Grow in light, very well-drained soil, in a warm, sunny site. Use a cloche or glass panel to protect from excessive winter wet and cold winds.*

☼ ◊ ✿✿   ↕90cm (36in) ↔60cm (24in)

Salvia cacaliifolia

Usually grown as an annual in cool climates, this upright and hairy, herbaceous perennial bears spikes of deep blue flowers in early summer, held above mid-green foliage. A distinctive sage with a brilliant flower colour, for bedding, summer infilling, or containers.

CULTIVATION: *Grow in light, moderately fertile, moist but well-drained soil enriched with organic matter. Choose a site in full sun to light dappled shade.*

☼ ☀ ◊ ◊ ♀ ✿   ↕90cm (36in) ↔30cm (12in)

Salvia coccinea 'Pseudococcinea'

A bushy, short-lived perennial often grown as an annual in frost-prone climates, with toothed, dark green, hairy leaves and loose, slender spikes of soft cherry-red flowers from summer to autumn. An exotic-looking addition to a summer display, and good in containers.

CULTIVATION: *Grow in light, moderately fertile well-drained soil, rich in organic matter. Site in full sun.*

☼ ◊ ✿ ✿ ‡60cm (24in) ↔30cm (12in)

Salvia discolor

This upright, herbaceous perennial, which is normally treated as a summer annual in cool climates, is valued for both its flowers and foliage. Its green leaves have a densely white-woolly surface, forming an unusual display in themselves until the long spikes of deep purplish-black flowers extend above them in late summer and early autumn.

CULTIVATION: *Thrives in light, moderately fertile, moist but well-drained, humus-rich soil. A position in full sun or light shade is best.*

☼ ☀ ◊ ♦ ❀ ‡45cm (18in) ↔30cm (12in)

Salvia fulgens

An upright, evergreen, summer-flowering subshrub bearing spikes of tubular, two-lipped red flowers. The oval, toothed, or notched leaves are rich green above and densely white-woolly beneath. Provides brilliant colour for bedding or containers. May not survive winter in frost-prone climates.

CULTIVATION: *Grow in light, moist but well-drained, moderately fertile, humus-rich soil. Site in full sun or semi-shade.*

☼ ☀ ◊ ♀ ❀ ❀ ‡15cm (6in) ↔20cm (8in)

Salvia guaranitica 'Blue Enigma'

A subshrubby perennial that grows well as a summer annual for bedding displays in cool climates. It is admired for its deep blue flowers, more fragrant than those of the species, *S. guaranitica*, which tower above the mid-green foliage from the end of summer until late autumn.

CULTIVATION: *Grow in light, moderately fertile, moist but well-drained soil enriched with organic matter. Choose a site in full sun to light dappled shade.*

☼ ☀ ◊ ♀ ❀❀ ‡1.5m (5ft) ↔90cm (36in)

Salvia x *jamensis* 'Hot Lips'

This eye-catching salvia has very striking red and white flowers that appear throughout summer. Unlike many salvias, it is reasonably frost-hardy, but in areas with cold winters choose a very sheltered position. Plant in early summer to allow the roots to establish well before winter. The aromatic leaves smell of mint when crushed.

CULTIVATION: *Grow in well-drained soil in full sun.*

☼ ◊ ❀ ❀
‡to 75cm (30in) ↔75cm (30in)

Salvia leucantha

The Mexican bush is a small evergreen shrub that must be grown under glass in cool climates for its winter flowers to be seen. The leaves are mid-green, white-downy beneath, and long spikes of white flowers with purple to lavender-blue calyces appear in winter and spring. Beautiful year-round in a greenhouse border or large container.

CULTIVATION: *Under glass, grow in potting compost in full light with shade from hot sun, and water moderately while in flower. Otherwise, grow in moist but well-drained, fertile soil in sun or partial shade.*

☼ ☀ ◊ ◗ ♡❀
‡60–100cm (24–39in) ↔40–90cm (16–36in)

Salvia microphylla 'Pink Blush'

An unusual flower colour for this shrubby perennial, with mid-green leaves and tall, slender spires of intense fuchsia-pink flowers. Lovely against a sunny, sheltered wall or fence. May not survive harsh winters. 'Kew Red' and 'Newby Hall' are also recommended, both with red flowers.

CULTIVATION: *Grow in light, moist but well-drained, moderately fertile soil that is rich in organic matter, in full sun. Plants damaged by frost may be trimmed, but do not cut into old wood.*

☼ ◊ ♀ ❀ ❀
‡90cm (3ft) ↔60cm (2ft)

Salvia officinalis 'Icterina'

A very attractive, yellow and green variegated form of sage that has a mound-forming, subshrubby habit. The aromatic, evergreen, velvety leaves can be used in cooking. Less significant spikes of small, lilac-blue flowers appear in early summer. Ideal for a herb or kitchen garden. 'Kew Gold' is a very similar plant, although its leaves are often completely yellow.

CULTIVATION: *Grow in moist but well-drained, fairly fertile, humus-rich soil. Site in full sun or partial shade.*

☼ ◊ ♀ ❀ ❀　‡to 80cm (32in) ↔1m (3ft)

Salvia officinalis Purpurascens Group

Purple sage is an upright, evergreen subshrub, suitable for a sunny border. Its red-purple young leaves and spikes of lilac-blue flowers are an attractive combination, the latter appearing during the first half of summer. The foliage is aromatic and can be used for culinary purposes. A useful plant to add colour to a herb garden.

CULTIVATION: *Grow in light, moderately fertile, humus-rich, moist but well-drained soil in full sun or light shade. Trim to shape each year after flowering.*

☀ ☀ ◊ ◑ ✿ ✿ ✿
‡to 80cm (32in) ↔1m (3ft)

Salvia officinalis 'Tricolor'

This variegated sage is an upright, evergreen perennial with grey-green woolly, aromatic leaves with cream and pink to beetroot purple marking. In early to midsummer, it carries spikes of lilac-blue flowers, attractive to butterflies. Less hardy than the species, so choose a warm, sheltered site, or grow for summer display.

CULTIVATION: *Grow in moist but well-drained, reasonably fertile soil enriched with organic matter, in full sun or light dappled shade. Shelter from winter wet and cold, drying winds, and trim back untidy growth each year after flowering.*

☀ ☀ ◊ ◑ ✿ ✿   ‡80cm (32in) ↔1m (3ft)

Salvia patens 'Cambridge Blue'

This lovely cultivar of *S. patens* is an upright perennial, with tall, loose spikes of pale blue flowers. It is a striking addition to a herbaceous or mixed border, or bedding and patio tubs. The flowers are borne during midsummer to mid-autumn, above the oval, hairy, mid-green leaves. In frost-prone climates, shelter at the base of a warm wall.

CULTIVATION: *Grow in well-drained soil, in full sun. Overwinter young plants in frost-free conditions.*

☼ ◊ ♀ ❀ ❀
‡45–60cm (18–24in) ↔ 45cm (18in)

Salvia pratensis Haematodes Group

A short-lived perennial, sometimes sold as *S. haematodes*, forming basal clumps of large, dark green leaves. In early and midsummer, spreading spikes of massed blue-violet flowers, with paler throats, emerge from the centre of the clumps. Provides colour for bedding, infilling in beds and borders, or containers. 'Indigo' is another attractive recommended cultivar.

CULTIVATION: *Grow in moist but well-drained, moderately fertile, humus-rich soil. Site in full sun or light shade.*

☼ ◊ ♀ ❀ ❀ ❀
‡to 90cm (36in) ↔ 30cm (12in)

Sarcococca hookeriana var. *digyna*

This perfumed, evergreen shrub is very similar to *S. confusa* (see facing page, below), but with a more compact and spreading habit. The tiny, fragrant white flowers, which are followed by small, black or blue-black fruits, have pink anthers; they are borne amid glossy leaves, more slender and pointed than those of *S. hookeriana*, in winter. The flowers are good for cutting.

CULTIVATION: *Grow in moist but well-drained, fertile, humus-rich soil, in shade. Dig up spreading roots in spring to confine to its allotted space.*

☼ ☀ ◐ ❀❀❀ ‡1.5m (5ft) ↔2m (6ft)

Satureja montana

Winter savory is an attractive herb that is not often seen in gardens. It forms a small semi-evergreen subshrub covered in small, dark green, slightly greyish leaves. Lavender-pink to purple, rosemary-like flowers clothe the bush all summer and are attractive to insects. The aromatic foliage has a spicy scent and can be used to flavour meat.

CULTIVATION: *Grow in reliably well-drained, neutral to slightly alkaline soil in full sun.*

☼ ◐ ❀❀❀ ‡40cm (16in) ↔20cm (8in)

Saxifraga 'Jenkinsiae'

This neat and slow-growing, evergreen perennial forms very dense cushions of grey-green foliage. It produces an abundance of solitary, cup-shaped, pale pink flowers with dark centres in early spring, on short, slender red stems. Good for rock gardens or troughs.

CULTIVATION: *Best in moist but sharply drained, moderately fertile, neutral to alkaline soil, in full sun. Provide shade from the hottest summer sun.*

☼ ◊ ❀ ❀ ❀ ‡5cm (2in) ↔20cm (8in)

Saxifraga Southside Seedling Group

A mat-forming, evergreen perennial that is suitable for a rock garden. Open sprays of small, cup-shaped white flowers, spotted heavily with red, are borne in late spring and early summer. The oblong to spoon-shaped, pale green leaves form large rosettes close to soil level.

CULTIVATION: *Grow in very sharply drained, moderately fertile, alkaline soil. Choose a position in full sun.*

☼ ◊ ❀ ❀ ❀ ‡30cm (12in) ↔20cm (8in)

Scabiosa caucasica 'Clive Greaves'

This delicate, perennial scabious, with solitary, lavender-blue flowerheads, is ideal for a cottage garden. The blooms have pincushion-like centres, and are borne above the clumps of grey-green leaves during mid- to late summer. The flowerheads cut well for indoor arrangements.

CULTIVATION: *Grow in well-drained, moderately fertile, neutral to slightly alkaline soil, in full sun. Deadhead to prolong flowering.*

☼ ◊ ♀ ❀❀❀　　　　　↔60cm (24in)

Scabiosa caucasica 'Miss Willmott'

A clump-forming perennial that is very similar to 'Clive Greaves' (above), but with white flowerheads. These are borne in mid- to late summer and are good for cutting. The lance-shaped leaves are grey-green and arranged around the base of the plant. Suits a cottage garden.

CULTIVATION: *Grow in well-drained, moderately fertile, neutral to slightly alkaline soil, in full sun. Deadhead to prolong flowering.*

☼ ◊ ♀ ❀❀❀
‡90cm (36in) ↔60cm (24in)

Schizostylis coccinea 'Sunrise'

This clump-forming, vigorous perennial bears upright spikes of salmon-pink flowers which open in autumn. The long leaves are sword-shaped and ribbed. Good in sheltered spots, for a border front, or above water level in a waterside planting. When cut, the flowers last well in water.

CULTIVATION: *Best in moist but well-drained, fertile soil, in sun. Naturally forms congested clumps, but these are easily lifted and divided in spring.*

☼ ◊ ❀❀❀
‡to 60cm (24in) ↔30cm (12in)

Sciadopitys verticillata

Japanese umbrella pine is a slow-growing, broad-needled evergreen tree that has a pyramidal growth habit when young and peeling red-brown bark. As it ages, it develops a loose habit with drooping, spreading branches. The distinctive whorls of needles at the end of the shoots make it an ideal accent tree or an interesting addition to a group of mixed evergreens.

CULTIVATION: *Grow in fertile, moist but well-drained, neutral to slightly acid soil, in full sun with some midday shade, or in partial shade.*

☼ ☼ ◊ ♀ ❀❀❀
‡10–20m (30–70ft) ↔6–8m (20–25ft)

Scilla bifolia

A small, bulbous perennial bearing
early-spring flowers which
naturalizes well under trees and
shrubs, or in grass. The slightly
one-sided spikes of many star-shaped,
blue to purple-blue flowers are
carried above the clumps of narrow,
basal leaves.

CULTIVATION: *Grow in well-drained,*
moderately fertile, humus-rich soil,
in full sun or partial shade.

☼ ◊ ♀ ❀❀❀
↕8–15cm (3–6in) ↔5cm (2in)

Scilla mischtschenkoana 'Tubergeniana'

This dwarf, bulbous perennial has
slightly earlier flowers than *S. bifolia*
(above), which are silvery-blue with
darker stripes. They are grouped
together in elongating spikes,
appearing at the same time as the
semi-upright, narrow, mid-green
leaves. Naturalizes in open grass.
Also known as *S. tubergeniana.*

CULTIVATION: *Grow in well-drained,*
moderately fertile soil that is rich in
well-rotted organic matter, in full sun.

☼ ◊ ♀ ❀❀❀
↕10–15cm (4–6in) ↔5cm (2in)

Sedum kamtschaticum 'Variegatum'

This clump-forming, semi-evergreen perennial has eye-catching, fleshy leaves, mid-green with pink tints and cream margins. During late summer, these contrast nicely with flat-topped clusters of small, star-shaped, yellow flowers which age to crimson later in the season. Suitable for rock gardens and borders.

CULTIVATION: *Grow in well-drained, gritty, fertile soil. Choose a site in full sun, but will tolerate light shade.*

☼ ◊ ♀ ✤✤✤ ‡10cm (4in) ↔25cm (10in)

Sedum 'Ruby Glow'

This low-growing perennial is an ideal choice for softening the front of a mixed border. It bears masses of small, star-shaped, ruby-red flowers from midsummer to early autumn, above clumps of fleshy, green-purple leaves. The nectar-rich flowers will attract bees, butterflies, and other beneficial insects.

CULTIVATION: *Grow in well-drained, fertile soil that has adequate moisture in summer. Position in full sun.*

☼ ◊ ♀ ✤✤✤
‡25cm (10in) ↔45cm (18in)

Sedum spathulifolium 'Cape Blanco'

A vigorous, evergreen perennial that forms a mat of silvery-green foliage, often tinted bronze-purple, with a heavy bloom of white powder over the innermost leaves. Small clusters of star-shaped, bright yellow flowers are borne just above the leaves in summer. A very attractive addition to a trough or raised bed.

CULTIVATION: *Grow in well-drained, moderately fertile, gritty soil. Position in full sun, but tolerates light shade.*

☀ ◊ ♀ ❁ ❁ ❁ ‡10cm (4in) ↔60cm (24in)

Sedum spathulifolium 'Purpureum'

This fast-growing, summer-flowering perennial forms tight, evergreen mats of purple-leaved rosettes; the central leaves are covered with a thick, silvery bloom. Flat clusters of small, star-shaped, bright yellow flowers appear throughout summer. Suitable for a rock garden, or the front of a sunny, well-drained border.

CULTIVATION: *Grow in gritty, moderately fertile soil with good drainage, in sun or partial shade. Trim occasionally to prevent encroachment on other plants.*

☀ ◑ ◊ ♀ ❁ ❁ ❁
‡10cm (4in) ↔60cm (24in)

Sedum spectabile 'Brilliant'

This cultivar of *S. spectabile*, the ice plant, is a clump-forming, deciduous perennial with brilliant pink flowerheads, excellent for the front of a border. The small, star-shaped flowers, packed into dense, flat clusters on fleshy stems, appear in late summer above the succulent, grey-green leaves. The flowerheads are attractive to bees and butterflies, and dry well on or off the plant.

CULTIVATION: *Grow in well-drained, fertile soil with adequate moisture during summer, in full sun.*

☼ ◊ ♥ ❀❀❀   ↕↔45cm (18in)

Sedum spurium 'Schorbuser Blut'

This vigorous, evergreen perennial forms mats of succulent, mid-green leaves which become purple-tinted when mature. Rounded clusters of star-shaped, deep pink flowers are borne during late summer. Suitable for a rock garden.

CULTIVATION: *Grow in well-drained, moderately fertile, neutral to slightly alkaline soil, in full sun. Tolerates light shade. To improve flowering, divide the clumps or mats every 3 or 4 years.*

☼ ☀ ◊ ♥ ❀❀❀
↕10cm (4in) ↔ 60cm (24in)

Sedum telephium Atropurpureum Group

This clump-forming, deciduous perennial is valued for its very dark purple foliage which contrasts well with other plants. During summer and early autumn, attractive pink flowers with orange-red centres are clustered above the oval, slightly scalloped leaves.

CULTIVATION: *Grow in well-drained, moderately fertile, neutral to slightly alkaline soil, in full sun. Divide clumps every 3 or 4 years to improve flowering.*

☼ ◊ ❀❀❀
‡45–60cm (18–24in) ↔30cm (12in)

Semiarundinaria fastuosa

Narihira bamboo makes an impressive plant, with very tall, straight and thick canes. These are shiny and mid-green with purplish stripes, particularly when young, and the leaves are glossy and green. Ideal for screening, as a windbreak, or as a specimen; it can form large clumps, but keep the roots confined using a strong root barrier to prevent the clump from spreading.

CULTIVATION: *Grow in moist but well-drained soil in sun or partial shade.*

☼ ☀ ◊ ◗ ♀ ❀❀❀
‡to 7m (22ft) ↔2m (6ft) or more

Sempervivum arachnoideum

The cobweb houseleek is a mat-forming, evergreen succulent, so-named because the foliage is webbed with white hairs. The small, fleshy, mid-green to red leaves are arranged in tight rosettes. In summer, flat clusters of star-shaped, reddish-pink flowers appear on leafy stems. Suitable for growing in a scree bed, wall crevice, or trough.

CULTIVATION: *Grow in gritty, sharply drained, poor to moderately fertile soil. Choose a position in full sun.*

☼ ◊ ♀ ❀❀❀ ‡8cm (3in) ↔30cm (12in)

Sempervivum ciliosum

A mat-forming, evergreen succulent carrying very hairy, dense rosettes of incurved, lance-shaped, grey-green leaves. It bears flat, compact heads of star-shaped, greenish-yellow flowers throughout summer. The rosettes of leaves die after flowering, but are rapidly replaced. Best in an alpine house in areas prone to very damp winters.

CULTIVATION: *Grow in gritty, sharply drained, poor to moderately fertile soil, in sun. Tolerates drought conditions, but dislikes winter wet or climates that are warm and humid.*

☼ ◊ ♀ ❀❀❀ ‡8cm (3in) ↔30cm (12in)

Sempervivum tectorum

The common houseleek is a vigorous, mat-forming, evergreen succulent with large, open rosettes of thick, oval, bristle-tipped, blue-green leaves, often suffused red-purple. In summer, dense clusters of star-shaped, red-purple flowers appear on upright, hairy stems. Very attractive growing on old roof tiles or among terracotta fragments.

CULTIVATION: *Grow in gritty, sharply drained, poor to moderately fertile soil. Choose a site in full sun.*

☼ ◊ ♀ ❀❀❀ ‡15cm (6in) ↔50cm (20in)

Sidalcea 'Elsie Heugh'

The prairie mallow bears spires of satiny, purple-pink flowers with prettily fringed petals over long periods from midsummer. These appear over a mound of lobed, glossy bright green leaves. Grow in a mixed or herbaceous border. The flowers are good for cutting; deadhead after flowering to encourage more blooms. Plants may need staking.

CULTIVATION: *Best in well-drained soil in sun.*

☼ ☀ ◊ ♀ ❀❀❀
↔90cm (36in) ↔45cm (18in)

Silene schafta

A clump-forming, spreading, semi-evergreen perennial with floppy stems bearing small, bright green leaves. Profuse sprays of long-tubed, deep magenta flowers with notched petals are borne from late summer to autumn. Suitable for a raised bed or rock garden.

CULTIVATION: *Grow in well-drained, neutral to slightly alkaline soil, in full sun or light dappled shade.*

☼ ☼ ◊ ♀ ✿✿✿
‡25cm (10in) ↔30cm (12in)

Silphium perfoliatum

The cup plant forms a big, imposing clump of resinous and aromatic dark green foliage above which branching heads of cup-shaped, yellow daisy flowers emerge from midsummer to autumn. They attract a variety of pollinating insects, which makes this plant ideal for wildlife gardens. It also tolerates poorly drained sites.

CULTIVATION: *Grow in sun or light shade, in a reliably moist soil.*

☼ ☼ ◊ ♀ ✿✿✿
‡2.5m (8ft) ↔1m (3ft)

Skimmia x *confusa* 'Kew Green' (male)

A compact, dome-shaped, evergreen shrub carrying aromatic, pointed, mid-green leaves. Conical spikes of fragrant, creamy-white flowers open in spring. There are no berries, but it will pollinate female skimmias if they are planted nearby. Good in a shrub border or woodland garden.

CULTIVATION: *Grow in moist but well-drained, moderately fertile, humus-rich soil. Tolerates full sun to deep shade, atmospheric pollution, and neglect. Requires little or no pruning.*

☼ ☀ ◐ ♡ ❀ ❀ ❀
‡0.5–3m (1¹/₂–10ft) ↔1.5m (5ft)

Skimmia japonica 'Rubella' (male)

This tough, dome-shaped, evergreen shrub bears dark red flower buds in autumn and winter, opening in spring as fragrant heads of white flowers. The oval leaves have red rims. No berries are produced, but it will pollinate female skimmias nearby. (Where space permits only one plant, 'Robert Fortune' produces both flowers and berries.) Tolerates pollution and coastal conditions.

CULTIVATION: *Grow in moist, fertile, neutral to slightly acid soil, in partial or full shade. Requires little pruning, but cut back any shoots that spoil the shape.*

☼ ☀ ◐ ♡ ❀ ❀ ❀   ‡↔to 6m (20ft)

Solanum crispum 'Glasnevin'

This long-flowering Chilean potato tree is a fast-growing, scrambling, woody-stemmed, evergreen climber. Fragrant, deep purple-blue flowers, borne in clusters at the tips of the stems during summer and autumn, are followed by small, yellow-white fruits. The leaves are oval and dark green. Slightly tender in cold areas, so grow on a warm, sunny wall.

CULTIVATION: *Grow in any moist but well-drained, moderately fertile soil, in full sun or semi-shade. Cut back weak and badly placed growth in spring. Tie to a support as growth proceeds.*

☼ ☀ ◊ ◖ ♀ ❀❀   ‡6m (20ft)

Solanum laxum 'Album'

This white-flowered potato vine is a scrambling, woody-stemmed, semi-evergreen climber. It produces broad clusters of fragrant, star-shaped, milk-white flowers, with prominent, lemon-yellow anthers, from summer into autumn. They are followed by black fruits. The leaves are dark green and oval. May not survive in cold areas.

CULTIVATION: *Grow in any moist but well-drained, fertile soil, in full sun or semi-shade. Thin out shoots in spring. The climbing stems need support.*

☼ ☀ ◊ ◖ ♀ ❀❀   ‡6m (20ft)

Solidago 'Goldenmosa'

This compact, vigorous goldenrod is a bushy perennial topped with bright golden-yellow flowerheads in late summer and early autumn. The leaves are wrinkled and mid-green. Valuable in a wild garden or for late summer colour; the flowers are good for cutting. Can be invasive.

CULTIVATION: *Grow in well-drained, poor to moderately fertile, preferably sandy soil, in full sun. Remove flowered stems to prevent self-seeding.*

☀ ◊ ♀ ❀❀❀
‡to 75cm (30in) ↔ 45cm (18in)

Sophora SUN KING ('Hilsop')

This shrubby, evergreen relative of the Japanese pagoda tree has dark green leaves divided into many small leaflets. It is valued for its deep yellow flowers that appear over a long period from late winter and into early spring, when few other plants are in flower. The display is particularly good after a long, hot summer. Grow as a specimen shrub or in a mixed border.

CULTIVATION: *Best in well-drained soil in sun.*

☀ ◊ ❀❀❀ ‡↔3m (10ft)

Sorbus aria 'Lutescens'

This compact whitebeam is a broadly columnar, deciduous tree, bearing oval and toothed, silvery-grey foliage which turns russet and gold in autumn. Clusters of white flowers appear in late spring, followed by brown-speckled, dark red berries. It makes a beautiful specimen tree, tolerating a wide range of conditions. 'Majestica' is similar but taller, with larger leaves.

CULTIVATION: *Grow in moist but well-drained, fertile soil, in sun. Tolerates heavy clay soils, semi-shade, urban pollution and exposed conditions. Remove any dead wood in summer.*

☼ ☀ ◊ ◊ ♀ ❀❀❀
‡10m (30ft) ↔8m (25ft)

Sorbus hupehensis var. *obtusa*

This variety of the Hubei rowan, *S. hupehensis*, is an open and spreading tree which gives a fine display of autumn colour. Broad clusters of white flowers in late spring are followed by round white berries; these ripen to dark pink later in the season. The blue-green leaves, divided into many leaflets, turn scarlet before they fall.

CULTIVATION: *Grow in any moist but well-drained soil, preferably in full sun, but tolerates light shade. Remove any dead or diseased wood in summer.*

☼ ☀ ◊ ◊ ♀ ❀❀❀   ‡↔to 8m (25ft)

Sorbus 'Joseph Rock'

This broadly columnar, upright, deciduous tree has bright green leaves which are divided into many sharply-toothed leaflets. These colour attractively to orange, red and purple in autumn. In late spring, white flowers appear in broad clusters, followed by round, pale yellow berries which ripen to orange-yellow.

CULTIVATION: *Grow in moist but well-drained, fertile soil, in sun. Very prone to fireblight, the main sign of which is blackened leaves; affected growth must be pruned back in summer to at least 60cm (24in) below the diseased area.*

☼ ◊ ◑ ❀❀❀   ‡10m (30ft) ↔7m (22ft)

Sorbus reducta

A deciduous shrub that forms a low thicket of upright branches. Much-valued for its ornamental, dark green foliage which turns a rich red in autumn. Small, open clusters of white flowers appear in late spring, followed by white, crimson-flushed berries. Tolerates pollution.

CULTIVATION: *Grow in well-drained, moderately fertile soil, in an open, sunny site. To thin congested plants, remove shoots that arise from the base while they are still young and soft.*

☼ ◊ ♡ ❀❀❀
‡1–1.5m (3–5ft) ↔2m (6ft)

Sorbus vilmorinii

A spreading shrub or small tree with elegant, arching branches bearing dark green leaves divided into many leaflets. The deciduous foliage gives a lovely display in autumn, turning orange- or bronze-red. Clusters of white flowers appear in late spring and early summer, followed later in the season by dark red berries which age to pink then white.

CULTIVATION: *Grow in well-drained, moderately fertile, humus-rich soil, in full sun or dappled shade. Remove any dead or diseased wood in summer.*

☼ ☀ ◊ ♀ ✿ ✿ ✿   ‡↔5m (15ft)

Spigelia marilandica

Indian pink is a pretty clump-forming woodland perennial. In early summer, red tubular flowers with canary yellow centres are borne just above the foliage. It is ideal for shaded borders or woodland gardens and it attracts butterflies.

CULTIVATION: *Thrives in moist, well drained, humus-rich soil in partial shade. Cut back after flowering to encourage a second flush of flowers in autumn. Toxic if ingested.*

☀ ◊ ✿ ✿ ✿
‡30–60cm (12–24in) ↔60cm (24in)

Spiraea japonica 'Anthony Waterer'

A compact, deciduous shrub that makes a good informal flowering hedge. The lance-shaped, dark green leaves, usually margined with creamy-white, are red when young. Dense heads of tiny pink flowers are borne amid the foliage in mid- to late summer.

CULTIVATION: *Grow in any well-drained, fairly fertile soil that does not dry out, in full sun. On planting, cut back stems to leave a framework 15cm (6in) high; prune back close to this every year in spring. Deadhead after flowering.*

☼ ◊ ❀ ❀ ❀   ↕↔to 1.5m (5ft)

Spiraea japonica 'Goldflame'

This compact, deciduous, flowering shrub bears pretty, bright yellow leaves which are bronze-red when young. Dense, flattened heads of tiny, dark pink flowers appear at the tips of slightly arching stems during mid- and late summer. Ideal for a rock garden. 'Nana' is even smaller, to 45cm (18in) tall.

CULTIVATION: *Grow in well-drained soil that does not dry out completely, in full sun. On planting, cut back stems to a framework 15cm (6in) high; prune back close to this each year in spring. Deadhead after flowering.*

☼ ◊ ❀ ❀ ❀   ↕↔75cm (30in)

Spiraea nipponica 'Snowmound'

This fast-growing and spreading, deciduous shrub has arching, reddish-green stems. The dense clusters of small white flowers in midsummer make an invaluable contribution to any shrub border. The rounded leaves are bright green when young, darkening as they age.

CULTIVATION: *Grow in any moderately fertile soil that does not dry out too much during the growing season, in full sun. Cut back flowered stems in autumn, and remove any weak growth.*

☼ ◊ ♀ ✿✿✿ ↕↔1.2–2.5m (4–8ft)

Spiraea x vanhouttei

Bridal wreath is a fast-growing, deciduous shrub, more compact in habit than *S. nipponica* 'Snowmound' (above), but with similar mounds of white flowers during early summer. The diamond-shaped leaves are dark green above with blue-green under-sides. Grow as an informal hedge, or in a mixed border.

CULTIVATION: *Grow in any well-drained, fertile soil that does not dry out, in sun. In autumn, cut back flowered stems, removing any weak or damaged growth.*

☼ ◊ ✿✿✿ ↕2m (6ft) ↔1.5m (5ft)

Stachyurus praecox

This spreading, deciduous shrub
bears oval, mid-green leaves on
arching, red-purple shoots. Hanging
spikes of tiny, bell-shaped, pale
yellow-green flowers appear on the
bare stems in late winter and early
spring. Suitable for a shrub border,
and lovely in a woodland garden.

CULTIVATION: *Grow in moist but well-
drained, humus-rich, fertile, neutral to
acid soil. Prefers partial shade, but will
tolerate full sun if soil is kept reliably
moist. Regular pruning is unnecessary.*

☀ ☀ ◊ ◊ ♀ ❀ ❀ ❀
↕1–4m (3–12ft) ↔3m (10ft)

Stipa gigantea

Golden oats is a fluffy, evergreen,
perennial grass forming dense
tufts of narrow, mid-green leaves.
In summer, these are topped by
silvery- to purplish-green
flowerheads which turn gold when
mature, and persist well into winter.
Makes an imposing feature at
the back of a border.

CULTIVATION: *Grow in well-drained, fertile
soil, in full sun. Remove dead leaves and
flowerheads in early spring.*

☀ ◊ ♀ ❀ ❀ ❀ ↕to 2.5m (8ft) ↔1.2m (4ft)

Styrax japonicus

Japanese snowbell is a gracefully spreading, deciduous tree bearing hanging clusters of fragrant, bell-shaped, dainty white flowers which are often tinged with pink. These appear during early to midsummer, amid oval, rich green leaves which turn yellow or red in autumn. Ideal for a woodland garden.

CULTIVATION: *Grow in moist but well-drained, neutral to acid soil, in full sun with shelter from cold, drying winds. Tolerates dappled shade. Leave to develop naturally without pruning.*

☼ ☀ ◊ ◑ ✽ ✽ ✽
‡10m (30ft) ↔ 8m (25ft)

Styrax obassia

The fragrant snowbell is a broadly columnar, deciduous tree bearing beautifully rounded, dark green leaves which turn yellow in autumn. Fragrant, bell-shaped white flowers are produced in long, spreading clusters in early and midsummer.

CULTIVATION: *Grow in moist but well-drained, fertile, humus-rich, neutral to acid soil, in full sun or partial shade. Shelter from cold, drying winds. Dislikes pruning; leave to develop naturally.*

☼ ☀ ◊ ◑ ✽ ✽ ✽
‡12m (40ft) ↔ 7m (22ft)

Symphytum x uplandicum 'Variegatum'

This upright, clump-forming, bristly perennial has large, lance-shaped, mid-green leaves with broad cream margins. Drooping clusters of pink-blue buds open to blue-purple flowers from late spring to late summer. Best suited to a wild garden or shady border. Less invasive than green-leaved types.

CULTIVATION: *Grow in any moist soil, in sun or partial shade. For the best foliage effect, remove flowering stems before they bloom. Liable to form plain green leaves if grown in poor or infertile soil.*

☼ ◑ ◊ ❀❀❀
‡90cm (36in) ↔60cm (24in)

Syringa meyeri 'Palibin'

This compact, slow-growing, deciduous shrub with a rounded shape is much valued for its abundant clusters of fragrant, lavender-pink flowers in late spring and early summer. The leaves are dark green and oval. Makes a bold contribution to any shrub border. Sometimes seen as *S. palibiniana*.

CULTIVATION: *Grow in deep, moist but well-drained, fertile, preferably alkaline soil, in full sun. Deadhead for the first few years until established. Prune out weak and damaged growth in winter.*

☼ ◊ ◊ ♀ ❀❀❀
‡1.5–2m (5–6ft) ↔1.5m (5ft)

Syringa pubescens subsp. *microphylla* 'Superba'

This upright to spreading, conical, deciduous shrub bears spikes of fragrant, rose-pink flowers at the tips of slender branches. First appearing in spring, they continue to open at irregular intervals until autumn. The oval, mid-green leaves are red-green when young. Makes a good screen or informal hedge. 'Miss Kim' is a smaller, more compact cultivar.

CULTIVATION: *Grow in moist but well-drained, fertile, humus-rich, neutral to alkaline soil, in full sun. Prune out any weak or damaged growth in winter.*

☼ ◊ ◑ ♀ ✸ ✸ ✸
‡to 6m (20ft) ↔ 6m (20ft)

Syringa vulgaris 'Charles Joly'

This dark-purple-flowered form of common lilac is a spreading shrub or small tree. Very fragrant, double flowers appear during late spring and early summer in dense, conical clusters. The deciduous leaves are heart-shaped to oval and dark green. Use as a backdrop in a shrub or mixed border.

CULTIVATION: *Grow in moist but well-drained, fertile, humus-rich, neutral to alkaline soil, in full sun. Young shrubs require minimal pruning; old, lanky stems can be cut back hard in winter.*

☼ ◊ ◑ ♀ ✸ ✸ ✸
‡↔ 7m (22ft)

Syringa vulgaris 'Katherine Havemeyer'

A spreading lilac, forming a large shrub or small tree, producing dense clusters of very fragrant, double, lavender-blue flowers. These open from purple buds in late spring and early summer. The deciduous, mid-green leaves are heart-shaped. 'Madame Antoine Buchner' is a very similar recommended lilac, with slightly pinker flowers.

CULTIVATION: *Grow in moist but well-drained, fertile, neutral to alkaline soil that is rich in humus, in sun. Mulch regularly. Little pruning is necessary, but tolerates hard pruning to renovate.*

☼ ◊ ◑ ♧ ❁ ❁ ❁   ↕↔7m (22ft)

Syringa vulgaris 'Madame Lemoine'

This lilac is very similar in form to 'Katherine Havemeyer' (above), but has compact spikes of large, very fragrant, double white flowers. These are borne in late spring and early summer, amid the deciduous, heart-shaped to oval, mid-green leaves. 'Vestale' is another recommended white lilac, with single flowers.

CULTIVATION: *Grow in deep, moist but well-drained, fertile, neutral to alkaline, humus-rich soil, in sun. Do not prune young plants, but older shrubs can be cut back hard to renovate in winter.*

☼ ◊ ◑ ♧ ❁ ❁ ❁   ↕↔7m (22ft)

Tamarix tetrandra

This large, arching shrub, with feathery foliage on purple-brown shoots, bears plumes of light pink flowers in mid- to late spring. The leaves are reduced to tiny, needle-like scales. Particularly useful on light, sandy soils; in mild coastal gardens, it makes a good windbreak or hedge. For a similar shrub that flowers in late summer, look for *T. ramossissima* 'Rubra'.

CULTIVATION: *Grow in well-drained soil, in full sun. Cut back young plants by almost half after planting. Prune each year after flowering, or the shrub may become top-heavy and unstable.*

☼ ◊ ♀ ❀❀❀ ↕↔3m (10ft)

Tanacetum coccineum 'Brenda'

A bushy, herbaceous perennial, grown for its daisy-like, magenta-pink, yellow-centred flowerheads in early summer, are borne on upright stems above aromatic, finely divided, grey-green foliage. The cut flowers last well in water. 'James Kelway' is very similar, while 'Eileen May Robinson' has pale pink flowers.

CULTIVATION: *Grow in well-drained, fertile, neutral to slightly acid soil, in an open, sunny site. Cut back after the first flush of flowers to encourage a second flowering later in the season.*

☼ ◊ ❀❀❀
↕70–80cm (28–32in) ↔45cm (18in)

Taxus baccata

Yew is a slow-growing, broadly conical, evergreen conifer with needle-like, dark green leaves. Male plants bear yellow cones in spring, and female plants produce cup-shaped, fleshy, bright red fruits in autumn. Excellent as a dense hedge, which can be clipped to shape, and as a backdrop to colourful plants. All parts are toxic.

CULTIVATION: *Grow in any well-drained, fertile soil, in sun to deep shade. Tolerates chalky or acidic soils. Plant both sexes together, for berries. Trim or cut back to renovate, in summer or early autumn.*

☼ ☀ ◊ ♀ ✿ ✿ ✿
‡to 20m (70ft) ↔10m (30ft)

Taxus baccata 'Dovastonii Aurea' (female)

This slow-growing, evergreen conifer has wide-spreading, horizontally tiered branches which weep at the tips. It is smaller than the common yew (above), and has yellow-margined to golden-yellow foliage. Fleshy, bright red fruits appear in autumn. All parts of this plant are poisonous if eaten.

CULTIVATION: *Grow in any well-drained, fertile soil, in sun or deep shade. Tolerates chalky or acidic conditions. Plant close to male yews for a reliable display of autumn berries. Trim or cut back to renovate, in summer or early autumn.*

☼ ☀ ◊ ✿ ✿ ✿
‡3–5m (10–15ft) ↔2m (6ft)

Taxus baccata 'Fastigiata' (female)

Irish yew is a dense, strongly upright, evergreen conifer which becomes columnar with age. The dark green leaves are not two-ranked like other yews, but stand out all around the shoots. All are female, bearing fleshy, berry-like, bright red fruits in late summer. All parts are toxic if eaten.

CULTIVATION: *Grow in any reliably moist soil, but tolerates most conditions including very dry, chalky soils, in full sun or deep shade. Plant with male yews for a reliable crop of berries. Trim or cut back to renovate, in summer or early autumn, if necessary.*

☼ ◐ ◊ ◊ ♀ ❀❀❀
‡to 10m (30ft) ↔ 4m (12ft)

Thalictrum delavayi 'Hewitt's Double'

An upright, clump-forming perennial, more floriferous than *T. delavayi*, that produces upright sprays of long-lasting, pompon-like, rich mauve flowers from midsummer to early autumn. Fnely divided, mid-green leaves are carried on slender stems shaded dark purple. An excellent foil in a herbaceous border to plants with bolder leaves and flowers.

CULTIVATION: *Grow in moist but well-drained, humus-rich soil, in sun or light shade. Divide clumps and replant every few years to maintain vigour.*

☼ ◐ ◊ ◊ ♀ ❀❀❀
‡1.2m (4ft) or more ↔ 60cm (24in)

Thalictrum flavum subsp. *glaucum*

This subspecies of yellow meadow rue is a summer-flowering, clump-forming perennial that bears large, upright heads of fragrant, sulphur-yellow flowers. These are carried above mid-green, divided leaves. Good at the margins of woodland.

CULTIVATION: *Best in moist, humus-rich soil, in partial shade. Tolerates sun and dry soil. Flower stems may need staking.*

☼ ☀ ◊ ◑ ♀ ❀❀❀
‡to 1m (3ft) ↔60cm (24in)

Thuja occidentalis 'Holmstrup'

This shrub-like form of white cedar is a slow-growing conifer with a conical shape. The dense, mid-green leaves are apple-scented and arranged in vertical sprays. Small oval cones appear amid the foliage. Plant alone as a specimen tree, or use as a hedge.

CULTIVATION: *Grow in deep, moist but well-drained soil, in full sun. Shelter from cold, drying winds. Trim as necessary in spring and late summer.*

☼ ◊ ◑ ♀ ❀❀❀
‡to 4m (12ft) ↔3–5m (10–15ft)

Thuja occidentalis 'Rheingold'

This bushy, spreading, slow-growing conifer is valued for its golden-yellow foliage, which is pink-tinted when young, and turns bronze in winter. Small, oval cones are carried amid the billowing sprays of apple-scented, scale-like leaves.
Good as a specimen tree.

CULTIVATION: *Grow in deep, moist but well-drained soil, in a sheltered, sunny site. Trim in spring and late summer, but be careful not to spoil the form.*

☼ ◊ ◊ ♀ ❀❀❀
‡1–2m (3–6ft) ↔3–5m (10–15ft)

Thuja orientalis 'Aurea Nana'

This dwarf Chinese thuja is an oval-shaped conifer with fibrous, red-brown bark. The yellow-green foliage, which fades to bronze over winter, is arranged in flat, vertical sprays. Flask-shaped cones are borne amid the foliage. Good in a rock garden.

CULTIVATION: *Grow in deep, moist but well-drained soil, in sun with shelter from cold, drying winds. Trim in spring and again in late summer as necessary.*

☼ ◊ ◊ ❀❀❀   ‡↔to 60cm (24in)

Thuja plicata 'Stoneham Gold'

This slow-growing, dwarf form of western red cedar is a conical conifer with fissured, red-brown bark and flattened, irregularly arranged sprays of bright gold, aromatic foliage; the tiny, scale-like leaves are very dark green within the bush. The cones are small and elliptic. Ideal for a rock garden.

CULTIVATION: *Grow in deep, moist but well-drained soil, in full sun with shelter from cold, drying winds. Trim in spring and again in late summer.*

☼ ◊ ♦ 🏆 ❄❄❄ ↕↔to 2m (6ft)

Thunbergia grandiflora

The blue trumpet vine is a vigorous, woody-stemmed, evergreen climber which can be grown as an annual in cold climates. Lavender- to violet-blue, sometimes white, trumpet-shaped flowers with yellow throats appear in hanging clusters during summer. The oval to heart-shaped, dark green leaves are softly hairy.

CULTIVATION: *Grow in moist but well-drained, fertile soil or compost, in sun. Provide shade during the hottest part of the day. Give the climbing stems support. Minimum temperature 10°C (50°F).*

☼ ◊ ♦ 🏆 ❀ ↕5–10m (15–30ft)

Thymus pulegioides '**Bertram Anderson**'

A low-growing, rounded, evergreen shrub carrying small, narrow, grey-green leaves, strongly suffused with yellow. They are aromatic and can be used in cooking. Heads of pale lavender-pink flowers are borne above the foliage in summer. Lovely in a herb garden. Sometimes sold as 'Anderson's Gold'.

CULTIVATION: *Best in well-drained, neutral to alkaline soil, in full sun. Trim after flowering, and remove sprigs for cooking as they are needed.*

☼ ◊ ♀ ❀ ❀ ❀
‡to 30cm (12in) ↔ to 25cm (10in)

Thymus serpyllum var. *coccineus*

A mat-forming, evergreen subshrub with finely hairy, trailing stems bearing tiny, aromatic, mid-green leaves. Crimson-pink flowers are borne in congested whorls during summer. Suitable for planting in paving crevices, where the foliage will release its fragrance when stepped on. May also be seen as *T. praecox* 'Coccineus'.

CULTIVATION: *Grow in well-drained, neutral to alkaline, gritty soil. Choose a position in full sun. Trim lightly after flowering to keep the plant neat.*

☼ ◊ ❀ ❀ ❀
‡25cm (10in) ↔ 45cm (18in)

Thymus 'Silver Queen'

This rounded, evergreen shrub is similar to 'Bertram Anderson' (see facing page, above), but with silver-white foliage. Masses of oblong, lavender-pink flowerheads are borne throughout summer. Plant in a herb garden; the aromatic leaves can be used in cooking.

CULTIVATION: *Grow in well-drained, neutral to alkaline soil, in full sun. Trim after flowering, and remove sprigs for cooking as needed.*

☼ ◊ ♀ ✤ ✤ ✤
‡to 30cm (12in) ↔ to 25cm (10in)

Tiarella cordifolia

This vigorous, summer-flowering, evergreen perennial is commonly known as foam flower, and gets its name from the tiny, star-shaped, creamy-white flowers. These are borne in a profusion of upright sprays, above lobed, pale green leaves which turn bronze-red in autumn. Ideal as ground cover in a woodland garden, as is the similar *T. wherryi*.

CULTIVATION: *Best in cool, moist soil that is rich in humus, in deep or light shade. Tolerates a wide range of soil types.*

☼ ☀ ◊ ♀ ✤ ✤ ✤
‡10–30cm (4–12in) ↔ to 30cm (12in)

Tilia 'Petiolaris'

The weeping silver lime makes a broad tree for a large garden. It has graceful branches and the heart-shaped leaves mature from fresh green in spring to a deeper green. They fade to yellow before they fall in autumn. Bees love the richly scented, summer flowers, which are however too small to be of great ornamental interest.

CULTIVATION: *Grow in moist but well-drained soil in sun or light shade.*

☀ ☀ ◊ ◊ ♀ ✿ ✿ ✿
‡30m (100ft) ↔20m (70ft)

Tolmiea menziesii 'Taff's Gold'

A spreading, clump-forming, semi-evergreen perennial carrying ivy-like, long-stalked, pale lime-green leaves, mottled with cream and pale yellow. An abundance of tiny, nodding, slightly scented, green and chocolate-brown flowers appear in slender, upright spikes during late spring and early summer. Plant in groups to cover the ground in a woodland garden.

CULTIVATION: *Grow in moist but well-drained, humus-rich soil, in partial or deep shade. Sun will scorch the leaves.*

☀ ☀ ◊ ◊ ✿ ✿ ✿
‡30–60cm (12–24in) ↔1m (3ft)

Trachelospermum jasminoides

Star jasmine is an evergreen, woody-stemmed climber with attractive, oval, glossy dark green leaves. The very fragrant flowers, creamy-white ageing to yellow, have five twisted petal lobes. They are borne during mid- to late summer, followed by long seed pods. In cold areas, grow in the shelter of a warm wall with a deep mulch around the base of the plant. The variegated-leaved form, 'Variegatum', is also recommended.

CULTIVATION: *Grow in any well-drained, moderately fertile soil, in full sun or partial shade. Tie in young growth.*

☼ ☀ ◊ ♀ ❀ ‡9m (28ft)

Trachycarpus fortunei

The Chusan palm is one of the hardiest palm trees. It has a single, upright stem with a head of fan-shaped, many-fingered, dark green leaves. Small yellow flowers appear in hanging clusters in early summer, followed by small, round black berries on female plants. It makes a good container tree for a warm, sheltered patio, but is also suitable for the open garden.

CULTIVATION: *Best in a sheltered spot in well-drained soil in sun.*

☼ ☀ ◊ ♀ ❀❀ ‡20m (70ft) ↔2.5cm (8ft)

Tradescantia x *andersoniana* 'J.C. Weguelin'

This tufted, clump-forming perennial bears large, pale blue flowers with three wide-open, triangular petals. These appear from early summer to early autumn in paired clusters at the tips of branching stems. The slightly fleshy, mid-green leaves are long, pointed and arching. Effective in a mixed or herbaceous border.

CULTIVATION: *Grow in moist, fertile soil, in sun or partial shade. Deadhead to encourage repeat flowering.*

☼ ☀ ◊ ❀❀❀
‡60cm (24in) ↔ 45cm (18in)

Tradescantia x *andersoniana* 'Osprey'

This clump-forming perennial bears clusters of large white flowers on the tips of the upright stems from early summer to early autumn. Each flower has three triangular petals, surrounded by two leaf-like bracts. The mid-green leaves are narrow and often purple-tinted. A long-flowering plant for a mixed or herbaceous border, lovely mixed with dark blue-flowered 'Isis'.

CULTIVATION: *Grow in moist but well-drained, fertile soil, in sun or partial shade. Deadhead to prevent self-seeding.*

☼ ☀ ◊ ◊ ❀❀❀
‡60cm (24in) ↔ 45cm (18in)

Tricyrtis formosana

An upright, herbaceous perennial grown for its white, purple-spotted, star-shaped flowers on zig-zagging, softly hairy stems. These appear in early autumn above lance-shaped, dark green leaves which clasp the stems. An unusual plant for a shady border or open woodland garden.

CULTIVATION: *Grow in moist soil that is rich in humus. Choose a sheltered site in deep or partial shade. In cold areas, provide a deep winter mulch where there is unlikely to be deep snow cover.*

☀ ☀ ◊ ❀❀❀
‡to 80cm (32in) ↔45cm (18in)

Trillium grandiflorum

Wake robin is a vigorous, clump-forming perennial grown for its large, three-petalled, pure white flowers which often fade to pink. These are carried during spring and summer on slender stems, above a whorl of three large, dark green, almost circular leaves. Effective in the company of hostas. The cultivar 'Flore Pleno' has double flowers.

CULTIVATION: *Grow in moist but well-drained, leafy, neutral to acid soil, in deep or light shade. Provide an annual mulch of leaf mould in autumn.*

☀ ☀ ◊ ◊ ♀ ❀❀❀
‡to 40cm (16in) ↔30cm (12in)

Trollius x *cultorum* 'Orange Princess'

This globeflower is a robust, clump-forming perennial with orange-gold flowers. These are held above the mid-green foliage in late spring and early summer. The leaves are deeply cut with five rounded lobes. Good for bright colour beside a pond or stream, or in a damp border. For bright yellow flowers, choose the otherwise similar 'Goldquelle'.

CULTIVATION: *Best in heavy, moist, fertile soil, in full sun or partial shade. Cut stems back hard after the first flush of flowers to encourage further blooms.*

☼ ☀ ◊ ♀ ❀ ❀ ❀
‡to 90cm (36in) ↔45cm (18in)

Tropaeolum majus 'Hermine Grashoff'

This double-flowered nasturtium is a strong-growing, often scrambling, annual climber. Long-spurred, bright red flowers appear during summer and autumn, above the light green, wavy-margined leaves. Excellent for hanging baskets and other containers.

CULTIVATION: *Grow in moist but well-drained, fairly poor soil, in full sun. The climbing stems need support. Minimum temperature 2˚C (36˚F).*

☼ ◊ ❀
‡1–3m (3–10ft) ↔1.5–5m (5–15ft)

Tropaeolum speciosum

The flame nasturtium is a slender, herbaceous climber, producing long-spurred, bright vermilion flowers throughout summer and autumn. These are followed by small, bright blue fruits. The mid-green leaves are divided into several leaflets. Effective growing through dark-leaved hedging plants, which contrast well with its flowers.

CULTIVATION: *Grow in moist, humus-rich, neutral to acid soil, in full sun or partial shade. Provide shade at the roots, and support the climbing stems.*

☼ ☀ ◊ ◗ ♀ ❀ ❀ ❀   ‡to 3m (10ft)

Tsuga canadensis 'Jeddeloh'

This dwarf form of the eastern hemlock is a small, vase-shaped conifer with deeply furrowed, purplish-grey bark. The bright green foliage is made up of needle-like leaves which are arranged in two ranks along the stems. An excellent small specimen tree for shady places; also popular for bonsai training.

CULTIVATION: *Grow in moist but well-drained, humus-rich soil, in full sun or partial shade. Provide shelter from cold, drying winds. Trim during summer.*

☼ ☀ ◊ ◗ ♀ ❀ ❀ ❀ ‡1.5m (5ft) ↔2m (6ft)

Tulipa clusiana
var. *chrysantha*

This yellow-flowered lady tulip is a
bulbous perennial which flowers in
early- to mid-spring. The bowl- to
star-shaped flowers, tinged red or
brownish-purple on the outsides,
are produced in clusters of up to
three per stem, above the linear,
grey-green leaves. Suitable for a
raised bed or rock garden.

CULTIVATION: *Grow in well-drained, fertile
soil, in full sun with shelter from strong
winds. Deadhead and remove any fallen
petals after flowering.*

☼ ◊ ♀ ❀❀❀   ‡30cm (12in)

Tulipa linifolia

This slender, variable, bulbous
perennial bears bowl-shaped red
flowers in early and mid-spring.
These are carried above the linear,
grey-green leaves with wavy red
margins. The petals have yellow
margins and black-purple marks at
the base. Good for a rock garden.

CULTIVATION: *Grow in sharply drained,
fertile soil, in full sun with shelter from
strong winds. Deadhead and remove any
fallen petals after flowering.*

☼ ◊ ♀ ❀❀❀   ‡20cm (8in)

Tulipa linifolia
Batalinii Group

Slender, bulbous perennials, often
sold as *T. batalinii*, bearing solitary,
bowl-shaped, pale yellow flowers with
dark yellow or bronze marks on the
insides. These appear from early to
mid-spring above linear, grey-green
leaves with wavy red margins.
Use in spring bedding; the
flowers are good for cutting.

CULTIVATION: *Grow in sharply drained,*
fertile soil, in full sun with shelter from
strong winds. Deadhead and remove
any fallen petals after flowering.

☼ ◊ ♀ ✿ ✿ ✿   ‡35cm (14in)

Tulipa turkestanica

This bulbous perennial produces up
to 12 star-shaped white flowers per
stem in early and mid-spring. They
are flushed with greenish-grey on the
outsides and have yellow or orange
centres. The linear, grey-green leaves
are arranged beneath the flowers.
Grow in a rock garden or sunny
border, away from paths or seating
areas as the flowers have an
unpleasant scent.

CULTIVATION: *Grow in well-drained, fertile*
soil, in full sun with shelter from strong
winds. Deadhead and remove any fallen
petals after flowering.

☼ ◊ ♀ ✿ ✿ ✿   ‡30cm (12in)

Tulipa cultivars

These cultivated varieties of tulip are spring-flowering, bulbous perennials, with a much wider range of flower colour than any other spring bulbs, from the buttercup-yellow 'Hamilton' through to the violet-purple 'Blue Heron' and the multi-coloured, red, white and blue 'Union Jack'. This diversity makes them invaluable for bringing variety into the garden, either massed together in large containers or beds, or planted in a mixed border. Flower shape is also varied; as well as the familiar cup-shaped blooms, as in 'Dreamland', there are also conical, goblet- and star-shaped forms. The flowers are good for cutting.

CULTIVATION: *Grow in well-drained, fertile soil, in sun with shelter from strong winds and excessive wet. Remove spent flowers. Lift bulbs once the leaves have died down, and store over summer in a cold greenhouse to ripen. Replant the largest bulbs in autumn.*

☼ ◊ ❀ ❀ ❀

‡15cm (6in)

2
‡60cm (24in)

3
‡50cm (20in)

4
‡40cm (16in)

5
‡50cm (20in)

6
‡30cm (12in)

1 *Tulipa* 'Ancilla' ♀ **2** *T.* 'Blue Heron' ♀ **3** *T.* 'China Pink' ♀
4 *T.* 'Don Quichotte' ♀ **5** *T.* 'Hamilton' **6** *T.* 'Oriental Splendour'

7 ‡60cm (24in)

8 ‡30cm (12in)

9 ‡35cm (14in)

10 ‡60cm (24in)

11 ‡20cm (8in)

12 ‡40cm (16in)

13 ‡60cm (24in)

14 ‡50cm (20in)

7 *T.* 'Dreamland' ♀ **8** *T.* 'Keizerskroon' **9** *T.* 'Prinses Irene' ♀
10 *T.* 'Queen of Sheba' **11** *T.* 'Red Riding Hood' ♀ **12** *T.* 'Spring Green' ♀
13 *T.* 'Union Jack' **14** *T.* 'West Point'

Uvularia grandiflora

Large merrybells is a slow-spreading, clump-forming perennial bearing solitary or paired, narrowly bell-shaped, sometimes green-tinted yellow flowers. They hang gracefully from slender, upright stems during mid- to late spring, above the downward-pointing, lance-shaped, mid-green leaves. Excellent for a shady border or woodland garden.

CULTIVATION: *Grow in moist but well-drained, fertile soil that is rich in humus, in deep or partial shade.*

☀ ☀ ◊ ◊ ♀ ❀❀❀
‡to 75cm (30in) ↔30cm (12in)

Vaccinium corymbosum

The highbush blueberry is a deciduous, acid-soil loving shrub with slightly arching shoots. The mid-green oval leaves turn yellow or red in autumn. Clusters of small, often pink-tinged, white flowers appear in late spring and early summer, followed by sweet, edible, blue-black berries. Best suited to a woodland garden on lime-free soil. Many cultivars are available that are suitable for growing in containers, such as 'Duke' and 'Spartan'.

CULTIVATION: *Grow in moist but well-drained acid soil, in sun or light shade. Plant two or three bushes to optimize pollination. Trim in winter.*

☀ ☀ ◊ ❀❀❀   ‡↔to 1.5m (5ft)

Vaccinium glaucoalbum

A mound-forming, dense, evergreen shrub bearing elliptic, leathery, dark green leaves with bright bluish-white undersides. Very small, pink-tinged white flowers appear in hanging clusters during late spring and early summer, followed by edible, white-bloomed, blue-black berries. Suits a lime-free woodland garden.

CULTIVATION: *Grow in open, moist but well-drained, peaty or sandy, acid soil, in sun or partial shade. Trim in spring.*

☼ ☀ ◊ ◑ ♀ ❀ ❀
‡50–120cm (20–48in) ↔1m (3ft)

Vaccinium vitis-idaea **Koralle Group**

These heavy-fruiting cowberries are creeping, evergreen shrubs with oval, glossy dark green leaves, shallowly notched at the tips. In late spring and early summer, small, bell-shaped, white to deep pink flowers appear in dense, nodding clusters. These are followed by a profusion of round, bright red berries which are edible, but taste acidic. Good ground cover for a lime-free, woodland garden.

CULTIVATION: *Best in peaty or sandy, moist but well-drained, acid soil, in full sun or partial shade. Trim in spring.*

☼ ☀ ◊ ◑ ♀ ❀ ❀ ❀
‡25cm (10in) ↔indefinite

Veltheimia bracteata

A bulbous perennial with basal rosettes of thick and waxy, strap-like, glossy, dark green leaves, from which upright flowering stems grow in spring, to be topped by a dense cluster of tubular, pink-purple flowers with yellow spots. An unusual house or conservatory plant where winter protection is needed.

CULTIVATION: *Plant bulbs in autumn with the neck just above soil level, in loam-based potting compost with added sharp sand. Site in full sun, reducing watering as the leaves fade. Keep the soil just moist during dormancy. Minimum temperature 2–5°C (36–41°F).*

☼ ◊ ♀☻ ‡45cm (18in) ↔30cm (12in)

Veratrum nigrum

An imposing, rhizomatous perennial that produces a tall, branching spike of many reddish-brown flowers from the centre of a basal rosette of pleated, mid-green leaves. The small, star-shaped flowers that open in late summer have an unpleasant scent, so choose a moist, shady site not too close to paths or patios.

CULTIVATION: *Grow in deep, fertile, moist but well-drained soil with added organic matter. If in full sun, make sure the soil remains moist. Shelter from cold, drying winds. Divide congested clumps in autumn or early spring.*

☼ ☀ ◊ ◊ ♀ ❋❋❋ ‡60–120cm (24–48in) ↔60cm (24in)

Verbascum bombyciferum

This mullein is a very tall perennial which forms basal rosettes of densely packed leaves covered in silky silver hairs. Short-lived, it dies back after the magnificent display of its tall, upright flower spike in summer, which is covered in silver hairs and sulphur-yellow flowers. For a large border, or may naturalize by self-seeding in a wild garden.

CULTIVATION: *Grow in alkaline, poor, well-drained soil in full sun. Support may be needed in fertile soil because of the resultant more vigorous growth. Divide plants in spring, if necessary.*

☼ ◊ ❀❀❀
‡1.8m (6ft) ↔ to 60cm (24in)

Verbascum 'Cotswold Beauty'

A tall, evergreen perennial that makes a bold addition to any large border. It bears upright spikes of saucer-shaped, peachy-pink flowers, with darker centres, over a long period from early to late summer. The flower spikes tower over the grey-green foliage, most of which is clumped near the base of the plant. It may naturalize in a wild garden.

CULTIVATION: *Best in well-drained, poor, alkaline soil in full sun. In fertile soil, it grows taller and will need support. Divide in spring, if necessary.*

☼ ◊ ❀❀❀
‡1.2m (4ft) ↔ 45cm (18in)

Verbascum dumulosum

This evergreen subshrub forms small, spreading domes of densely felted, grey or grey-green leaves on white-downy stems. In late spring and early summer, clusters of small, saucer-shaped yellow flowers with red-purple eyes appear amid the foliage. In cold areas, grow in small crevices of a warm, sunny wall.

CULTIVATION: *Best in gritty, sharply drained, moderately fertile, preferably alkaline soil, in full sun. Shelter from excessive winter wet.*

☼ ◊ ♀ ❀
‡to 25cm (10in) ↔to 40cm (16in)

Verbascum 'Gainsborough'

This short-lived, semi-evergreen perennial is valued for its spires of saucer-shaped, soft yellow flowers, borne throughout summer. Most of the oval and grey-green leaves are arranged in rosettes around the base of the stems. A very beautiful, long-flowering plant for a herbaceous or mixed border.

CULTIVATION: *Grow in well-drained, fertile soil, in an open, sunny site. Often short-lived, but easily propagated by root cuttings taken in winter.*

☼ ◊ ♀ ❀❀❀
‡to 1.2m (4ft) ↔30cm (12in)

Verbascum 'Letitia'

This dense, rounded, evergreen subshrub produces a continuous abundance of small, clear yellow flowers with reddish-purple centres throughout summer. The lance-shaped, irregularly toothed, grey-green leaves are carried beneath the clustered flowers. Suitable for a raised bed or rock garden. In cold climates, grow in crevices of a warm and protected drystone wall.

CULTIVATION: *Best in sharply drained, fairly fertile, alkaline soil, in sun. Choose a site not prone to excessive winter wet.*

☼ ◊ ♥ ❀ ❀
‡to 25cm (10in) ↔ to 30cm (12in)

Verbena bonariensis

This is a very useful and easy border plant that can effortlessly add height to a garden. Its tall, wiry stems rise above most other perennials in midsummer, without obscuring them, and bear small heads of vibrant purple-violet flowers. These remain well into autumn and are attractive to bees and butterflies.

CULTIVATION: *Grow in moist but well-drained soil in sun.*

☼ ◊ ◊ ♥ ❀ ❀ ❀
‡1.5m (5ft) ↔ 50cm (20in)

Verbena 'Sissinghurst'

A mat-forming perennial bearing rounded heads of small, brilliant magenta-pink flowers, which appear from late spring to autumn – most prolifically in summer. The dark green leaves are cut and toothed. Excellent for edging a path or growing in a tub. In cold areas, overwinter under glass.

CULTIVATION: *Grow in moist but well-drained, moderately fertile soil or compost. Choose a site in full sun.*

☼ ◊ ♀ ❀
‡to 20cm (8in) ↔to 1m (3ft)

Veronica gentianoides

This early-summer-flowering, mat-forming perennial bears shallowly cup-shaped, pale blue or white flowers. These are carried in upright spikes which arise from rosettes of glossy, broadly lance-shaped, dark green leaves at the base of the plant. Excellent for the edge of a border.

CULTIVATION: *Grow in moist but well-drained, moderately fertile soil, in full sun or light shade.*

☼ ◒ ◊ ❀❀❀
‡↔45cm (18in)

Veronica prostrata

The prostrate speedwell is a dense, mat-forming perennial. In early summer, it produces upright spikes of saucer-shaped, pale to deep blue flowers at the tips of sprawling stems. The small, bright to mid-green leaves are narrow and toothed. Grow in a rock garden. The cultivar 'Spode Blue' is also recommended.

CULTIVATION: *Best in moist but well-drained, poor to moderately fertile soil. Choose a position in full sun.*

☼ ◊ ♀ ❀ ❀ ❀
‡to 15cm (6in) ↔ to 40cm (16in)

Veronica spicata subsp. *incana*

The silver speedwell is an entirely silver-hairy, mat-forming perennial with upright flowering stems. These are tall spikes of star-shaped, purple-blue flowers, borne from early to late summer. The narrow leaves are silver-hairy and toothed. Ideal for a rock garden. The cultivar 'Wendy', with bright blue flowers, is also recommended.

CULTIVATION: *Grow in well-drained, poor to moderately fertile soil, in full sun. Protect from winter wet.*

☼ ◊ ♀ ❀ ❀ ❀ ‡↔30cm (12in)

Viburnum x *bodnantense* '**Dawn**'

A strongly upright, deciduous shrub carrying toothed, dark green leaves, bronze when young. From late autumn to spring, when branches are bare, small, tubular, heavily scented, dark pink flowers, which age to white, are borne in clustered heads, followed by small, blue-black fruits. 'Charles Lamont' is another lovely cultivar of this shrub, as is 'Deben', with almost white flowers.

CULTIVATION: *Grow in any deep, moist but well-drained, fertile soil, in full sun. On old crowded plants, cut the oldest stems back to the base after flowering.*

☀ ◊ ◊ ♀ ❁ ❁ ❁   ‡3m (10ft) ↔2m (6ft)

Viburnum x *carlcephalum*

This vigorous, rounded, deciduous shrub has broadly heart-shaped, irregularly toothed, dark green leaves which turn red in autumn. Rounded heads of small, fragrant white flowers appear amid the foliage during late spring. Suits a shrub border or a woodland garden.

CULTIVATION: *Grow in deep, moisture-retentive, fertile soil. Tolerates sun or semi-shade. Little pruning is necesary.*

☀ ☼ ◊ ♀ ❁ ❁ ❁   ‡↔3m (10ft)

Viburnum davidii

A compact, evergreen shrub that forms a dome of dark green foliage consisting of oval leaves with three distinct veins. In late spring, tiny, tubular white flowers appear in flattened heads; female plants bear tiny but decorative, oval, metallic-blue fruits, later in the season. Looks good planted in groups.

CULTIVATION: *Best in deep, moist but well-drained, fertile soil, in sun or semi-shade. Grow male and female shrubs together for reliable fruiting. Keep neat, if desired, by cutting wayward stems back to strong shoots, or to the base of the plant, in spring.*

☼ ☀ ◊ ◐ ♀ ❀❀❀   ↕↔1–1.5m (3–5ft)

Viburnum farreri

This strongly upright, deciduous shrub has oval, toothed, dark green leaves which are bronze when young and turn red-purple in autumn. During mild periods in winter and early spring, small, fragrant, white or pink-tinged flowers are borne in dense clusters on the bare stems. These are occasionally followed by tiny, bright red fruits.

CULTIVATION: *Grow in any reliably moist but well-drained, fertile soil, in full sun or partial shade. Thin out old shoots after flowering.*

☼ ☀ ◊ ◐ ♀ ❀❀❀
↕3m (10ft) ↔2.5m (8ft)

Viburnum opulus 'Xanthocarpum'

A vigorous, deciduous shrub bearing maple-like, lobed, mid-green leaves which turn yellow in autumn. Flat heads of showy white flowers are produced in late spring and early summer, followed by large bunches of spherical, bright yellow berries. 'Notcutt's Variety' and 'Compactum' are recommended red-berried cultivars, the latter much smaller, to only 1.5m (5ft) tall.

CULTIVATION: *Grow in any moist but well-drained soil, in sun or semi-shade. Cut out older stems after flowering to relieve overcrowding.*

☼ ☀ ◊ ◊ ♀ ❀ ❀ ❀
‡5m (15ft) ↔ 4m (12ft)

Viburnum plicatum 'Mariesii'

A spreading, deciduous shrub bearing distinctly tiered branches. These are clothed in heart-shaped, toothed, dark green leaves which turn red-purple in autumn. In late spring, saucer-shaped white flowers appear in spherical, lacecap-like heads. Few berries are produced; for a heavily fruiting cultivar, look for 'Rowallane'.

CULTIVATION: *Grow in any well-drained, fairly fertile soil, in sun or semi-shade. Requires little pruning, other than to remove damaged wood after flowering; be careful not to spoil the form.*

☼ ☀ ◊ ♀ ❀ ❀ ❀
‡3m (10ft) ↔ 4m (12ft)

Viburnum tinus 'Eve Price'

This very compact, evergreen shrub has dense, dark green foliage. Over a long period from late winter to spring, pink flower buds open to tiny, star-shaped white flowers, carried in flattened heads; they are followed by small, dark blue-black fruits. Can be grown as an informal hedge.

CULTIVATION: *Grow in any moist but well-drained, moderately fertile soil, in sun or partial shade. Train or clip after flowering to maintain desired shape.*

☼ ☀ ◊ ♦ ♀ ❉ ❉ ❉　　　　↔3m (10ft)

Vinca major 'Variegata'

This variegated form of the greater periwinkle, also known as 'Elegantissima', is an evergreen subshrub with long, slender shoots bearing oval, dark green leaves which have creamy-white margins. Dark violet flowers are produced over a long period from mid-spring to autumn. Useful as ground cover for a shady bank, but may be invasive.

CULTIVATION: *Grow in any moist but well-drained soil. Tolerates deep shade, but flowers best with part-day sun.*

☼ ☀ ◊ ♦ ♀ ❉ ❉ ❉
‡45cm (18in) ↔indefinite

Vinca minor 'Atropurpurea'

This dark-flowered lesser periwinkle is a mat-forming, ground cover shrub with long trailing shoots. Dark plum-purple flowers are produced over a long period from mid-spring to autumn, amid the oval, dark green leaves. For light blue flowers, choose 'Gertrude Jekyll'.

CULTIVATION: *Grow in any but very dry soil, in full sun for best flowering, but tolerates partial shade. Restrict growth by cutting back hard in early spring.*

☼ ☀ ◊ ◑ ♀ ❀❀❀
‡10–20cm (4–8in) ↔indefinite

Viola cornuta

The horned violet is a spreading, evergreen perennial which produces an abundance of slightly scented, spurred, violet to lilac-blue flowers; the petals are widely separated, with white markings on the lower ones. The flowers are borne amid the mid-green, oval leaves from spring to summer. Suitable for a rock garden. The Alba Group have white flowers.

CULTIVATION: *Grow in moist but well-drained, poor to moderately fertile soil, in sun or partial shade. Cut back after flowering to keep compact.*

☼ ☀ ◊ ◑ ♀ ❀❀❀
‡to 15cm (6in) ↔to 40cm (16in)

Viola 'Jackanapes'

A robust, clump-forming, evergreen perennial bearing spreading stems with oval and toothed, bright green leaves. Spurred, golden-yellow flowers with purple streaks appear in late spring and summer; the upper petals are deep brownish-purple. Good for containers or summer bedding.

CULTIVATION: *Grow in moist but well-drained, fairly fertile soil, in full sun or semi-shade. Often short-lived, but easily grown from seed sown in spring.*

☼ ☀ ◊ ◔ ♀ ❀ ❀ ❀
‡to 12cm (5in) ↔ to 30cm (12in)

Viola 'Nellie Britton'

This clump-forming, evergreen perennial with spreading stems produces an abundance of spurred, pinkish-mauve flowers over long periods in summer. The oval, mid-green leaves are toothed and glossy. Suitable for the front of a border.

CULTIVATION: *Best in well-drained but moist, moderately fertile soil, in full sun or partial shade. Deadhead frequently to prolong flowering.*

☼ ☀ ◊ ◔ ♀ ❀ ❀ ❀
‡to 15cm (6in) ↔ to 30cm (12in)

Vitis coignetiae

The crimson glory vine is a fast-growing, deciduous climber with large, heart-shaped, shallowly lobed, dark green leaves which turn bright red in autumn. Small, blue-black grape-like fruits appear in autumn. Train against a wall or over a trellis.

CULTIVATION: *Grow in well-drained, neutral to alkaline soil, in sun or semi-shade. Autumn colour is best on poor soils. Pinch out the growing tips after planting and allow the strongest shoots to form a permanent framework. Prune back to this each year in mid-winter.*

☼ ☀ ◊ ♀ ❀ ❀ ❀   ↕15m (50ft)

Vitis vinifera 'Purpurea'

This purple-leaved grape vine is a woody, deciduous climber bearing rounded, lobed, toothed leaves; these are white-hairy when young, turning plum-purple, then dark purple, before they fall. Tiny, pale green summer flowers are followed by small purple grapes in autumn. Grow over a robust fence or pergola, or through a large shrub or tree.

CULTIVATION: *Grow in well-drained, slightly alkaline soil, in sun or semi-shade. Autumn colour is best on poor soils. Prune back to an established framework each year in mid-winter.*

☼ ☀ ◊ ♀ ❀ ❀ ❀   ↕7m (22ft)

Weigela florida 'Foliis Purpureis'

A compact, deciduous shrub with arching shoots that produces clusters of funnel-shaped, dark pink flowers with pale insides in late spring and early summer. The bronze-green foliage is made up of oval, tapered leaves. Pollution-tolerant, so is ideal for urban gardens.

CULTIVATION: *Best in well-drained, fertile, humus-rich soil, in full sun. Prune out some older branches at ground level each year after flowering.*

☼ ◊ ❀ ❀ ❀   ‡1m (3ft) ↔1.5m (5ft)

Weigela 'Florida Variegata'

A dense, deciduous shrub that produces abundant clusters of funnel-shaped, dark pink flowers with pale insides. These are borne in late spring and early summer amid attractive, grey-green leaves with white margins. Suitable for a mixed border or open wood-land. Tolerates urban pollution. 'Praecox Variegata' is another pretty variegated weigela.

CULTIVATION: *Grow in any well-drained, fertile, humus-rich soil, in full sun. Prune out some of the oldest branches each year after flowering.*

☼ ◊ ♀ ❀ ❀ ❀   ‡↔2–2.5m (6–8ft)

Wisteria floribunda 'Alba'

This white-flowered Japanese wisteria is a fast-growing, woody climber with bright green, divided leaves. The fragrant, pea-like flowers appear during early summer in very long, drooping spikes; bean-like, velvety-green seed pods usually follow. Train against a wall, up into a tree or over a well-built arch.

CULTIVATION: *Grow in moist but well-drained, fertile soil, in sun or partial shade. Prune back new growth in summer and in late winter to control spread and promote flowering.*

☼ ☀ ◊ ◊ ♀ ✳ ✳ ✳   ↕9m (28ft) or more

Wisteria frutescens 'Amethyst Falls'

This American native wisteria grows much less rampantly than Asian wisterias, making it ideal for small gardens. The dangling clusters of fragrant lilac flowers bloom in late spring and sometimes again in summer. Grow against a sheltered wall, along a sunny fence, or in a large container. Provide strong support.

CULTIVATION: *Grow in moist, well-drained, fertile soil full sun or partial shade. Water regularly until established. Prune in late winter, as needed.*

☼ ☀ ◊ ◊ ✳ ✳ ✳
↕3–6m (10–20ft) ↔1–2m (3–6ft)

Wisteria sinensis

Chinese wisteria is a vigorous, deciduous climber that produces long, hanging spikes of fragrant, pea-like, lilac-blue to white flowers. These appear amid the bright green, divided leaves in late spring and early summer, usually followed by bean-like, velvety-green seed pods. 'Alba' is a white-flowered form.

CULTIVATION: *Grow in moist but well-drained, fertile soil. Provide a sheltered site in sun or partial shade. Cut back long shoots to 2 or 3 buds, in late winter.*

☼ ☀ ◊ ◑ ✿ ✿ ✿ ↕9m (28ft) or more

Yucca filamentosa '**Bright Edge**'

An almost stemless, clump-forming shrub with basal rosettes of rigid, lance-shaped, dark green leaves, to 75cm (30in) long, with broad yellow margins. Tall spikes, to 2m (6ft) or more, of nodding, bell-shaped white flowers, tinged with green or cream, are borne in mid- to late summer. An architectural specimen for a border or courtyard. The leaves of 'Variegata' are edged white.

CULTIVATION: *Grow in any well-drained soil, in full sun. Remove old flowers at the end of the season. In cold or frost-prone areas, mulch over winter.*

☼ ◊ ♀ ✿ ✿ ↕75cm (30in) ↔1.5m (5ft)

Yucca flaccida 'Ivory'

An almost stemless, evergreen shrub that forms a dense, basal clump of sword-like leaves. Tall spikes, to 1.5m (5ft) or more, of nodding, bell-shaped white flowers are borne in mid- and late summer. The lance-shaped, dark blue-green leaves, fringed with curly or straight threads, are arranged in basal rosettes. Thrives in coastal gardens and on sandy soils.

CULTIVATION: *Grow in any well-drained soil, but needs a hot, dry position in full sun to flower well. In cold or frost-prone areas, mulch over winter.*

☼ ◊ ♀ ❀ ❀　‡55cm (22in) ↔1.5m (5ft)

Zantedeschia aethiopica

This relatively small-flowered arum lily is a clump-forming perennial. Almost upright, cream-yellow flowers are borne in succession from late spring to midsummer, followed by long, arrow-shaped leaves, evergreen in mild areas. May be used as a waterside plant in shallow water. May not survive in cold areas.

CULTIVATION: *Best in moist but well-drained, fertile soil that is rich in humus. Choose a site in full sun or partial shade. Mulch deeply over winter in cold, frost-prone areas.*

☼ ☀ ◊ ◖ ❀ ❀　‡90cm (36in) ↔60cm (24in)

Zantedeschia aethiopica 'Green Goddess'

This green-flowered arum lily is a clump-forming, robust perennial which is evergreen in mild climates. Upright and white-centred flowers appear from late spring to mid-summer, above the dull green, arrow-shaped leaves. Good in shallow water, in a site sheltered from heavy frosts.

CULTIVATION: *Grow in moist, humus-rich, fertile soil. Choose a site in full sun or partial shade. Mulch deeply over winter in cold, frost-prone areas.*

☼ ☀ ◊ ♦ ♀ ✿✿
‡90cm (36in) ↔ 60cm (24in)

Zauschneria californica 'Dublin'

This deciduous Californian fuchsia, sometimes called 'Glasnevin', is a clump-forming perennial bearing a profusion of tubular, bright red flowers during late summer and early autumn. The grey-green leaves are lance-shaped and hairy. Provides spectacular, late-season colour for a dry stone wall or border. In climates with cold winters, grow at the base of a warm wall.

CULTIVATION: *Best in well-drained, moderately fertile soil, in full sun. Provide shelter from cold, drying winds.*

☼ ◊ ♀ ✿✿
‡to 25cm (10in) ↔ to 30cm (12in)

THE PLANTING GUIDE

Berrying Plants

If birds do not enjoy the feast as soon as they ripen, berries can enhance the garden for many months. Red berries are most common, but there are black, white, yellow, pink, blue and even purple berries, to be included in almost any garden. Pale berries look best against a dark background such as an evergreen hedge, and red berries are attractive against a wintery, sunny sky.

Actaea alba
Perennial Clump-forming plant with divided leaves and fluffy flowers followed by pearly, white berries with black eyes.
‡90cm (36in) ↔45–60cm (18–24in)

Arbutus unedo f. rubra
♀ **Evergreen tree** The pink flowers and red, globular fruits are both at their best in autumn.
‡↔8m (25ft)

Arum italicum 'Marmoratum'
♀ **Perennial** Evergreen, marbled foliage and spikes of red berries in autumn when the leaves die down.
‡30cm (12in) ↔15cm (6in)

Berberis dictyophylla
Shrub This deciduous shrub is at its best in winter when the white shoots are studded with red berries.
‡2m (6ft) ↔1.5m (5ft)

Berberis x stenophylla 'Corallina Compacta'
♀ **Evergreen shrub**   page 85

Berberis verruculosa
♀ **Evergreen shrub**   page 87

Callicarpa bodinieri var. giraldii 'Profusion'
♀ **Shrub**   page 96

Celastrus orbiculatus
Climber Strong-growing climber with yellow autumn colour and yellow fruits opening to reveal red seeds.
‡14m (46ft)

Clerodendrum trichotomum var. fargesii
♀ **Shrub** Fast-growing plant with fragrant white flowers and turquoise berries set against red calyces.
‡↔5m (15ft)

Coriaria terminalis var. anthocarpa
Shrub, not fully hardy Slightly tender, arching subshrub with small leaves and clusters of translucent yellow berries.
‡1m (3ft) ↔2m (6ft)

Cornus 'Norman Hadden'
♀ **Evergreen tree** Some leaves turn yellow and drop each autumn, when the cream and pink flowers are followed by large red fruits.
‡↔8m (25ft)

Cotoneaster conspicuus 'Decorus'
♀ **Evergreen shrub**   page 146

Cotoneaster 'Rothschildianus'
♀ **Evergreen shrub** An arching shrub with golden yellow berries in autumn after white flowers in summer.
‡↔5m (15ft)

Euonymus planipes
Shrub The foliage is bright red in autumn and falls to reveal red capsules containing orange seeds.
‡↔3m (10ft)

Gaultheria mucronata 'Wintertime'
♀ **Evergreen shrub**   page 220

Hippophae rhamnoides
Shrub   page 257

Hypericum kouytchense
♀ Shrub   page **270**

Ilex aquifolium & cultivars
Evergreen shrubs   pages **272–273**

Ilex x meserveae 'Blue Angel'
Evergreen shrub Compact, slow-growing shrub with glossy, dark, bluish-green leaves and red berries.
‡4m (12ft) ↔2m (6ft)

Ilex verticillata 'Winter Red'
Shrub Deciduous shrub with white flowers in spring and masses of small red berries in winter.
‡2.5–3m (8–10ft) ↔3m (10ft)

Leycesteria formosa
Shrub Tall, arching stems tipped with white flowers within maroon bracts, followed by purple berries.
‡↔2m (6ft)

Lonicera nitida 'Baggesen's Gold'
♀ Evergreen shrub   page **323**

Lonicera periclymenum
'Graham Thomas'
♀ Climber   page **324**

Physalis alkekengi
♀ **Perennial** Creeping plant with upright stems: orange lanterns containing orange berries follow white flowers.
‡60–75cm (24–30in) ↔90cm (36in)

Pyracantha 'Cadrou'
Evergreen shrub Spiny shrub with white flowers and red berries, resistant to scab.
‡↔2m (6ft)

Rosa 'Fru Dagmar Hastrup'
♀ Shrub rose   page **464**

Rosa 'Geranium'
♀ **Shrub rose** Arching, prickly stems with neat red flowers and large, long red hips.
‡2.5m (8ft) ↔1.5m (5ft)

Rosa 'Scharlachglut'
Shrub rose Vigorous long-stemmed rose that can be trained as a climber, with showy scarlet flowers and bright scarlet hips.
‡3m (10ft) ↔2m (6ft)

Sambucus racemosa
'Plumosa Aurea'
Shrub Divided yellow leaves in summer and clusters of small, red berries.
‡↔3m (10ft)

Skimmia japonica 'Fructu Albo'
Evergreen shrub Neat evergreen with white flowers and bright, white fruits.
‡60cm (24in) ↔1m (3ft)

Sorbus aria 'Lutescens'
♀ Tree   page **498**

Sorbus hupehensis var. *obtusa*
♀ Tree   page **498**

Tropaeolum speciosum
♀ Perennial climber   page **521**

Viburnum davidii
♀ Evergreen shrub   page **535**

Viburnum opulus 'Xanthocarpum'
♀ Shrub   page **536**

Plants with Scented Flowers

Fragrance is too often forgotten when planting a garden. Yet there are as many shades of fragrance as there are of colours: they can affect mood, take you back to your childhood, or whisk you off to a far-off land with a single sniff. The most strongly scented flowers are often white or insignificant in appearance, but have evolved to make their presence felt in other ways.

Abelia chinensis
Shrub Spreading, with heads of small pale pink flowers in late summer.
↕1.5m (5ft) ↔2.5m (8ft)

Buddleja alternifolia
♀ **Shrub**   page **92**

Camellia 'Inspiration'
♀ **Evergreen shrub**   page **99**

Camellia japonica 'Adolphe Audusson'
♀ **Evergreen shrub**   page **100**

Camellia japonica 'Elegans'
♀ **Evergreen shrub**   page **100**

Chimonanthus praecox 'Grandiflorus'
♀ **Shrub**   page **120**

Chimonanthus praecox 'Luteus'
♀ **Shrub** After the large leaves fall, pale yellow flowers scent the winter air.
↕4m (12ft) ↔3m (10ft)

Choisya ternata
♀ **Evergreen shrub**   page **121**

Choisya x dewitteana 'Aztec Pearl'
♀ **Evergreen shrub** Narrowly divided, deep green leaves and white, pink-tinged flowers in spring and autumn.
↕↔2.5m (8ft)

Clematis montana var. *grandiflora*
♀ **Climber**   page **128**

Daphne bholua 'Gurkha'
Shrub   page **164**

Daphne tangutica Retusa Group
♀ **Evergreen shrubs**   page **164**

Dianthus 'Doris'
♀ **Perennial**   page **172**

Erica erigena 'Golden Lady'
Evergreen shrub   page **187**

Hamamelis (Witch Hazels)
Shrubs   pages **238–239**

Hosta 'Honeybells'
Perennial   page **260**

Jasminum officinale
Climber A strong, twining climber with white flowers in summer which have an intense, sweet scent.
↕12m (40ft)

Jasminum officinale 'Argenteovariegatum'
♀ **Climber**   page **292**

Lilium Pink Perfection Group
♀ **Bulb**   page **317**

Lonicera caprifolium
Climber The Italian honeysuckle has pink and cream, fragrant flowers in summer.
↕6m (20ft)

Lonicera periclymenum 'Belgica'
Climber In early summer this honeysuckle produces creamy-yellow flowers streaked with maroon.
↕7m (22ft)

Lonicera periclymenum
'Graham Thomas'
♀ Climber page 324

Magnolia grandiflora 'Goliath'
Evergreen tree page 331

Mahonia x media 'Charity'
Evergreen shrub page 335

Philadelphus 'Beauclerk'
♀ Shrub page 384

Philadelphus 'Belle Etoile'
♀ Shrub page 384

Phlox paniculata 'White Admiral'
♀ **Perennial** Large heads of pure white flowers with a sweet, peppery scent, borne in summer.
↕1m (3ft)

Pittosporum tenuifolium
Evergreen shrub page 405

Primula florindae
♀ Perennial page 414

Rosa 'Albertine'
♀ Rambler rose page 454

Rosa 'Arthur Bell'
♀ Cluster-flowered bush rose page 450

Rosa 'Blessings'
Large-flowered bush rose page 448

Rosa 'Climbing Iceberg'
♀ **Climbing rose** This fine rose has many pure white flowers all summer.
↕2.5m (8ft)

Rosa 'Compassion'
♀ Climbing rose page 452

Rosa GRAHAM THOMAS
♀ Modern shrub rose page 463

Rosa 'Just Joey'
♀ Large-flowered bush rose page 449

Rosa MARGARET MERRIL
Cluster-flowered bush rose page 451

Rosa PEACE
♀ Large-flowered bush rose page 449

Rosa 'Penelope'
♀ Modern shrub rose page 463

Rosa REMEMBER ME
♀ Large-flowered bush rose page 449

Sarcococca hookeriana var. *digyna*
Evergreen shrub page 483

Skimmia japonica 'Rubella'
♀ Evergreen shrub page 495

Syringa meyeri 'Palibin'
♀ Shrub page 505

Syringa vulgaris 'Madame Lemoine'
♀ Shrub page 507

Ulex europaeus 'Flore Pleno'
♀ **Evergreen shrub** Spiny bush that has a few of its double flowers, scented of coconut, open almost all year.
↕2.5m (8ft) ↔2m (6ft)

Viburnum x burkwoodii 'Park Farm Hybrid'
♀ **Evergreen shrub** Upright shrub with bronze new leaves and deep pink, scented flowers in late spring.
↕3m (10ft) ↔2m (6ft)

Viburnum carlesii 'Aurora'
♀ **Shrub** Bushy shrub with pink flowers in late spring, opening from red buds.
↕↔2m (6ft)

Architectural Plants

Every garden needs plants that are larger than life, with the sort of shape or texture that cannot be ignored. Too many of these plants will give a sub-tropical effect, with bold leaves and spiky shapes, but if carefully placed among less extraordinary plants, they become the focus of a view, or a "full stop" in the border. Make use of light and shade to emphasise bold silhouettes.

Acanthus spinosus
Perennial page 31

Aesculus parviflora
♀ **Shrub** page 46

Agave americana
♀ **Succulent** Viciously spiny plant with steel-grey leaves that curve to make a magnificent rosette.
‡2m (6ft) ↔3m (10ft)

Ailanthus altissima
Tree This can be a tall tree but will produce divided leaves 1.2m (4ft) long if cut back hard every year.
‡25m (80ft) ↔15m (50ft)

Araucaria heterophylla
♀ **Conifer** page 64

Betula nigra
Tree page 89

Betula pendula 'Youngii'
Tree page 89

Catalpa bignonioides 'Aurea'
♀ **Tree** Spreading tree that can be pruned hard annually for large, gold leaves.
‡↔10m (30ft)

Chamaecyparis nootkatensis 'Pendula'
Conifer page 118

Chamaecyparis pisifera 'Filifera Aurea'
♀ **Conifer** A broad, arching shrub with whip-like, golden shoots.
‡12m (40ft) ↔5m (15ft)

Corylus avellana 'Contorta'
♀ **Shrub** page 143

Crocosmia masoniorum
♀ **Perennial** page 152

Eryngium giganteum
♀ **Biennial** Rosettes of deep green leaves, and spiny, white stems and flowerheads in the second year.
‡90cm (36in) ↔30cm (12in)

Eucalyptus pauciflora subsp. *niphophila*
♀ **Evergreen tree** page 197

Fagus sylvatica 'Pendula'
♀ **Tree** A tree of huge proportions with horizontal and arching branches cascading to the ground.
‡15m (50ft) ↔20m (70ft)

Fargesia nitida
Ornamental grass page 205

Gunnera manicata
♀ **Perennial** page 233

Helianthus 'Monarch'
♀ **Perennial** page 250

Kniphofia caulescens
Perennial page 297

Macleaya x *kewensis* 'Kelway's Coral Plume'
♀ **Perennial** page 330

Melianthus major
♀ **Perennial, not hardy** page 340

Paeonia delavayi
Shrub   page 363

Phormium cookianum subsp.
hookeri 'Cream Delight'
♀ Perennial   page 392

Phormium tenax
Perennial   page 393

Phormium tenax
Purpureum Group
♀ Perennials   page 394

Phyllostachys aurea
♀ **Bamboo** The golden bamboo,
with yellow-brown canes and yellow-
green leaves.
‡2–10m (6–30ft) ↔indefinite

Phyllostachys aureosulcata
f. *aureocaulis*
Bamboo Large bamboo with bright,
golden canes and narrow leaves.
‡3–6m (10–20ft) ↔indefinite

Phyllostachys nigra
♀ **Bamboo**   page 398

Phyllostachys nigra
f. *henonis*
♀ **Bamboo**   page 399

Pleioblastus variegatus
♀ **Bamboo**   page 407

Prunus 'Amanogawa'
♀ **Tree** Very slender, upright growth,
resembling a Lombardy poplar, with
semi-double pink flowers in spring.
‡8m (25ft) ↔4m (12ft)

Prunus 'Kiku-shidare-zakura'
Tree   page 422

Rodgersia aesculifolia
♀ **Perennial** Creeping rhizomes produce
clumps of large leaves like those of horse
chestnuts, and pink flowers.
‡2m (6ft) ↔1m (3ft)

Sorbaria tomentosa var. *angustifolia*
♀ **Shrub** A spreading shrub with feathery
leaves, red stems and fluffy, white
flowerheads.
‡↔3m (10ft)

Stipa gigantea
♀ **Ornamental grass**   page 503

Trachycarpus fortunei
♀ **Hardy palm** Slow-growing but
cold-tolerant palm with fan-shaped leaves
and a furry trunk with age.
‡20m (70ft) ↔2.5m (8ft)

Viburnum plicatum 'Pink Beauty'
Shrub A spreading shrub with horizontal
tiers of branches, covered with white
flowers turning to pink.
‡3m (10ft) ↔4m (12ft)

Woodwardia radicans
♀ **Hardy fern** A large, evergreen fern
with huge, arching fronds.
‡2m (6ft) ↔3m (10ft)

Yucca filamentosa
Evergreen shrub A clump-forming,
stemless plant with soft leaves and spires
of creamy flowers.
‡75cm (30in) ↔1.5m (5ft)

Yucca filamentosa 'Bright Edge'
♀ **Shrub**   page 543

Yucca flaccida 'Ivory'
♀ **Shrub**   page 544

Yucca gloriosa
♀ **Evergreen shrub** Erect trunks with
grey-green narrow, sharp-tipped leaves,
and large clusters of white flowers.
‡↔2m (6ft)

Plants to Attract Garden Birds

Native and visiting birds will visit gardens to feed on a wide variety of plants, especially those bearing berries and seeds (see also page 548). Unfortunately, their feeding necessarily means that the food source – and the attractive autumn display – does not last long, so it is worth offering them a variety of plants, and providing extra food on a regular basis.

Atriplex hortensis
var. *rubra*
Annual Vigorous plant with deep red leaves that contrast well with other plants; its seeds are loved by birds.
↕1.2m (4ft) ↔30cm (12in)

Berberis thunbergii
Shrub Green leaves and small, yellow flowers in summer become red leaves and berries in autumn.
↕2m (6ft) ↔2.5m (8ft)

Cortaderia selloana 'Pumila'
♀ **Ornamental grass** A compact pampas grass with short flower spikes.
↕1.5m (5ft) ↔1.2m (4ft)

Cotoneaster lacteus
♀ **Evergreen shrub**   page 147

Cotoneaster simonsii
♀ **Shrub**   page 147

Crataegus x *lavalleei*
'Carrierei'
♀ **Tree**   page 150

Cynara cardunculus
♀ **Perennial**   page 159

Daphne mezereum
Shrub Upright branches, fragrant pink flowers in spring, red berries in autumn.
↕1.2m (4ft) ↔1m (3ft)

Hedera helix & **cultivars**
Evergreen climbers pages 246–247 When ivy reaches its flowering stage the black berries are attractive to many birds; ivy also provides valuable nesting sites.

Helianthus annuus
'Music Box'
Annual Multi-coloured sunflowers that produce heads of seeds that may be harvested to feed birds in later months.
↕70cm (28in) ↔60cm (24in)

Ilex aquifolium 'Handsworth
New Silver'
♀ **Evergreen shrub**   page 273

Lonicera periclymenum 'Serotina'
♀ **Climber**   page 324

Mahonia aquifolium
Evergreen shrub A suckering shrub with gently spiny leaves, and yellow flowers followed by black berries.
↕1m (3ft) ↔1.5m (5ft)

Malus x *zumi* 'Golden Hornet'
Tree   page 337

Miscanthus sinensis
Ornamental grass This grass forms clumps of long, arching leaves and silver or pink flowerheads in late summer.
↕2.5m (8ft) ↔1.2m (4ft)

Onopordum nervosum
♀ **Biennial** A large, silvery, prickly plant with thistle-like purple flowers.
↕2.5m (8ft) ↔1m (3ft)

Papaver somniferum
'White Cloud'
Annual White, double flowers and pretty seedheads attractive to birds.
↕1m (3ft) ↔30cm (12in)

Prunus padus
Tree A spreading tree with pendent spikes of small white flowers followed by black berries.
↕15m (50ft) ↔10m (30ft)

Pyracantha 'Mohave'
Evergreen shrub Dense, spiny growth with dark green leaves and bright red berries.
↕4m (12ft) ↔5m (15ft)

Ribes odoratum
Shrub Weakly-branched shrub with fresh green leaves, yellow scented flowers in spring and black berries in late summer.
↕↔2m (6ft)

Rosa filipes 'Kiftsgate'
♀ **Climbing rose**   page **452**

Rosa pimpinellifolia
Species shrub rose This very spiny bush has single white flowers followed by purplish-black hips.
↕1m (3ft) ↔1.2m (4ft)

Rosa 'Scabrosa'
♀ **Rugosa shrub rose** Deep pink flowers and bright red hips on a mounded bush with deeply veined foliage.
↕↔1.7m (5½ft)

Sambucus nigra 'Aureomarginata'
Shrub Fast-growing plant for any soil, with yellow-edged leaves, white flowers and heads of black elderberries.
↕↔6m (20ft)

Silybum marianum
Biennial Rosettes of spiny leaves, veined with white; prickly mauve seedheads and thistle seeds.
↕1.5m (5ft) ↔60–90cm (24–36in)

Sorbus aucuparia 'Fastigiata'
Tree Upright, narrow tree with red berries in late summer after white spring flowers.
↕8m (25ft) ↔5m (15ft)

Viburnum betulifolium
Shrub Spectacular displays of red berries follow white flowers in summer when several plants are grown together.
↕↔3m (10ft)

Viburnum opulus
Shrub Strong-growing shrub with white flowers in summer, bright autumn colour and red berries.
↕5m (15ft) ↔4m (12ft)

Vitis vinifera 'Purpurea'
♀ **Climber**   page **540**

Flowers for Cutting

It is useful to be able to cut flowers from the garden, either to use on their own or to add to bought flowers. Many annuals are grown especially for cutting, but other garden plants can supply flowers for the house without spoiling the display. To produce many smaller stems for cutting, pinch out the shoots of free-branching plants such as asters and delphiniums in early summer.

Achillea 'Coronation Gold'
♀ Perennial   page 39

Aconitum 'Bressingham Spire'
♀ Perennial   page 42

Aster 'Little Carlow'
♀ Perennial   page 74

Aster pilosus var. pringlei 'Monte Cassino'
Perennial Thin stems of narrow, upright habit, forming a dense bush with needle-like leaves and small white flowers.
‡1m (3ft) ↔30cm (12in)

Astilbe 'Fanal'
♀ Perennial   page 75

Astrantia major 'Shaggy'
♀ Perennial The bracts around the clusters of flowers are longer than usual.
‡30–90cm (12–36in) ↔45cm (18in)

Baptisia australis
♀ Perennial   page 79

Campanula lactiflora 'Prichard's Variety'
♀ Perennial Compact cultivar with heads of violet-blue flowers in midsummer.
‡75cm (30in) ↔45cm (18in)

Campanula persicifolia 'Chettle Charm'
Perennial Mats of deep green foliage and thin stems with white, blue-tinted flowers.
‡1m (3ft) ↔30cm (12in)

Chrysanthemums
Perennials, not hardy   pages 122–125

Clematis 'Vyvyan Pennell'
Climber   page 109

Crocosmia x crocosmiiflora 'Solfatare'
♀ Perennial   page 151

Delphinium 'Bellamosum'
Perennial Well-branched stems with thin spikes of deep blue flowers for a long period.
‡1–1.2m (3–4ft) ↔45cm (18in)

Delphinium 'Bruce'
♀ Perennial   page 166

Delphinium 'Sungleam'
♀ Perennial   page 169

Border Carnations (Dianthus)
Perennials   page 171

Dianthus 'Coronation Ruby'
♀ Perennial A laced pink with pink and ruby-red flowers with a clove scent.
‡38cm (15in) ↔30cm (12in)

Eryngium x tripartitum
♀ Perennial   page 192

Geum 'Mrs J. Bradshaw'
♀ Perennial Hairy basal leaves, wiry branched stems and double scarlet flowers.
‡40–60cm (16–24in) ↔60cm (24in)

Kniphofia 'Royal Standard'
♀ Perennial   page 298

Lathyrus odoratus (Sweet Peas)
Annual climbers   page 303

Leucanthemum x superbum 'Wirral Supreme'
♀ Perennial   page 310

Narcissus 'Merlin'
♀ Bulb   page 349

Narcissus 'White Lion'
♀ Bulb Double, white flowers too heavy
to stand up in the garden; best when cut.
‡40cm (16in)

Osteospermum 'Whirlygig'
Subshrub, not hardy   page 361

Paeonia lactiflora
'Sarah Bernhardt'
♀ Perennial   page 364

Phlox maculata 'Alpha'
♀ Perennial   page 389

Phlox maculata 'Omega'
♀ Perennial Conical heads of fragrant,
small white flowers with a deep pink eye.
‡90cm (3ft) ↔45cm (18in)

Physostegia virginiana 'Vivid'
♀ Perennial   page 400

Rosa 'Alexander'
Large-flowered bush rose   page 448

Rosa 'Iceberg'
Cluster-flowered bush rose   page 451

Rosa 'Royal William'
Large-flowered bush rose   page 449

Rosa 'Silver Jubilee'
Large-flowered bush rose   page 449

Roses, Modern Shrub
Deciduous shrubs   pages 462–463

Rudbeckia fulgida var. *sullivantii*
'Goldsturm'
♀ Perennial   page 468

Rudbeckia laciniata 'Goldquelle'
Perennial   page 468

Scabiosa caucasica 'Clive Greaves'
♀ Perennial   page 485

Schizostylis coccinea 'Sunrise'
Perennial   page 486

Solidago 'Goldenmosa'
♀ Perennial   page 497

Solidago x *luteus* 'Lemore'
♀ Perennial This generic hybrid
produces heads of yellow daisy-like
flowers.
‡90cm (3ft) ↔30cm (12in)

Tanacetum coccineum 'Brenda'
Perennial   page 508

Tulipa 'Sorbet'
♀ Bulb Late blooms are pale pink with
carmine streaks and flashes.
‡60cm (24in)

Veronica spicata subsp. *incana*
♀ Perennial   page 533

ACKNOWLEDGMENTS

The publisher would like to thank the following for their kind permission to reproduce the photographs:

Key: a-above; b-below/bottom; l-left; r-right; c-centre; ca-centre above; cl-centre left; cr-centre right; br- below/bottom right; bc-below/bottom centre; t-top; tc-top centre; tr-top right

6 Marianne Majerus Garden Images. **12 GAP Photos:** Marcus Harpur. **28 GAP Photos:** Martin Hughes-Jones (b). **31 GAP Photos:** Elke Borkowski (t). **32 Marianne Majerus Garden Images** (t) (b). **47 GAP Photos:** Richard Bloom (b); J.S. Sira (t). **48 Alamy Images:** Steffen Hauser/botanikfoto (t). **53 Getty Images:** Nacivet (b). **54 Alamy Images:** Holmes Garden Photos (b). **55 GAP Photos:** Christina Bollen (t). **56 GAP Photos:** Richard Bloom (t). **Marianne Majerus Garden Images** (b). **65 GAP Photos:** Maayke de Ridder (b). **68 GAP Photos:** Geoff Kidd (t). **72 GAP Photos:** Martin Hughes-Jones (t). **77 Garden World Images:** Francoise Davis (b). **Getty Images:** Jayme Thornton (t). **78 GAP Photos:** Jonathan Buckley (t). **82 GAP Photos:** Visions (tr). **83 Garden World Images:** Gilles Delacroix (tc). **97 Garden World Images:** Gilles Delacroix (t). **100 GAP Photos:** S&O (tr). **110 GAP Photos:** Mark Bolton (b). **113 GAP Photos:** Dave Bevan (b). **114 Garden World Images:** Gilles Delacroix (b). **117 The Garden Collection:** Pedro Silmon (bl). **118 The Garden Collection:** Derek Harris (b). **119 GAP Photos** (t). **138 Garden World Images:** MAP/ Arnaud Descat (b). **144 GAP Photos:** Marcus Harpur (b); **The Garden Collection:** Liz Eddison (t). **148 Derek Gould** (bl). **149 GAP Photos:** Gerald Majumdar (b). **150 Harpur Garden Library** (t). **155 GAP Photos:** Martin Hughes-Jones (b). **156 Garden World Images:** MAP/Frédéric Didillon (t). **162 Caroline Reed** (tl). **163 Andrew Lawson** (ca). **170 Photolibrary:** Andrea Jones (t). **178 Getty Images:** Rob Whitworth (b); **Elaine Hewson** (t). **181 Caroline Reed** (t). **193 Clive Nichols** (t). **198 The Garden Collection:** Torie Chugg (t). **200 GAP Photos:** Dave Bevan (t). **Garden World Images:** Gilles Delacroix (b). **205 Garden Exposures Photo Library** (t). **206 GAP Photos:** Visions (b). **212 A–Z Botanical:** Geoff Kidd (br). **219 GAP Photos:** Geoff Kidd (b). **224 Garden World Images:** Glenn Harper (t). **227 Marianne Majerus Garden Images** (bc). **229 Marianne Majerus Garden Images** (cr). **234 GAP Photos:** J.S. Sira (t). **237 Photolibrary:** J.S. Sira (t). **247 Caroline Reed** (t). **251 GAP Photos:** John Glover (b). **252 The Garden Collection:** Nicola Stocken Tomkins (b). **253 GAP Photos:** Martin Hughes-Jones (b). **262 Garden World Images:** MAP/Frédéric Tournay (b). **264 GAP Photos:** Howard Rice (b). **271 Clive Nichols** (tr). **275 Garden World Images:** Trevor Sims (tr). **279 GAP Photos:** J.S. Sira (tc). **289 GAP Photos:** J.S. Sira (t). **300 Garden World Images:** Gilles Delacroix (b). **304 GAP Photos:** Lee Avison (t). **312 GAP Photos:** Friedrich Strauss (b). **319 Photolibrary:** Paroli Galperti (b). **325 Garden World Images:** Gilles Delacroix (t). **326 Garden World Images:** Lee Thomas (t). **337**

Marianne Majerus Garden Images (b).
338 GAP Photos: John Glover (t). 339
Marianne Majerus Garden Images (b).
340 GAP Photos: Ron Evans (b). 341
Getty Images: Richard Bloom (b).
342 GAP Photos: Jo Whitworth (t).
347 Harpur Garden Library (br). 348
Marianne Majerus Garden Images (cl).
358 Photolibrary: Chris Burrows (b).
365 GAP Photos: Jo Whitworth (b). 367
Clive Nichols (b). 371 Alamy Images:
William Tait. 384 GAP Photos: Martin
Hughes-Jones (b). 391 GAP Photos:
Adrian Bloom (t). 394 GAP Photos:
Neil Homes (t). 399 GAP Photos: Martin
Hughes-Jones (b). 402 GAP Photos:
J.S. Sira (t); Garden Picture Library:
J.S. Sira (tl). 412 Marianne Majerus
Garden Images (b). 420 GAP Photos:
Visions (b). 422 GAP Photos: Pernilla
Bergdahl (t). 429 GAP Photos: Martin
Hughes-Jones (t). 432 GAP Photos:
Martin Hughes-Jones (b). 448

Photolibrary: Dennis Davis (tr).
452 GAP Photos: Dave Bevan (l).
466 GAP Photos: Neil Holmes (t). 471
Marianne Majerus Garden Images (b).
476 GAP Photos: Rob Whitworth (t).
481 GAP Photos: Jo Whitworth (t).
Marianne Majerus Garden Images (b).
483 Garden World Images: Philip Smith
(b). 494 GAP Photos: Martin Hughes-
Jones (b). 508 Marianne Majerus
Garden Images (t). 509 GAP Photos:
Richard Bloom (t). 511 Photolibrary:
John Glover (b). 515 GAP Photos:
Visions (t). 516 GAP Photos: Howard
Rice (t). 526 Garden World Images:
Ashley Biddle (b). 538 GAP Photos:
Martin Hughes-Jones (b). 542 Alamy
Images: Frank Paul (b). 529 Harry
Smith Collection (br).

All other images © Dorling Kindersley
For further information see:
www.dkimages.com

PUBLISHER'S ACKNOWLEDGMENTS

Dorling Kindersley would also like to thank:

Text contributors and editorial assistance Geoff Stebbings, Candida Frith-Macdonald, Simon Maughan, Andrew Mikolajski, Sarah Wilde, Tanis Smith and James Nugent; at the Royal Horticultural Society, Vincent Square: Susanne Mitchell, Karen Wilson and Barbara Haynes
Design assistance Wendy Bartlet, Ann Thompson
DTP design assistance Louise Paddick
Additional picture research Charlotte Oster, Sean Hunter; special thanks also to Diana Miller, Keeper of the Herbarium at RHS Wisley
Index Ella Skene

First Edition
Managing Art Editor Lee Griffiths
DTP Design Sonia Charbonnier
Production Mandy Inness
Picture Research Sam Ruston, Neale Chamberlain

Second Edition
RHS Editor Simon Maughan
RHS Consultant Leigh Hunt
Picture Research Sarah Hopper
Editor Fiona Wild

Revised Edition
RHS Editor Simon Maughan
RHS Consultant Leigh Hunt
Picture Research Sarah Hopper
Editors Helen Fewster, Caroline Reed, Nidhilekha Mathur
Art Editors Joanne Doran, Elaine Hewson, Nitu Singh
Production Editors Joanna Byrne, Luca Frassinetti
DTP Designers Pushpak Tyagi, Anurag Trivedi

THE ROYAL HORTICULTURAL SOCIETY

For over 200 years, the Royal Horticultural Society, Britain's premier gardening charity, has been promoting horticultural excellence by providing inspiration through its shows, gardens, and expertise. Membership of the RHS brings many benefits to anyone interested in gardening, whatever their level of skill, and membership subscriptions represent a vital element of the Society's funding.

To find out more about becoming a member, contact:

**RHS Membership Department
PO Box 313, London, SW1P 2PE**

0845 062 1111

www.rhs.org.uk